# DEVON CLOCKS AND CLOCKMAKERS

# DEVON CLOCKS AND CLOCKMAKERS

CLIVE N. PONSFORD

DAVID & CHARLES
Newton Abbot London North Pomfret (Vt)

**British Library Cataloguing in Publication Data**

Ponsford, Clive N.
Devon clocks and clockmakers.
1. Clock and watch makers – England – Devon – History
I. Title
681.1'13'0922 TS543.G7

ISBN 0-7153-8332-9



Typeset by Typesetters (Birmingham) Ltd,
Smethwick, West Midlands
and printed in Great Britain
by Redwood Burn Limited, Trowbridge, Wilts
for David & Charles (Publishers) Limited
Brunel House Newton Abbot Devon

Published in the United States of America
by David & Charles Inc
North Pomfret Vermont 05053 USA

# CONTENTS

# LIST OF ILLUSTRATIONS

PLATES

FIGURES

# PREFACE

This book, covering the period from the fourteenth century to about 1900, gives details of more than 1,750 former clock and watchmakers who lived and worked in one of England's largest counties. It deals with both church clocks and house clocks and with watches.

The original project was to revise the late J. K. Bellchambers's *Devonshire Clockmakers* (1962) and to enlarge it by including details of nineteenth-century makers. But after numerous old trade directories, archives and early newspapers had been searched through, it became evident that revision would be so extensive that an entirely new and very much larger volume would be necessary. However, some of the excellent illustrations from Mr Bellchambers's book have been retained, and his pioneer research on William Stumbels and the other Totnes makers has been included.

I have drawn on material gathered for my earlier volume *Time in Exeter* (1978) and have included in Chapter 2 the results of a survey of the county's turret clocks undertaken with the Reverend Prebendary J. G. M. Scott of Bampton, the Exeter diocesan adviser on church towers, bells and clocks. Mr Günther von Waskowski, of Newton Abbot, placed at my disposal his collection of superb photographs of William Stumbels clocks, a good selection of which adorn the following pages. I wish to express my sincere thanks to Prebendary Scott and to Mr von Waskowski for their help. I also owe a debt to Mrs Lorna Bellchambers and her son James for making available notes and correspondence relating to *Devonshire Clockmakers*.

In addition, I am deeply grateful to Mr Roger Ellis for lending me the unpublished Memoirs of his great-great-grandfather, the Exeter watchmaker Henry Ellis, and for permission to publish extracts from these; to Mr Ian Stoyle, of Thorverton, for providing additional material on mid nineteenth-century makers; and to Mr Geoffrey Buggins for permission to publish extracts from the business records of

Thwaites & Reed. I am also greatly indebted to Mr William Upjohn Shipley of Chestnut Hill, Massachusetts, Mr Roland Cleak, Mr John Leyman, Mr Reginald Norman, Mr A. F. Lilley, Mr Robert Sherlock and Mr Charles Hadfield for help in various ways; and to the staffs of record offices and libraries in Exeter, Plymouth and London, especially Mrs Margery M. Rowe and the late Mr Trevor Falla, of the Devon Record Office.

Finally, I must thank my wife Pam and my children for putting up with so much during the years when this book was in preparation.

CLIVE N. PONSFORD

1

# THE CLOCK AND WATCH TRADE

## EARLY MECHANICAL TIMEKEEPERS

The word 'clock' so spelt first appears in Devon in a document recording expenditure at Exeter Cathedral in 1376–7. At Barnstaple a few years later, in 1389, a timekeeper described as 'le cloc' was set up at the parish church. St Petrock's in Exeter's High Street also possessed a clock at an early date. Its churchwardens' accounts survive on parchment from 1425, and these record a payment of 2½d for a ladder (*scala*) for the clock in 1430–1. Four or five years later a cord was bought for it at 1d, and in 1447–8 'a man of Wynkelegh' [Winkleigh, north Devon] was paid 10s for repairing and amending the clock. A new clock (*novi orilogii*) was made for the church in 1470–1 at a cost of 33s 4d, an additional 16d being spent on wire for the chime (*le wyre pro le chyme*) and 4d on a weight (*payse*). The maker's name was not recorded but the next year John Fylberd of Taunton, Somerset, received 22d in expenses for coming to mend it.

At Ashburton a payment of 17½d was made in 1479–80 to Roger Torryng 'for a hammer bought for le clocke'. Expenditure on a new clock and a set of chimes was itemised in 1536–7 as follows:

> £9 payd unto John Clockmaker makyn of the clock and the chyme; 10s ffor wyer ffor the clocke and the chyme; 22d ffor plate ffor the clocke; Item payd 21s 8d ffor 3 scheffers [wooden pulleys for the weight lines] weyyn 65lb ffor the clock and the chyme; 11s 2d ffor castyn of the peysys [weights] ffor the clocke and chyme; 3s 4d ffor 60lb of the same lyd [lead]; 2s 10d ffor naylys ffor the bell whelys and to the clockmaker; 3s 4d ffor makyn of hamers for the chyme (Alison Hanham, *Churchwardens' Accounts of Ashburton, 1479–1580*, Devon & Cornwall Record Society, 1970).

Church records show that clocks were also installed at St John's on Fore Street Hill, Exeter, in 1536–7, at Chudleigh *c*1573 – it replaced an earlier one – and at Sandford *c*1595. Among other Devon parishes which possessed clocks before the year 1600 were: Braunton,

Broadclyst, Chagford, Crediton, Dawlish, Halberton, Holy Trinity, Exeter, North Molton, South Tawton, Talaton, Tavistock and Winkleigh. Clocks were also to be found in some public buildings such as Plymouth Guildhall (1526–7) and the Quay Hall, Barnstaple (pre-1591).

Exeter's Elizabethan historian John Hooker, in his unpublished Commonplace Book (Devon Record Office), has a delightful story about a house clock owned by Hugh Oldham who was bishop of Exeter from 1504 until his death in 1519. He relates that the bishop was very temperate in all his ways and liked to dine punctually at eleven every morning and to sup at five every afternoon; to ensure that these hours were observed he kept a clock in his house and appointed a special man to look after it. Important business, however, often prevented the bishop from coming to the table at the appointed hour, and on those occasions the clock-keeper would delay the clock's striking until his master was ready, 'for how so ever the days went the clocke must stryke XI or fyve accordinge as my lord was to dyne or supp'. Other members of the household naturally found this confusing; but it amused Bishop Oldham who frequently made a point of asking what was the clock, only to be told by his clock-keeper: 'As pleaseth your lord; for if you be redy to go to dinner the clock will be XI'. On hearing this the bishop 'would smyle and go his waye'.

The rector of St Edmund's, the Reverend John Williams, was another sixteenth-century Exeter clock-owner. Remarkably, he also owned a printing press and a set of plumber's tools (for working in lead). In his will dated 6 May 1567 (proved 9 December 1572) he bequeathed to Gregory Dodds, the then dean of Exeter, 'my little clock which I had of Sir William Hearne, clerk, deceased late parson of St Petrocks'. Privately owned clocks were, however, very rare at that time and most people relied upon public clocks and chimes, and on bells that were rung at fixed times. The Devon & Cornwall Record Society has published a book of 266 household inventories of the period 1531–1699, and clocks occur in only eight. They ranged in value from 10s to £2 10s and were to be found on staircases and in halls and galleries.

## NOTABLE CRAFTSMEN

The first Devon clockmakers were workers in iron – blacksmiths and clocksmiths. Among those who repaired the Exeter Cathedral clock in the fourteenth and fifteenth centuries were John Gifford, Henry Boteler, Roger the Clockmaker, Toker of Ashburton, and John Sare of Stokeinteignhead. In 1483, Thomas, clockmaker of Exeter, was paid 10s 4d for mending the church clock at Ottery St Mary, receiving 8d per day for his work, plus board, and fodder for his horse. During Queen Elizabeth I's reign, John Crowder of East Budleigh – the village in east Devon near which Sir Walter Raleigh was born – worked as a church clock repairer over a wide area of the county from *c*1560 to the 1590s. In Exeter, John Haydon, by trade a lantern-maker, kept the clock in repair at St John's Church from *c*1585; the man who succeeded him as keeper, in 1609, Matthew Hoppin, was a locksmith, as were many of the craftsmen who made and repaired turret clocks in the seventeenth century. John Savage was the first clockmaker so described to be admitted as a freeman of Exeter. He leased a shop in the High Street, next to St Stephen's Church, and about 1620 made a clock for Plymouth Guildhall. His son Peter continued the business and, as will be seen in Chapter 3, repaired a hall clock from a Somerset country house and also mended watches.

Makers of brass lantern clocks included the Quakers Abell Cottey (1655–1711) of Crediton and Arthur Davis of Tiverton, Westleigh, Cullompton and Kentisbeare; and John Michell of Chardstock on the Dorset border. The last named also made longcase clocks, one of which, in a fine walnut case with a particularly well-engraved dial, is illustrated in Plate 4. In north Devon, John Morcombe who made the town clock of 1622–3 at Hartland, and Phineas Parker who died in 1695, were early Barnstaple makers. Ephraim Dyer, of Bideford, made longcase clocks and can be traced as a repairer of church clocks between 1689 and 1723. In the south of the county, William Clement, a son or namesake of the anchor-escapement pioneer, established a reputation for fine work at Totnes and was followed there by the great William Stumbels. An Edward Clement worked at Exeter, where he was one of ten or more clockmakers in business around the year 1700. At Tiverton, Simon Thorne (ii) was a prolific eighteenth-century maker of longcase clocks. Thomas Land was described as a watchmaker of St Saviour's parish, Dartmouth, when he married in 1705; he then worked at Tiverton and from there moved on to Honiton where later makers included Francis Pile (Pyle) who flourished between *c*1731 and 1763, and Henry Deeme who married in 1756 and died about 1786.

John Peream was a gunsmith and clockmaker at Ottery St Mary in east Devon. His will, proved on 21 May 1756 by his widow Esther, contains some interesting bequests:

> To my son William Peream all my working tools, all my gun stocks and stocking timber, all the new guns, secondhand or other guns, brass work, gun locks and all the clock bells I now have by me for his own use . . .
>
> To my grandson William two large silver spoons with the Apostles heads thereon, a gold ring and the new clock and case with my name on it.

At a more rural level, Edmund Huxtable operated as a clockmaker, locksmith and agricultural dealer in the Chittlehampton area of north Devon. One of his account books has been preserved in the Devon Record Office (2603M/Bl); it contains many entries relating to farming and a few concerning clockmaking. In September 1777 he supplied a customer, John Gould, with a new clock costing £2 15s; he also supplied him with reap hooks, cheese and bacon. In August 1778 he charged the Reverend Peter Beavis a shilling 'for Righting & Cleaning ye Cookow Clock'. Other entries include:

> Mr William Lake, October ye 16, 1776, for Righting a Clock 1s
> 1778: Oct ye 19 for a pendulum of a clock 3s
> 1788: a new key to ye lock of the Clock case 8d

Among Devon clockmaking families which established themselves in the eighteenth century and continued on into the nineteenth were the Upjohns of Exeter, Plymouth and Bideford; the Pollards of Crediton, Exeter and Plymouth Dock; the Tuckers of Exeter and Tiverton; the Bucknells of Crediton; and the Murches of Honiton.

## CLOCKS AND CLOCK CASES

### *Types of clock*

Most of the standard types of clock were either made or sold in Devon. The longcase, or grandfather, clock was popular and in old advertisements was usually referred to as an eight-day clock, the less expensive yet numerous thirty-hour variety seldom getting a mention. An auction at Plymouth in 1804 included 'a capital eight-day clock by Ellicott'; and a sale in 1798 at a house on the Strand at Topsham, near Exeter, featured 'an eight-day clock, mahogany case, neat front, with the moon, and High Water at Topsham Bar, nicely calculated, on an improved plan'. The *Sherborne Mercury* of 22 February 1768 announced a sale 'at Dolish, in Luppitt, near Honiton, lately the dwelling house of Doctor Farrer, now settled in London', where the household furniture included 'an elegant eight-day chyme clock, in a neat, mahogany case, quite new'. Earlier descriptions include 'long swing pendulum clock' and, in an inventory of 1710, 'the pendulum clock that stands in the great stare case'. The term 'cottage clock' was used to describe humbler versions of these tall-case clocks in the nineteenth century. Bracket clocks were referred to either as 'spring clocks' or as 'table clocks'. The weight-driven wall clock of the 'Act of Parliament' variety was a 'large dial'; the smaller, spring-driven wall clock was a 'spring dial'. An inventory made in 1674 of the goods of Thomas Russell of Crediton, woolcomber, included 'a brass clocke' valued at £1 10s – presumably a lantern clock. It is related that Samuel Good, a clock and watch repairer at Seaton in the later Victorian years, used to go to sales after he retired to buy old lantern clocks to convert 'into grandfathers for which he would get as much as £20' (Tapley-Soper Index, Westcountry Studies Library, Exeter).

Longcase clocks with simple tidal dials, such as the one mentioned above showing 'High Water at Topsham Bar', were made by a number of Devon makers, mostly during the second half of the eighteenth century. These clocks have arch dials and show the age of the moon and the corresponding time of high water at the particular place for which they were made. 'Topsham Bar' clocks were a speciality of the Exeter makers; the old entrance to the city's ship canal was just above Topsham, on the Exe, and the time of high water was important to Exeter merchants as the river was only navigable at certain states of the tide. Various other Devon examples show high water at such places as Plymouth Dock, Torbay, Totnes, Bideford and Sidmouth.

A good quality longcase clock by Angel Sparke of Plymouth, typical of the thousands made in the county between *c*1680 and 1860, is illustrated in Plates 1 and 2. It has a lively maritime scene in the dial arch, and dates from the 1780–90 period.

### *Fashions in cases*

Contemporary references to clock cases are rare. It is known that in London there were specialist clock-case makers, but in Devon their manufacture seems to have been part of the trade of the joiner and cabinet maker, judging by an advertisement in *The Protestant Mercury: or The Exeter Post Boy* of 18 May 1716:

> Nicholas Williams, Joyner in Southgate-street, EXON, sells all Sorts of Chests of Drawers Hanging-Presses Clock-Cases Cabinets Scrutores Commode-Tables Desks Book-Cases and Looking Glasses, of the Newest Fashion and best Fineer'd Work in Walnut-Tree, also Japan'd Work.

The well-matched, veneered cases of some of Jacob Lovelace's longcase clocks – probably made by his cabinet-maker neighbours in Exeter, the Channon family – and the tidal clock by William Stumbels of Totnes (Plate 30) are good examples of the 'best Fineer'd Work in Walnut-Tree', and Devon clocks are also to be found in 'Japan'd Work', or lacquer, cases (Plates 10, 12, 45, 46 and 47). Walnut continued in favour with the case-makers in the middle years of the eighteenth century, but then made way for mahogany. A late instance of the use of walnut is provided by a longcase clock of *c*1770 made by Thomas Upjohn, of Exeter.

An early mention of mahogany – sometimes spelt 'mohogany' – occurs in the St Sidwell's, Exeter, churchwardens' accounts in 1737:

> To Thomas Marshall for ye Mohogany rails Communion Table & wineScott [wainscot] floor £29 15s 8d.

The wood was all the fashion in the 1760s, being imported at West Country ports and offered for sale in newspapers of the period; a cabinet-maker in Exeter was in business at 'The Sign of the Mohogany Tree'. The high ceilings of Georgian houses provided the county's case-makers with an opportunity to supply mahogany longcase clocks of pleasing slender proportions with lofty arch dials. Oak and painted and varnished deal were used to case the less costly clocks, especially those of thirty-hour duration.

Devon's case-makers were conservative to the end and continued the eighteenth-century longcase styles well into the following century. Broad cases of pronounced Midlands character make an occasional appearance on Devon clocks after the spread of the railways in the middle years of the nineteenth century. A white-dial longcase clock by Spiegelhalder of Exeter has such a case, as does an example by John Sharland of Tiverton. Another, by William Huxtable, South Molton, has a label indicating that it was sent to him by rail via Barnstaple.

## ORGANISATION OF CLOCKMAKING TRADE

Clockmaking does not appear to have been highly organised in Devon. Tools and patterns were obtained from London and, by the early nineteenth century, travellers from Bristol and the Midlands were regular callers at Devon watch and clockmaking shops, encouraging expansion into jewellery, plated goods and other retail lines. Certain clockmakers, in Exeter and elsewhere, acted as wholesale dealers, offering a wide range of components and materials and supplying watch parts.

A typical clockmaker's establishment in a Devon city or town at the end of the eighteenth century would have consisted of the master, his family, a journeyman, and possibly two or three living-in apprentices. Because maximum light was a prime consideration, watch repairs were usually carried out on a bench by the front window of the shop; the man or boy making up the clocks was often banished to a back workshop.

In the larger city retail businesses it was not unusual for a jeweller to sell ready-made clocks and watches, or to employ working watchmakers and silversmiths, and vice versa; and the names of jewellers, goldsmiths and silversmiths will be found on watches and on the dials of clocks. Some of these names are included in the List of Makers, but a more comprehensive list will be found in Exeter Museum's published *Catalogue of Exeter Silver.* This gives details of gold and silversmiths from a wide area of the South West who sent their wares to the Exeter Assay Office (closed 1883).

In market towns and more rural areas clockmakers often mixed their trade to a remarkable extent, some also being gunsmiths and ironmongers and, in the nineteenth century, insurance agents, stamp distributors, beer retailers and auctioneers. In at least two Devon instances, they were also taxidermists. Many of the town craftsmen termed themselves watchmakers rather than clockmakers, and it is clear that the repair of watches formed the major part of their business. But in country areas watches probably played a lesser part in the trade, the staple product in the eighteenth century being the thirty-hour longcase clock.

### *Apprenticeship*

Boys showing an aptitude for the trade were usually taken on for a month's trial and then indentured as apprentices on or about their fourteenth birthday. Their fathers would pay a premium which, in the eighteenth century, could range from a few pounds to as much as £60, depending on the prospects and the master's standing in the trade.

These premiums were originally intended as a kind of present for the tradesman's wife, to engage her to be kind to the apprentice and take motherly care of him. Later, however, they became a source of capital and were especially useful when a master was first setting out in business. Apprenticeship was for seven years in most instances, after which the apprentice was free to 'journey' about to find employment as a journeyman and, eventually, if he had sufficient ambition and capital, to set up in business on his own account. R. Campbell in *The London Tradesman*, 1747, wrote:

> It requires no great strength nor much education to make a practical watchmaker, but a man who intends to be master of the theory ought to have a tolerable education and should have some smattering of mechanics and mathematics . . . The trade is not much over-stocked in Town, and no trade has better encouragement in our plantations, or in any other part of Europe. If he understands his business, he may have bread almost anywhere.

Some apprentices ran away before completing their full term. Others were abandoned by their masters. One such was Joseph Purchase, an Exeter woolcomber's son who was apprenticed on 1 January 1699/1700 for seven years to John Blackmore of Exeter, watchmaker. In March 1701/2 he was discharged at Exeter Quarter Sessions from his apprenticeship, the court being told that John Blackmore 'hath deserted his dwelling house and left his said apprentice wholly destitute of maintenance and made noe provision for the instructing of his said apprentice'. He was then bound apprentice to John Lamby of St Edmund's parish, Exeter, to learn the 'Art trade and mistery of a Clockmaker & Jackmaker' for the remainder of his seven-year term. Bad luck dogged young Purchase, however, and at the Quarter Sessions in July 1705 it was stated that 'John Lamby hath for some time absconded himselfe & left his habitation and taken noe care of or made any provision for the maintenance of the sd Joseph Purchase'. And once again the court discharged him from his apprenticeship.

### *Brassfounders and ironmongers*

Brassfounders played a principal role in the clock trade, casting wheels, plates, pillars and dial parts which they then supplied in the rough to makers. But some brassfounders also made complete clocks; for example Richard Luscombe who worked at Totnes in the first half of the eighteenth century. Luscombe was one of the Totnes parish church bellringers and he and his companions made a gift of a brass chandelier, now hung in the south chapel of the church, inscribed 'for the ringers use for ever'. The names of the ringers are engraved on it and against that of Richard Luscombe the word 'fecit' indicates that he was its maker. The churchwardens' accounts of Ashburton record that in 1729 Luscombe entered into a twenty-year agreement to keep the church clock and chimes in repair.

It was a brassfounder, William Howard, who added the minute dial to the Exeter Cathedral clock in 1759. Howard, a freeman of Exeter, had a foundry on the old Exe Bridge near St Edmund's Church, and was extensively employed on church clock repairs. He also made brass chandeliers, or 'branches' as they were then called; one dated 1746 is at Hickleton Church, Yorkshire. Howard died in 1768, but the business was continued by his widow Elizabeth until her death in September 1781. In the *Exeter Flying Post* in April the next year it was announced that a two-day sale would take place of the stock in trade 'at the house of the late Mrs Howard on the Old Bridge, Exeter'. This consisted of 'all sorts of braziery, foundery, and pewterer's goods' and included clocks and roasting jacks. Clocks were also made at that period at brass foundries in North Street, Exeter, by Robert Harrison and James Hine. The latter came from Bridgwater, Somerset, a town noted for its brass-foundrywork.

Ironmongers, too, played their part, one of them advertising for workmen in January 1790 as follows: 'Wanted immediately, five or six Journeymen Clock or Jackmakers. Apply to Mr Samuel Kingdon, Ironmonger, Exeter'.

### *Blind watchmakers*

In E. J. Wood's *Curiosities of Clocks and Watches* (London, 1866), it was stated that in the earlier part of that century there was a sign over a door in Barnstaple denoting that clocks and watches were repaired there by William Huntley, a blind man. He was born blind and was brought up to the business by his father, who was himself a clock and watchmaker. The son had plenty of employment, being considered by the inhabitants to be very efficient in his profession. He repaired musical clocks and watches, and seldom met with any difficulty in repairing the most complicated; in fact when others had failed to correct faulty working of a clock or watch, he often discovered and set right the defect. The *Exeter Flying Post* of 26 October 1826 had carried a reference to Huntley, saying that he had been blind from his birth and describing him as an ingenious watchmaker. The paper added that a few years earlier he had tendered to maintain the town clock at Barnstaple and keep it in repair.

Other watchmakers, although not blind, suffered from failing eyesight. The work was exacting, and long hours spent at the bench on delicate and tiny mechanisms took their toll. The nineteenth-century Tiverton watchmaker, John Walter Tothill Tucker, went totally blind; and another sufferer was the Exeter watch and chronometer maker Daniel Ross. When in 1856 the latter published a little volume of poems entitled *Serious Thoughts*, it was stated that any profits resulting from its sale would be donated to the West of England Institution for the Blind.

### *Clocks from London and Bristol*

Some Devon clockmakers such as John Skinner of Exeter and his son John, and Ralph Banks and William Lancaster, both of Plymouth Dock (now Devonport), were customers of John Thwaites of Clerkenwell, London. They obtained from him many high-grade clocks, especially bracket clocks – which, as already mentioned, were then referred to as 'spring clocks' – and sold them with their names on the dials, although Thwaites usually signed and numbered the movements. Entries in Thwaites's massive ledgers and day books in Guildhall Library, London, show that clocks for Devon makers were supplied to order either singly or in batches of two or more. The following excerpts from the day books concerning Ralph Banks, William Lancaster and John Skinner are especially valuable for the descriptions of clocks then in vogue, such as spring clocks with compass-head, round-top, bell-top and cove-top cases.

8 July 1796: Mr R. Banks, Plymouth Dock. A new plain spring clock with 8in dial plate silvered, flat pendulum, day of the month by a hand under the centre, in a mahogany compass head case. Packing case & package.

12 July 1797 (Plymouth Dock from the Bell Inn, Friday Street): Mr R. Banks. Two new spring clocks, one with an 8in & one with a 7in arch dial plate silvered, with days of the month by a hand under the centre, in mahogany cases with brass handles & wood open works &c.

11 Aug 1797: Mr R. Banks. Two new spring clocks.

6 October 1797: Plymouth Dock/Mr Lancaster. A spring repeater with the quarters on 6 bells and a 7in silvered plate, flat pendulum in a mahogany case, compass head, and 3 pannells with brass mouldings and open works. A packing case with proper package.

14 September 1798: Mr R. Banks. A spring clock with 7in round Japanned dial plate, flat pendulum & metal hands gilt in a satin wood case. A spring clock with 7in Japanned dial plate, flat pendulum &c, metal hands gilt in a mahogany round top case.

30 January 1799: Mr R. Banks. A spring clock with 7in Japanned dial plate, flat pendulum & metal hands gilt in a satin wood case, quarter columns.

20 April 1799: Mr Banks. A spring clock with 7in silvered dial plate, flat pendulum, in a mahogany case & wood open works at the side.

June 1799: Mr Banks. A large spring dial with going ratchett & heavy bob & large barrell & fuzee & steel hands in a mahogany case with Japann'd board of 2 feet diameter & brass ring & convex glass.

The day books record many transactions over the years with John Skinner. In July 1803 he ordered five spring clocks and, the following September, a further two. The July order was as follows, except that the prices have been omitted as they are in code:

London July 1803
Mr Skinner

To a spring clock with 8in Japann$^{d}$ dial plate flat pendulum &c in a mahogany cove top case brass ring & convex glass.

To 2 spring clocks with 7in Japann$^{d}$ plates flat pendulums &c in mahogany round top cases & pannell brass ring & convex glasses.

To a spring clock with 7in silvered dial plate flat pendulum &c in a mahogany bell top case brass fretts.

To a D[itt]o. D[itt]o. in a mahogany round top case.

To 3 packing cases for the above with package &c.

Sometimes Skinner would send an order for clock parts only. In February 1804, for example, he ordered a 'spring to a barrel', a new crown wheel and pinion, and a verge for a spring clock. A best eight-day clock was ordered in October 1808 'with 13in silvered dial plate, dead scapement, going ratchett, & weight shells'. The ledgers give information on prices. In 1803 John Skinner paid £8, £9 9s, and £10 each for spring clocks. Spring dials (wall clocks) were less expensive, one in 1809 costing £7 8s. In the same year an eight-day clock and case (a longcase clock) was charged to Skinner at £14 0s 6d. In 1810 he bought two eight-day turret clocks from Thwaites, one of which is in the pediment of Exeter Prison and is still in use. It carries the signature 'John Skinner Exeter 1810' on the setting dial and also on the end of one of the winding barrels.

A large brass-cased drum clock examined by the author has two white-painted 380mm (15in) copper dials with the date 1829 scratched on the back of each. The clock was made by Thwaites's successors, Thwaites & Reed, for John Skinner (ii). The backplate of the very solidly made six-pillar, rack-striking movement has the signature 'Skinner EXETER', but the front plate is signed 'T & R 5707'. This number is repeated on the top edge of the plate, and another signature, that of 'A & I Thwaites', appears on the underside of a mechanism on the top of the movement for regulating the pendulum. This refers to the eighteenth-century partnership of Ainesworth & John Thwaites and reveals that this particular part of the clock had been in stock for many years before being fitted to this movement.

Devon makers were also supplied with ready-made clock movements by the Bristol firm of Thomas Hale & Sons, known until about 1842 as Wasbrough Hale & Co and originally as Wasbrough – a name found on the handles of numerous old bronze skillets (cooking pans). Established in 1726 and based in Narrow Wine Street, this firm built up an extensive trade in the making of clocks, wind dials, weathercocks and machinery. Around the year 1858 Thomas Hale & Sons published a list of the various church and other clocks it had made, and stated that the number of house clocks made up to December that year was 39,861. Many of these clocks carried the names of retailers. Devon examples include eight-day longcase clocks signed 'John Tickell Kingsbridge' and 'Ellis & Co Exeter'. The Ellis clock, dating from *c*1830, has the initials 'W H & Co' on the top of the bell. These stand for Wasbrough Hale & Co and should not be confused with 'W & H', standing for Walker & Hughes, the Birmingham dial makers.

## WATCHES

The seventeenth-century records of Exeter Quarter Sessions include several cases alleging the theft of watches. In October 1624, for instance, one Avis Skelton was accused of stealing a watch and purse belonging to Henry Gould during a sermon in Exeter Cathedral at six o'clock in the morning. And in 1693 John Adams, mariner, of Churchill, Somerset, testified that 'a certayne watch now found on him, being broken in pieces, was by him bought of a boy in Bristoll for three pence'.

The following references to watches occur in Devon wills:

1617 John Trosse, of Exeter, gentleman: 'to William Trosse my son £100 and my howre watch'.

1673 Richard Inglett, of Plymouth, physician: 'to my son Theodore . . . my watch'.

1733 John Elston, of Exeter, silversmith: 'to my son John Elston my silver watch, also such tools &c as I have in or belonging to the shop'.

1738 Edward Whiteway, of Ashburton, clothier: 'to my beloved wife my silver watch with the deaths head upon it'.

The earliest surviving Devon watches known to the author date from the early years of the eighteenth century. An example by John Richards of Honiton, *c*1710, has a silver champlevé dial with a calendar aperture. A similar watch, signed and dated 1718 by the Exeter maker Jacob Lovelace, also has this latter feature, which occurs again, much later, on a 1762 watch by Francis Pile of Honiton. This last watch also has a silver champlevé dial and was made at a time when the fashion generally was for white-enamel dials. A watch hallmarked 1799, made or supplied by William Gard of Exeter, has the name of the owner, ELIAS TREMLETT, around the dial in place of the hour numbers, but as there was a letter too many the final TTs are used to indicate 9 o'clock. Samuel Pearce, of Honiton, also supplied watches of this type; one was marked with the name Robert Troke on the dial plate and another with the name John Woolcott. 'Stop seconds' watches were also fashionable towards the end of the eighteenth century. One such is signed 'Abr$^{\underline{m}}$ Daniel, Plymouth'. It has a centre-sweep seconds hand, above which is a secondary dial for the hours and minutes; the movement can be stopped and started by means of a lever that projects through the side of the inner case.

In an advertisement in *The Exeter Mercury* of 28 June and 5 July 1765, George Flashman, clock and watchmaker, announcing that he had moved to premises opposite Exeter Guildhall, stated that he had for sale silver, pinchbeck, shagreen and tortoise-shell watches, from 3 guineas to 10, and clocks and gold watches in proportion.

It is not possible to say to what extent watches were actually made in Devon. The trade was a complex and subdivided one, the manufacture of a watch representing the labour of a number of component makers and craftsmen and women (see 1813 description of watchmaking, Chapter 5). Some of the cases of eighteenth-century Devon watches have London hallmarks, and some of the nineteenth-century ones bear the dagger and three wheatsheaves of the Chester Assay Office. At least one craftsman, however, made watch cases in Devon. His name was Francis Bailey and he was also a maker of spurs for fighting cocks. *Brice's Weekly Journal* of 12 April 1728 has this advertisement:

> FRANCIS BAILEY. Late of Exeter. Now living in Bridgewater [sic]. After many years longer Practice than any Pretender to it in the Country, still makes and sells Silver and Steel Cock-spurs, which on Experience have ever given all Gentlemen the best Satisfaction. He sells the Silver ones at 4s a Pair and those of Steel at 3s. He also makes and sells Barbers Cards [combs], with Steel or Iron pins, either for mixing or Back-drawing; together with Gold and Silver Watch Cases of plain or O-gee Joints, according to the newest Fashions.

### *Clerkenwell watches*

Contemporary newspapers provide evidence that complete watches were sent from Clerkenwell to Devonshire bearing the names of the local watchmakers who were to sell them. A notice from the *Exeter Flying Post* of 15 January 1795 will serve as illustration:

> LOST out of the Western Mercury coach, from London to Exeter, on the 19th of December last, a box, directed to R. Banks, watchmaker, Plymouth Dock; containing, one gold watch named Wm. Lancaster, Plymouth Dock; also one metal watch, cap't and jewel'd, and seconds, named R. Banks, Plymouth Dock, No. 25,434; also three silver watches named R. Banks, Plymouth Dock, No. 111, 112, and No. 113; also a pair of gold cases; three large gold seals, one of which is engraved with the letter S, and crest over; also a plain gold button. Whoever will give information to Mr Baily, watchmaker, No. 29 Red Lion Street [Clerkenwell], or to R. Banks, watchmaker, Catharine Street, Plymouth Dock, so that the above articles may be recovered, shall receive twelve guineas reward.

A similar award was offered in 1791 for a box of watches that was lost either from the warehouse of the Swan with Two Necks in Lad Lane, London, or on the way by mail coach to A. Joseph & Son, Plymouth. The box was being sent by Baily & Upjohn of Clerkenwell, but the watches themselves bore the names of a curious assortment of watchmakers: E. Poole, London; T. Manser, London; Roger Staine, 'Bruxtian'; and Charles Straune, London. There were seventeen plain silver watches, plus one with an engraved edge, and three gilt-metal ones. Oddly enough, watch No 765 by T. Manser was followed by No 766 by Roger Staine. One wonders if these were, in fact, trade names, Baily & Upjohn signing only their better quality watches, or those that they sold themselves.

In the early nineteenth century, the Clerkenwell watch manufacturer William J. Upjohn used to travel in the West of England with his watches. He also undertook the more difficult types of repair, and parcels of watches were regularly sent by coach to his workshops by Devon makers.

## SIGNATURES ON CLOCKS AND WATCHES

In the early days clock and watchmakers would sign their work out of pride, but it was not compulsory. Later, as commerce developed and the fame of British watchmakers spread abroad, great quantities of empty cases and dial plates were exported. Many of these were then fitted with poor quality movements and signed with counterfeit names or those of eminent London craftsmen. In an attempt to prevent this – 'and the like ill practices in England' – Parliament, in 1697–8, banned the export of dials and metal cases without movements, and enacted that all clocks and watches should be signed with their maker's name and place of abode or freedom.

Many makers, particularly in the late-seventeenth and early-eighteenth centuries, added the word 'fecit' to their signatures on clocks. The church clock at Dulverton, for example, carries the date '28 June 1708' and 'Lewes Pridham Sandford Fecit', meaning that Lewes Pridham, of Sandford, made or executed this. The word was sometimes misspelt 'facit'.

In clockmaker's signatures Jno was used as an abbreviation for John (not Jonathan), and a capital 'J' was sometimes rendered as 'I'. But clock signatures are not always what they appear, for there was a tendency in the nineteenth century for repairers to substitute their names for those of the original makers. For example, early this century a correspondent from Dawlish reported: 'We have a clock which has been in our family for many generations. A great number of years ago it needed repairing and was sent to Mr Strowbridge. When it came back his name "H. Strowbridge, Dawlish", was engraved upon the dial.'

The old ecclesiastical names of Barum for Barnstaple and Exon for Exeter will be found on many clocks and watches made before, say, 1775, and occasionally thereafter. Devonport did not become known as such until 1 January 1824; clocks made there before that date were signed with the place-name Plymouth Dock, or possibly Hamoaze Dock.

# 2

# TURRET CLOCKS AND CHIMES: CLOCK TOWERS

20 Nov$^{r}$ – 1915
Stop'd the striking 'Police Order'
because of the Zepps.
Started again
Armistice Day, 11 Nov$^{r}$ 1918.

*(A note by the clock at the old police station at Cullompton, referring to German Zeppelin airships.)*

The story of the weight-driven turret clock in Devon begins at Exeter Cathedral in the 1370s and continues to the present day. During this period of more than 600 years there have been countless renewals and repairs and some churches have had a succession of two, three, or even four clocks. Where clock movements of obvious antiquity have survived as at Exeter Cathedral, Ottery St Mary and Hartland, there is evidence of substantial alterations and rebuilding. Occasionally, clocks were moved from one church to another. For example, the present clock at Exminster was once at Crediton; Sandford's came from Kenton and Bishop's Tawton's from Barnstaple; while Kentisbeare acquired secondhand a curious machine made up apparently of parts from other clocks and using bell-clappers as internal counterweights for the hands.

## SIX NOTABLE PUBLIC CLOCKS

### *Exeter Cathedral*

Exeter Cathedral has the largest of the four famous West of England 'astronomical' dials – the others are at Wells Cathedral, Wimborne Minster, and Ottery St Mary parish church. The Exeter dial is within the cathedral; it shows the supposed orbit of the sun around the earth in twenty-four hours and the phases and age of the moon, and is constructed of blue-painted oak boards in a painted medieval stonework surround. The present clock movement was made in 1885, but its predecessor, containing workmanship of many centuries, is preserved as a working exhibit on the floor of the north transept, beneath the dial. In its present form the old clock is a jumble; it was originally two separate units but was reassembled as one by J. J. Hall in 1910. The oldest mechanical part is the vertically framed strike train (Plate 5), originally mounted on a post high up in the north tower; it has flail-locking and was designed to strike on the massive Peter bell cast in 1484. The going train – formerly behind the dial – has a four-post frame and was made after the invention of the pendulum, possibly by the celebrated Exeter clockmaker Jacob Lovelace who was paid £21 17s 6d in 1713–14 for 'a bill about ye Clocke'. A little quarter-striking unit, perched on top of the going train, looks like a modified sixteenth- or early seventeenth-century iron-framed domestic clock. The brass escape wheel is inscribed 'JK 1841', and the present hammer that strikes the Peter bell is dated 1831.

The earliest reference to a time-measuring device at Exeter Cathedral occurs in the Patent Rolls of Edward II in which mention is made of a writing of 1284 by Bishop Peter Quinil (or Quivil), concerning the Cathedral bells, musical instruments and horologe.

> 1317–18. January 6, Westminster. Inspeximus and confirmation of a writing of Peter, bishop of Exeter, granting with the consent of the chapter of Exeter to Roger de Repford the bellfounder (*campanistario*), and Agnes his wife, and Walter their son, and their heirs, the tenement, which Nicholas de Peynton, the bishop's bondman (*nativus*) held in Peyngton, subject to a yearly rent of 1d for every secular service, and the said Roger, Agnes, and Walter and their heirs shall make or cause to be made at the costs of the chapter of Exeter the bells (*campanas*) of the church, and as often as need shall be they shall repair or cause to be repaired the musical instruments (*organa*) and clock (*orilogium*); while so employed all necessaries of food and drink shall be supplied to them: dated in the chapter of Exeter II Ides of July, 1284 (*Calendar of the Patent Rolls of Edward II*, vol 3, p72).

The fabric account rolls preserved at the Cathedral provide few clues as to the nature of this mechanism. Indeed, there are so few references before the 1370s that one is inclined to think that the horologe was not a mechanical clock in the modern sense. In 1329–30 there was a payment: 'In the salary to the smith for making hinges for the clock in the church 8d' (*In stipend' fabri pro vertivell' fac' pro horlog' in ecclesia viij d*). Then in 1372–3 there were expenses about the casting of a clock bell by Thomas Karoun, followed by an important series of entries in the account roll for 1376–7, recording the construction of a room in the north tower for the new horologe called the clock (*pro horalogio quod vocatur clock de novo construend'*). Examination of the document confirms that the actual word used is *clock* and not *clokke*, as some writers have stated. The total expense of the room newly-built was £10 6s 5½d. The clock placed there was put in order by John Gifford (*Johannis Gyfford*). Later accounts mention weights and cords.

In 1423–4 a payment of 73s 4d was made to 'John Budde the painter of Exeter for painting the new clock made in the Cathedral church'. This is probably the date of the stonework surrounding the dial, which has cresting, carved grapes resembling fircones, and other architectural details similar to those decorating the nearby stone screen of St Paul's Chapel in the north transept.

The cathedral clock was for centuries Exeter's standard of time and it had to be kept in good running order, hence the large number of repairs and alterations. In July 1742 the Dean and Chapter ordered that it should be set every day at 5 o'clock in the afternoon by a regulator, and in 1759 a minute dial was added. But about 1817 certain manufacturers complained about the timekeeping and submitted a petition, suggesting the old clock was worn out and should be replaced. Various repairs were then carried out. In 1854, in an unusual experiment, the old striking part was linked with one of Charles Shepherd's galvano-magnetic regulators which had been placed at Exeter Guildhall. On 14 September that year the *Exeter Flying Post* reported:

> The electric clock at the Guildhall is now in connection with an apparatus fitted up under the superintendence of Messrs Ellis Brothers, of this city, by which means the great bell at the Cathedral strikes the hour indicated by the Guildhall clock, thus rendering the old clock comparatively useless, unless for the regulation of the services of the church. In olden time the bell of St Peter was the regulation to the whole city; the improvements or alteration that has been made is calculated to sustain its reputation by the combination of modern science with mechanism.

The original regulator used is believed to be a mahogany-cased one now in Exeter Museum.

### *Matthew the Miller*

The Matthew the Miller clock at St Mary Steps parish church, Exeter, takes its name from the three automata (Plate 3) that ornament the exterior of the tower opposite the site of the city's former West Gate. Mounted beneath a canopy above the dial they feature a central figure – seated, wearing perhaps a breastplate, certainly with plumed helmet and sceptre – and two flanking pikemen with plumed headgear, pikes, swords and hammers. According to local legend they represent Matthew, a miller of old whose comings and goings were a byword for punctuality, and his two sons. They were set up in the reign of King James I and have been providing a round-the-clock sideshow as one of the sights of Exeter ever since.

To mark the passing of each quarter of an hour the jacks, Matthew's sons, turn their heads, raise hammers, and smite the bells mounted beneath their feet. But the full performance takes place only on the hour. First the jacks sound the quarters and then, when the hours are struck on the tenor bell in the tower, Matthew the Miller takes over, beating time with his sceptre and nodding his head.

A popular rhyme, dating back at least to the eighteenth century, runs as follows:

> Matthew the Miller's alive,
> And Matthew the Miller's dead:
> Every hour, in Westgate Tower,
> Matthew the Miller nods his head.

The origins of the clock only recently came to light among the papers of the Exeter Consistory Court. These give details of a dispute over expenditure by the churchwardens, and reveal that the clock originally carried the date 1621 and that the seated timber figure was known from the outset as Matthew the Miller. There was also a carving, since removed, showing a laden packhorse and trees growing in front of a millhouse. The maker was Matthew Hoppin, an Exeter locksmith who was extensively employed in Devon on turret-clock repairs; and the clock, dial and jacks were set up over a period of a year or so between 1619 and 1621. The present movement is a 1725 replacement, and the Gothic-style canopy above the figures dates from

1835. The whole clock was extensively restored in 1980 when it was found necessary to replace the decayed body and limbs of Matthew and also the dial, which has a central revolving disc with a sun to indicate the hours and a hand to represent the minutes. Four quaintly carved classical figures, including Apollo and Ceres, decorate the dial corners. Alexander Jenkins in his 1806 *History of Exeter* states:

> This Matthew was an opulent miller who resided nearby at Cricklepit; he was remarkable for his integrity, and regular course of life; and his punctuality of going at one hour for, and returning from the city with, his grist, occasioned him to be so much noticed by the neighbourhood that they knew exactly the hour of the day by the time of his passing; and from this circumstance the statue received the name.

It is difficult, however, to reconcile the military aspect of the three figures with the carving of the scene by the millhouse. Do they really represent the miller and his sons, or were they set up on the tower and then identified with the legend? It is possible that they were originally intended for Exeter Cathedral where, in 1615, it was proposed unsuccessfully that a dial should be set up on the outside of the north tower 'with bells and quarter-smiters as it is at Westminster' (Cathedral archives: Petition touching ye bell metal).

### *Hartland town clock*

The illustration (Plate 6) may give a misleading impression of size, for this is a small clock with a 'fieldgate' frame about 585mm (23in) long and 535mm (21in) high. The capstan-winding bars are mere handgrips. Empty holes in the top rail above the going train indicate that it once had an overslung foliot, but the entire movement appears to have been re-wheeled in brass and now has an anchor escapement. The pendulum beats fifty-four to the minute and, as befits the town's proximity to Devon's most rugged coastline, was formerly suspended by a piece of whalebone. The clock is well documented in the Borough of Harton (Hartland) portreeve's accounts (Devon Record Office), which record that it replaced an earlier clock in 1622–3 and was made by John Morcombe of Barnstaple at a cost of 33s. He subsequently kept it in repair, regularly visiting the town at the time of the Easter fair and being provided with food and drink. In 1657–8 a heavier clock bell was cast by 'Mr Pennyngton' and Morcombe was paid £1 for 'new makeing' the clock; in 1716–17 the clock was taken to 'Bytheford' (Bideford) to be 'righted' by the clockmaker Ephraim Dyer at a cost of £2, but there is an earlier gap in the accounts so it is not exactly certain when the movement was altered to its present form. The clock now strikes on a nineteenth-century bell; the dial is modern.

*'Grandisson's clock', Ottery St Mary*

The parish church of Ottery St Mary has two working clocks and a set of chimes. The newer one (by Gillett & Bland of Croydon, 1874–5) is in the tower, but the older clock is on a gallery in the south transept and to reach it one must climb a ruggedly hewn timber stairway of great antiquity. The old clock is famous for its astronomical dial, similar to the one in Exeter Cathedral but smaller, and differing in not having a stonework surround. The dial together with the movement was restored in 1907 by J. J. Hall of Exeter, whose claims that it was a fourteenth-century clock installed by John de Grandisson, Bishop of Exeter, were later hotly disputed. The movement has been much rebuilt and modified. In its present form it cannot be much older than *c*1700, but it does contain parts of an earlier clock or clocks. The going and strike trains are mounted end to end in a cage frame with flat top and bottom rails and roughly shaped octagonal corner-posts. Simple crook finials are attached to the nuts at the corners. The main pivot bars are exceptionally broad, suggesting that they may once have been part of a fieldgate-type frame similar to the one in which the quarter train is still carried above the strike. The strike locking-plate is especially interesting, as it is mounted on an internally notched plate which must originally have formed part of a flail-locking clock. Three pins on the locking-plate were fitted to let off a chime barrel every four hours. The pendulum bob is bell-shaped, and the older iron wheels behind the dial have triangular teeth.

*Ipplepen: St Andrew's Church*

The unlit third stage of the church tower houses one of Devon's most remarkable working clocks, undocumented but reputedly more than 300 years old. The clock strikes the hours only, has no dial, and is the only one in the county with an oak four-post frame. The corner-posts have chamfered mid-sections and are topped by round finials; the pivot-bars and wheels are of iron except for the second and escape wheels of the going train, which are brass. A long pendulum with a bell-shaped bob and a horizontal link to the pallets is mounted above the movement, and swings back and forth across the front. The clock

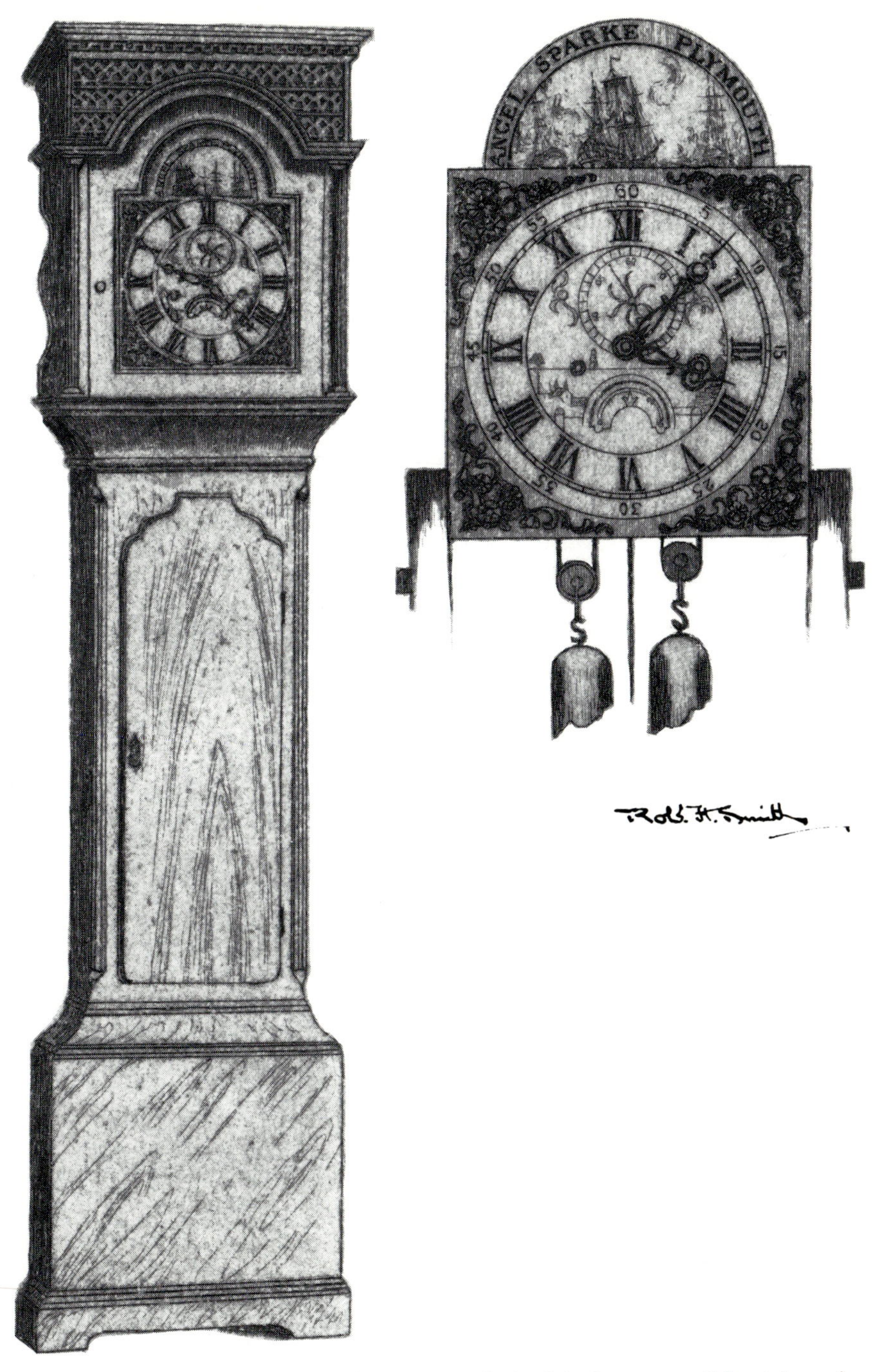


Plates 1 & 2 A typical Devon arch dial longcase clock of the later years of the eighteenth century. The maker was Angel Sparke, Plymouth. The figure on the right shows the dial of the same clock. Etchings by Robt. H. Smith

Plate 3 (*left*) St Mary Steps Church, Exeter. The quarter-jacks of 1621 flanking the central figure of Matthew the Miller on the outside of the tower. They were set up by Matthew Hoppin, an Exeter locksmith who specialised in turret clock-work

Plate 4 (*below left*) 'John Michell of Chardstock fecit'. Hood and dial of eight-day walnut long-case clock, *c*1710, with five-pillar movement, inside count-wheel, small escape wheel and equally small seconds dial. The half-hour markings are of an unusual pattern. An exceptional clock for a village maker

Plate 5 (*below right*) Strike train of the old clock at Exeter Cathedral as shown in an 1854 print when it was converted experimentally for electro magnetic control. The octagonal lantern pinions and capstan-winding bars were subsequently removed. This flail-looking train was probably made for the Peter bell cast in 1484 and was mounted high up in the tower, many feet above the going train – behind the dial in the north transept – to which it was linked by a wire. An early Tudor clock in the chapel at Cotehele House, Cornwall, has a similar vertical wrought-iron frame

has neither capstan bars nor winding squares, but is wound twice a day from a windlass at the foot of the tower using a system of counter-wound ropes. An interesting detail is provided by the 78-tooth iron mainwheel of the going train; this has spaced notches on its inner rim, indicating that it was originally the locking-plate of an earlier clock (cf Ottery St Mary). The Ipplepen clock in its present form probably dates from the period 1680 to 1700.

### *A Vulliamy masterpiece*

The royal clockmaker Benjamin Lewis Vulliamy made the clock in the central tower at the Royal William Victualling Yard, Plymouth. Dated 1831, it carries the number 1100 together with an inscription which describes its maker as clockmaker to His Majesty King William IV. The clock, said to have been composed originally of 1,393 pieces, has three black-painted cast-iron horizontal box frames upon which the working parts are mounted; its total width is just over 2.4m (8ft). The workmanship is superb, the finish being more akin to watchwork than to turret clockmaking. Vulliamy described the clock in 1847 as follows:

> This eight-day clock, which strikes the quarters as well as hours, was made under the immediate direction of Sir John Rennie, and, though somewhat limited for space, is in every respect a very perfect piece of work. The time is shown upon four very large faces, and there are probably few clocks to be found in so exposed a situation. This clock may be said to face the Atlantic, and to brave the storms of the Western Ocean. It stands at the head of Plymouth Sound; and a straight line drawn from the clock across the centre of the breakwater would cross the ocean, and only terminate on the coast of America. All persons conversant with the south coast of this island are aware of the violence of the gales of wind which occasionally visit Plymouth Harbour. Though much confined for room, the general arrangements of this clock are everything that I could wish; it has a two-seconds pendulum, with a bob weighing 258 lbs, yet the weight of the maintaining power, calculated with a single line, is only 26½ lbs. The 'scape-wheel of the clock carries a seconds-hand to show time within, and a degree plate is employed to indicate the arc of vibration of the pendulum; consequently there is not any difficulty in determining exactly the deviations of the clock from mean time. The mode employed for communicating the motion from the clock to the hands is simple and convenient, and has not on any occasion got out of order. This clock, notwithstanding its exposed situation, has kept time so correctly as to

become the standard for regulating time in all Government establishments at Plymouth, Devonport, Stonehouse, and all other public clocks in the neighbourhood.

The clock is believed to have cost around £650 and in its early days was kept in order by Lancaster's, the Devonport clockmakers. It is not only the finest turret clock in the county, but one of the finest in the world.

## CHURCH CLOCKS WITH FOUR-POST ('BIRDCAGE') WROUGHT-IRON FRAMES

**Ashreigney:** Brass plate affixed to frame reads 'Edmund Tout & John Babbage wardens, Pollard Crediton, Fecit 1813'.

**Beer:** Old clock preserved in church; early eighteenth century, iron wheels, modified escapement.

**Berry Pomeroy:** Late eighteenth century, probably by William Pike, Totnes; restored 1911.

**Bishop's Tawton:** Early eighteenth century, but not installed until *c*1826; knob finials on swan-necked stems, believed to be former Barnstaple parish church clock.

**Bradninch:** Disused clock mounted in tower alongside its successor; 1761, by Jno. Pyke, Wiveliscombe (brass plate).

**Brampford Speke:** Eighteenth century, dial originally single handed, but minute hand added in 1881.

**Brixham, St Mary's:** Old clock on display; 1740, by William Stumbels, Totnes (brass plate).

**Broadhembury:** Old clock privately preserved; 1733–4, by David Evans, Exeter (churchwarden's accounts), scroll finials.

**Buckerell:** Early to mid eighteenth century, some iron wheels, dial originally single handed.

**Chudleigh:** Old clock preserved in church; early to mid eighteenth century, re-wheeled by James Pike, Newton Abbot, *c*1756 (churchwarden's accounts).

**Cockington:** 1724 – 'For a new Clocke in ye Church by Agreement

£8, for two Rops for ye Clock 4s 6d' (churchwarden's accounts); no dial.

**Colyton:** Early eighteenth century, possibly by Lewis Pridham, Sandford, 'reconstructed and repaired' by J. J. Hall, Exeter, in 1921, separate iron-wheeled quarter train.

**Combe Martin:** Old clock at Ilfracombe Museum; 1734, by John Cole, Barnstaple (churchwarden's accounts), scroll finials, weights composed of stones in iron-bound wooden buckets.

**Dartmouth, St Saviour's:** Old clock preserved in church; 1705, by Ambrose Hawkins, Exeter (signature on frame).

**Dawlish, St Gregory's:** Old clock *c*1715 at Dawlish Museum; possibly by David Evans, Exeter; scroll finials.

**Denbury:** Eighteenth century with counter-wound rope winding, restored 1948 but now in disuse.

**Drewsteignton:** Old clock replaced 1893 and preserved in church; possibly 1680–1 by Halstaffe, Exeter (churchwarden's accounts), but more probably fifty years later in date.

**Dulverton, Somerset:** 1708, by Lewis Pridham, Sandford (brass plate).

**East Budleigh:** 1732, by Edward Upjohn, Topsham (churchwarden's accounts), rebuilt 1828 by James Blackmore, Sidmouth, now auto-wound.

**Exeter, St Mary Steps:** 1725, by Joseph Robertson, Exeter; three-train, with external automaton figure (Matthew the Miller) and quarter jacks originally *c*1621 by Matthew Hoppin, Exeter; extensively restored 1980 and now auto-wound.

**Exminster:** 1748, by John Tickell, Crediton, and originally at Crediton parish church; installed *c*1840 together with chimes (since removed); modified.

**Gittisham:** Old clock in store at Bristol Museum; 1729–30, by John Legg, Axminster, cost £8 10s originally (churchwarden's accounts).

**Hemyock:** 1747–8 (churchwarden's accounts); modified in 1920s by Thwaites & Reed of Clerkenwell and J. J. Hall, Exeter.

**High Bickington:** Early eighteenth century, possibly by John Cole, Barnstaple; modified, now auto-wound.

**Jacobstowe:** Early eighteenth century, attributed to Lewis Pridham, Sandford.

**Kentisbeare:** Installed in the nineteenth century, but composed of parts from two or more older clocks, two dials, restored 1979.

**Kingsteignton:** 1811, by Henry Pyke.

**Modbury:** Old clock preserved in church; 1705, by Ambrose Hawkins, Exeter (signature on frame).

**Musbury:** 1729, by Richard Bannister, Axminster; unusual strike train with two barrels, restored 1979.

**Newton St Cyres:** 1711, by Lewis Pridham, Sandford (brass plate); cost £16 with dial (churchwarden's accounts), restored 1971, now auto-wound.

**North Molton:** Old clock displayed in church as working exhibit; early to mid eighteenth century, scroll finials.

**North Tawton:** Early eighteenth century, possibly by Lewis Pridham, Sandford; rebuilt 1899 by C. H. Cornish, Plymouth.

**Pilton, Barnstaple:** Reputedly 1713, by Richard Webber, Pilton; re-wheeled with new pinwheel escapement by North Devon horologist Inkerman Rogers, 1925 (name and date engraved on wheels); no dial, clock bell in pinnacle on tower.

**Plymtree:** 1792–3, by William Upjohn (ii), Exeter (brass plate), modified, cost £27 6s originally (churchwarden's accounts).

**Sandford:** Formerly at Kenton, erected by J. J. Hall, Exeter, as war memorial in 1920; similar to Kingsteignton parish church clock; early 1800s.

**South Brent:** Old clock preserved privately; eighteenth century, no dial.

**South Tawton:** Old clock in Exeter Museum collection; early eighteenth century, 'new made' in 1741, frame stamped with name 'Markes'.

**Stoke Canon:** Eighteenth century, modified, dates '1731' and '1732' inscribed on clock-house lintel, restored 1979.

**Talaton:** Old clock abandoned in clock room in 1925; mid-eighteenth century with passing quarter strike.

**Thorverton:** Three-train, by John Tickell, Crediton, 1751 (brass plate), outer dial added in 1903; separate chime barrel, *c*1700 or earlier.

**Tiverton, St George's:** Old clock now in Tiverton Museum; 1737, possibly by Simon Thorne, Tiverton; hands also preserved.

**Torrington, Great:** Early eighteenth century, modified, frame badly distorted, knob finials on curved stems; many similarities with the Bishop's Tawton clock, probably by the same maker.

**Ugborough:** 1780–1 – 'To Mr Pike in Exchange of the old clock for a new one £21' (churchwarden's accounts); probably William Pike of Totnes, as the clock was taken there for attention in 1783–4.

**Upottery:** 1794, by Adam Cleak, Bridport, the gift of Henry Addington, later Viscount Sidmouth (brass plate), modified with gravity escapement and additional Westminster quarters, now auto-wound.

**West Down:** 1714–16, reputedly the gift of Sir Nicholas Hooper, diamond-shaped dial.

The church clock at Loddiswell is similar to clocks of this group, but has a wrought-iron frame of plate and spacer type. It is stamped with the name 'Wills' – possibly one of the Truro makers. A vertically framed early eighteenth-century quarter clock, similar to the one at Colyton, survives in disuse at Otterton.

### CHURCH CLOCKS WITH FOUR-POST ('BIRDCAGE') CAST-IRON FRAMES

**Ashprington:** 1840, by Wasbrough Hale & Co, Bristol, the gift of Richard Durant; two-second pendulum.

**Bishopsteignton:** 1850, by W. Harner, Colyford.

**Bradninch:** 1852, by John Moore & Sons, Clerkenwell; two-second pendulum; formerly at St David's 'Pepperpot' Church, Exeter (demolished and replaced by present church).

**Broadclyst:** *c*1830, by T. & J. Pollard, Exeter (signed but not dated).

**Clyst St Mary:** 1843, by John Moore & Sons, Clerkenwell, 'given by Mr Golsworthy'.

**Combeinteignhead:** 1852, by T. C. Pollard, manufacturer, London; provided under the will of Robert Crowther and installed at a cost of about £100.

**Cornworthy:** 1796, by William Dorrell, Bridge Water Square, London (inscription).

**Crediton:** Three-train, 1838, by Thwaites & Reed, Clerkenwell; cost £150 (governors' accounts), separate cast-iron framed chime barrel.

**Dalwood:** Old clock in tower, *c*1800, inscribed 'Murch Honiton'.

**Dunsford:** 1830, by T. & J. Pollard, Exeter.

**Exeter, Holy Trinity:** 1820, by John Upjohn, Exeter; installed originally at a cost of £125, removed to Exeter Museum in 1976.

**Newton Ferrers:** 1847, by John Moore & Sons, Clerkenwell; two-second pendulum.

**Plymouth, Stoke Damerel:** 1811, by John Pollard, Plymouth Dock; 2.4m (8ft) blue dial with gilt lettering.

**Shillingford St George:** 1812–13, by T. & J. Pollard, Exeter; restored 1980.

**Sowton:** 1846, by Matthew Murch, Honiton.

**Talaton:** 1817, by T. & J. Pollard, Exeter; formerly at Haldon House, installed secondhand by J. J. Hall, Exeter, in 1925; brass urn finials.

**Torquay, St Saviour's, Torre:** 1852, by John Moore & Sons, Clerkenwell.

**Uffculme:** 1848, by John Moore & Sons, Clerkenwell; two-second pendulum.

**Uplyme:** 1846, by Matthew Murch, Honiton; modified.

**Yealmpton:** 1851, by John Moore & Sons, Clerkenwell.

The church clock at Sampford Courtenay, made by Biddle of Clerkenwell, has a cast-iron plate and spacer frame, as has the clock at Colebrooke, inscribed 'Fairer, Maker to the Queen, Anno 1861, London, No. 1227'. The clock at Zeal Monachorum is unusual for Devon; it was made in the nineteenth century by Whitehurst of Derby, and has extended barrels. Smith's of Derby installed it secondhand in 1912.

CHURCH CLOCKS WITH CAST-IRON HORIZONTAL (FLATBED) FRAMES

**Honiton, St Paul's:** 1851, by Matthew Murch, Honiton; three-train.

**Tavistock:** 1849, by E. J. Dent, London; three-train, with four dials and separate chime barrel. The Duke of Bedford, then owner of the Tavistock Abbey estate, was a large contributor towards the original cost of nearly £500. In a letter from London, dated 24 March 1849, to Arthur H. D. Acland Esq, Dent wrote: 'I hope soon to be ready to put up the large clock at Tavistock, then I can attend to business and enhale the pure Devonshire air' (Devon Record Office).

**Woodbury:** 1846, by E. J. Dent, London; three-train with compensation pendulum; installed at a cost of £125.

Many Devon tower clocks date from after 1860. These are mostly of flatbed type and were made by specialist firms, although some carry the names of local clockmakers through whom, perhaps, the original orders for manufacture were placed. The standard of workmanship is invariably excellent. The church clock of 1868 at Bovey Tracey has an inscription: 'J. W. Benson, Ludgate Hill, London, Clockmaker to HRH the Prince of Wales'. Other church clocks by Benson include Payhembury (1896), Sidbury (1905) and those at Awliscombe, Bishop's Nympton, Clyst Honiton, Hatherleigh, Stoke Fleming and Withycombe Raleigh, Exmouth. The Derby firm of J. Smith & Sons is well represented in Devon towers, with clocks dating from the late Victorian period to the present day. Smith's made the 1889 quarter-striking clock at St Michael's, Teignmouth, and the Kingswear clock of 1897. Further examples are at Bampton, Cheriton Fitzpaine, Chudleigh, Combe Martin, Cullompton, Dartmouth, Exbourne,

Huntsham, Milton Abbot, Morchard Bishop, South Molton, Stockland and Stoke Gabriel.

Gillett & Bland, of Croydon, installed clocks with an unusual fifteen-legged gravity escapement at Ottery St Mary (1874–5) and Black Torrington. This firm (by then trading as Gillett & Co) won the contract for the 1885 clock at Exeter Cathedral which, like its medieval predecessor, is in two parts, with the strike train high up in the north tower and the going and quarter trains behind the dial in the transept below. There are at least twenty other Gillett clocks in Devon towers (the later ones signed Gillett & Johnston). They include St Peter's, Tiverton, and Lympstone, both 1883; Axminster (1887); St Peter's, Barnstaple (1913); Babbacombe, Torquay, Brixton, Coldridge, Collaton St Mary, Harberton, Ilsington, Kenn, Kenton, Paignton, St Marychurch, Torquay, Willand; and Widecombe in the Moor (1979), one of the last gravity escapement clocks to be set up in Britain. The clocks at Bideford (1878) and St Giles in the Wood (1879), although clearly made by Gillett, bear the name of the local supplier, R. Squire & Son, Bideford.

An 1864 clock at St Paul's, Tiverton, has a four-legged gravity escapement, as has the timepiece at Tiverton Town Hall (1868). Both were made by J. B. Joyce of Whitchurch, Shropshire. This firm also made the church clocks at Holsworthy (1869), St Mary's, Brixham (1929), and the town hall clock at Kingsbridge (1875). Another North Country manufacturer, William Potts & Sons of Leeds, is represented at Poltimore (1868), Otterton (1891), and Petrockstowe (1919). In north Devon, John Gaydon of High Street, Barnstaple, installed a number of clocks, as did Charles Ford, also of Barnstaple (see List of Makers). The clock of *c*1891 at Chulmleigh was set up by the village's clockmaker, William Passmore.

J. & H. Jump of London made the church clocks at Revelstoke (1882) and Holbeton. F. Templer Depree of Exeter, and the firm's successors, are represented at Heavitree, Exeter (1910), Stokeinteignhead (1907), South Zeal and Upton Pyne. W. F. Evans of Birmingham made the clock at Rockbeare, and probably the one at Topsham. At Combe Raleigh (1877) the maker was G. Wadham of Bath; at Winkleigh (1872), Edward Funnell of Brighton; at Starcross (1866) and Halberton, Henry Weight of Malmesbury; and at South Tawton (1893), Leeson & Son of Coleshill, Birmingham. C. Price of Clerkenwell made the clock at Northam, and Thwaites & Reed, also of

Clerkenwell, the one at Alphington near Exeter. Finally, in east Devon, there is the clock at Sidmouth parish church, dated 1876; it is in the style of Benson but carries an inscription, 'Chas Frodsham & Co, Clockmakers to the Queen, No 1560, 84 Strand, London'.

The tower at Torre Abbey, Torquay, houses an exceptionally good, early eighteenth-century clock with scroll-top corner-posts and iron wheels. Various modifications were made to it by W. Carleton of Torquay in 1875; the name of the original maker has not been recorded. James Pike, of Newton Abbot, made the castle clock at Powderham in 1773. Other makers of country-house turret clocks included Lewis Pridham of Sandford, William Stumbels of Aveton Gifford and Totnes, and Richard Hillson of Plympton. Daniel Ross, of Exeter, supplied the 1842 clock at Old Blundell's School, Tiverton. The stables clock at Saltram House was made in 1762 by William Smith of Upper Moorfields, London; another London clock, by Grignion & Son of Covent Garden, dated 1776, is at the Royal Naval Hospital, Stonehouse, Plymouth; and a third, 1818, by Dutton & Sons, is at Plymouth Custom House. Gillett & Johnston, of Croydon, made a clock for the Britannia Royal Naval College, Dartmouth; it strikes ships-bells, including the dog-watches.

## FINANCE AND FEATURES

### *Time and money*

Church rates and subscription lists were the usual ways of financing parish clocks except where they were provided as gifts by persons living, or under the terms of a will. A small contribution 'received from certain parishioners for the making of the clock' was recorded in the 1537–8 churchwardens' accounts of St John's, Exeter. Later, when the city was under siege in the Prayer Book Rebellion of 1549, the St John's parishioners used money received for an item of church plate 'for the relief of the poor . . . and for mendying of ye clock'. At Drewsteignton in 1680–1, the parish officers ordered that rent received for the parish lands should go towards 'ye Makeing & Setting up of the Clock, & for Halstaff ye ClockMaker & his Men, their Diett & Lodging'. During Sir Jonathan Trelawny's time as bishop of Exeter (1689–1707), some thirty-five leading parishioners of St Petrock's, Exeter, signed a petition seeking his consent for them to levy a parish rate to pay for a new clock, as their old one was defective. A subscription list at Dawlish in 1715 raised 35s towards the £12 needed for a clock. Heading the list of 'The names of those which gives towardes the Church Clock' was the vicar, Mr Humphry Harvey, with 10s; the other twenty-five contributors averaged a shilling each. At Kingswear in 1767 the minister, churchwardens and principal inhabitants of the parish petitioned successfully to the bishop for a faculty to sell a bell to pay for their new clock. In the nineteenth century, with the ending of church rates, the power of churchwardens declined and new clocks were provided either as gifts or by public subscription. A great many were set up throughout the country to commemorate Queen Victoria's jubilees of 1887 and 1897, and to a lesser degree after World War I as memorials to the fallen. Among the last Devon church clocks to be partly paid for with a church rate was one set up at St Sidwell's, Exeter, in 1837.

### *Dials*

The provision of a gilded dial, motion work and 'rodding' could in some instances double the cost of installing a clock, and was not always considered necessary. Indeed, there is a revealing letter written in 1914 by the Thorverton village clockmaker, James Richard Cummings, to the incumbent at Rewe nearby concerning the church clock:

> It's difficult to advise you for the best. You must abandon the idea of a dial. Of course, it can be done, but you'll find it a failure, as the old clock would never control a pair of hands & dial-work in a storm. It was never constructed to do so . . . If a new clock is under consideration I'd aim at one striking quarters, rather than a dial – I always find it more serviceable, except in a thickly populated place where illuminated dials can be fixed.

In old accounts, dials were sometimes referred to as 'dyall tables' and were constructed from wooden planks set in a frame. The diamond-shaped dial as at West Down in north Devon was common, and so too were octagonal and hexagonal ones; but by 1800 these had been superseded by the round, slightly convex, metal dial. At Exeter an illuminated dial, perhaps the first of its kind in Devon, was installed in 1835 at a school fronting the High Street (see Chapter 5); and later in the century the gilt, skeleton dial became popular. Although church clocks with three or even four dials were not unusual in the Victorian era, two was about the maximum in the eighteenth century. The great majority of clocks had just one dial and one hand.

The church clock at Buckland-in-the-Moor has a unique dial with the chapters marked in Gothic lettering and spelling out 'My Dear Mother':

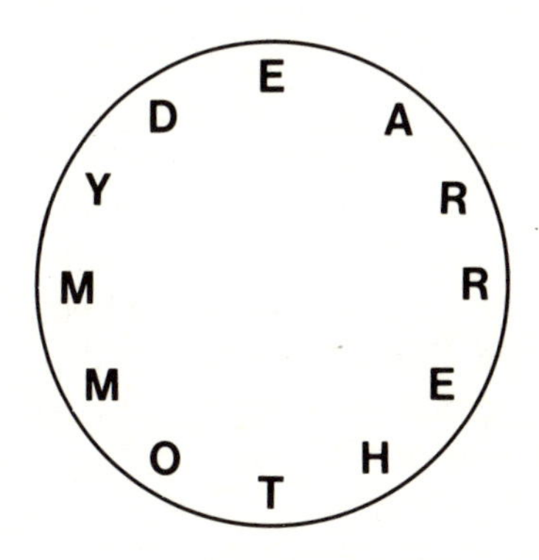


It and three bells were presented in 1930 by William Whitley, of Welstor near Ashburton, in memory of his mother. He owned property in Buckland parish, and also caused the Ten Commandments to be cut into the granite face on the top of Buckland Beacon.

### *Chime barrels and tunes*

Chime barrels, some nearly three hundred years old, survive in a number of Devon churches. They operate in the manner of a musical box, and by means of wires linked to hammers which play simple tunes on the bells. Some are hand-cranked, as at Branscombe and Colyton, but the larger ones are weight driven, being released at three- or four-hourly intervals by pins on the hour locking-plate of the clock. The tower at Otterton in east Devon houses a long-abandoned, 'fieldgate'-framed chime barrel, dating perhaps from *c*1700. Another, standing on end and minus its frame, was photographed by the author in St Michael's churchyard, Honiton; and may have been brought there from the demolished Allhallows chapel in the town where, prior to 1835, the chimes played the 149th Psalm and 'Britons, Strike Home'. An old chime barrel of *c*1700 or earlier at Thorverton, restored once again to working order, plays the hymn tune 'York'. A later example, made in 1767 by Thomas Bilbie, the Cullompton bellfounder, is at Beaminster parish church in Dorset.

There are references to chimes in old accounts at St Petrock's,

Exeter, in 1470–1 and at Ashburton in 1536–7, as noted in Chapter I, at Tavistock in 1574–5 (a new set) and at Chudleigh *c*1582. Philip Wyot, Barnstaple's town clerk from 1586 to 1608, was less than enthusiastic about the chimes at his parish church. In 1593 he recorded in his journal: 'The Chaims now going, w[ch] coste besides the bell that was had before £25 – a great charge to small effect.'

In 1727 at Axminster in east Devon, Joseph Elford, a clockmaker there, agreed with the parish to make a new clock with quarters and chimes for £22, 'ye Chimes to goe to the Tune of ye One hundred & Fiftyeth Psalm & Britons strike home Six times round every time it goes which Chimes are to play Every Four houres'. After being silent for nineteen years, the chimes were repaired in 1822 and set to the tune of 'God Save the King'. This they played but imperfectly, as there were not bells enough! At Exeter an eighteenth-century writer noted that 'at 8, 12, and 4 o'Clock, St Petrock's chimes play Sternhold's queer old tune of the Fourth Psalm'. At Tavistock the present chimes play a different tune each weekday (at 9am, 12 noon, 6pm and 9pm). On Mondays the tune is 'Auld Lang Syne', on Tuesdays 'The Minstrel Boy', on Wednesdays 'The Blue Bells of Scotland', on Thursdays 'Ye Watchers and Ye Holy Ones', on Fridays 'The Church's One Foundation', and on Saturdays 'Home Sweet Home'.

The chimes at Holsworthy include a tune written for them by Samuel Sebastian Wesley, and those at All Saints Church, Brixham (1929), play three of the Reverend H. F. Lyte's hymns; he was the parish's first incumbent and wrote 'Abide With Me'. Sir George Grove in his *Dictionary of Music* relates that John Davy (1765–1824), the composer of 'The Bay of Biscay' and other songs who was born at Upton Helions in Devon, took some horseshoes from a smithy when he was aged about six and amused himself by imitating upon them the chimes of the neighbouring church of Crediton. The tower at Crediton still houses a massive, cast-iron framed chime barrel, contemporary perhaps with its 1838 church clock; the tunes are 'Duke Street' and 'The Sicilian Mariners' Hymn' (known locally as the 'Honiton Psalm'). In Davy's day the tunes may have been different.

## CLOCK TOWERS

Devon's Victorian and Edwardian clock towers recall a period of vanished opulence. The larger ones have four illuminated dials and sound the hours and sometimes the quarters. They are topped by weather vanes and some of them were originally provided with ornamental lamps at each corner, drinking fountains and, as at Exeter, drinking troughs for cattle and horses. Most of the towers are four-sided, but some of the later ones, such as those at Torquay and Tiverton, have three sides and three dials.

Interior space is generally at a premium; and weights suspended overhead can pose a hazard. At Seaton the movement is mounted at ground level with a long rod taking the drive to the dials. In contrast the movement – by F. T. Depree – at Exeter Clock Tower is on a level with the dials and it is necessary to climb several iron ladders and platforms to reach it. In some towers, iron rungs and ladders set vertically against the inside walls were considered sufficient to meet the clock-winder's needs. The movement at Tiverton – by Smith of Derby – is mounted half-way up the tower, with a separate quarter-striking train placed immediately below it because of lack of space.

Barnstaple's Clock Tower (1862) is a memorial to Prince Albert; those at North Tawton and Seaton (1887) and Exmouth (1897), were set up to commemorate respectively the fiftieth and sixtieth years of the reign of Queen Victoria. The approximate dates of the others are Plymouth (1862), Lympstone (1885), Exeter (1898), Torquay (1903), and Tiverton (1908). The old St Leonard's tower in Newton Abbot houses a Golden Jubilee clock of 1887, made by Gillett of Croydon.

## FIRE IN THE CLOCKROOM

For centuries, fire was the scourge of Devon's thatched towns and villages and, on one occasion, careless workmen in the clockroom were nearly responsible for the destruction of the beautiful church tower at Cullompton. A note in one of the parish registers records the incident:

> On Wednesday the Eighth day of December Anno Dom. 1725, there happened a sudden (& was likely to be a very dreadfull Fire in the Clock chamber of the Tower of this Town, which was discovered about Eight a Clock in the evening, through the Window of the Belfrey, the Town was alaurmed, the wind being very high, & evening very light, the people were amazed, but after the Keys of the Church were brought, some few entered just as the fire began to Flame, & found the Planching on Fire, a bourd or two thereof, being three inch Plank, was burnt thro', & one of the Beams burnt down two or three inches deep. So that had it not been

> discover'd in that instant of time, the Tower, Church & whole Town would in all Probability have been burnt down, but, God be thanked, it was happily discover'd & extinguish'd without doing any more hurt, then burning the Planks & Beam as abovesd. It was Occasioned by the Carelessness of a Carpenter & Locksmiths leaving fire unextinguished in an Iron pan, cover'd with a board, in the sd Room, some time the Same day; tho' they tho't it had been clean extinguished, having Piss'd in the sd Fire to extinguish it.

On the subject of fire, it is worth noting that clockmakers sometimes resorted to 'burning the clock'. The Braunton church-wardens' accounts for 1689–90 show a payment of one shilling 'to John May for Wood for the Clocke to burn off the oyle and dirt thereof'. Payment of a similar amount was recorded at Gittisham in east Devon in 1725–6 for 'Boiling ye Clock and Brasses', a further entry indicating that the clock was first taken apart.

# 3

# NOTABLE MAKERS: EXETER

## MATTHEW HOPPIN

Matthew Hoppin made the original clock in the tower of St Mary Steps Church, Exeter, with its automata of Matthew the Miller and his two sons (see Chapter 2). He kept at least three other church clocks in repair and also the town clock at Totnes, more than twenty-two miles away to the south-west. Details of an agreement entered into by Hoppin are recorded in the earliest extant parish register of Holy Trinity, Exeter:

> Memoranda the 27 of Februarie 1607[8]. Mathewe Hopping in consideration hee hathe mended the clocke of the Holie Trinitie in Exon & hath received 6s for his labour. and the sayd Mathewe Hopping doth promisse covenant & agree with those of the sayd parish whose names are here under subscribed to keepe the sayd clocke & mend the faults for two shillings a yeare as long as he liveth to bee payd everie halfe yeare 12d. and to beginne their pay at our ladie day next being the 25 of March. 1608. Provided alwayes unlesse the clocke take any hurt by extraordinarie meanes.

The signatures of several parishioners are appended, some signing with their mark, as did 'Mathewe Hoppinge' whose symbol – a circle with a stroke through it, resembling the earth on its axis – was singularly appropriate for a man specialising in time measurement.

Hoppin, the son of a St Sidwell's, Exeter, locksmith, married Jone (Joan) Pere on 6 October 1608. They lived in the parish of St Mary Major and it was there that he was buried, near the west front of Exeter Cathedral, on 28 October 1625. A bond in £100 to administer his estate was entered into by his widow and by William Hoppin, who was described as a blacksmith.

WILLIAM HOPPIN

Matthew Hoppin was succeeded by 'his brother' William who had served his apprenticeship with him as a locksmith. William, the son of Charles Hoppin, had been baptised at St Sidwell's on 11 February 1603 and was admitted as an Exeter freeman in 1627. Among the Totnes borough archives is an agreement that he made concerning the town clock and chimes.

William Hoppin Clockmaker his Covenant to repaire the Cheames; and to have yearly at Mich – £2/ 1627.
– : It is agreed between the town of Totnes and Wm. Hoppin of the Citty of Exon clockmaker, that hee the sayd Wm. Hoppin shall at all tymes needfull upon warning given him, repayre and mayntayn of the clock-watch and cheames of the sayd town, as in a deed wrytten and made between the sayd town and Mathew Hoppin of the sayd Citty deceased appeareth. And in consideration hereof the sayd town is to pay him yearly, as formerly agreed on with the sayd deceased Mathew Hoppin and as in the forsayd wryttin and appeareth. And the tyme of his yearly paye to begin at michellmas next anno 1628. In wittnes our hands the 27th day of January 1627 –

Wittnes Barnard
Hatch
Willam Hoppin

the 27 Jan 1627 was paid him by Mr. Lea: Churchwarden for new mending the cheames – £2.

The Hoppin business prospered under William. He was a leading parishioner in St Mary Arches, serving as churchwarden in 1637 and as a constable of the city. In 1641–2 he repaired the clock in the now ruined church of St Edmund's, on the medieval Exe bridge, and he also looked after the clocks of St Mary Steps and St Kerrian's – a church which formerly stood in North Street, Exeter. The church-wardens' accounts of the latter have this item:

1637: £1 9s 10d paid unto Hoppinge the Lockier for makinge of newe wheeles for the clocke and mending the rest and for brasse wyer . . . And of 1s paide for the paire of newe jemies [hinges] for the clocke doore.

In the same year, 1637, he entered into an agreement with the parish of Chudleigh, and the church clock and chimes were repaired in his workshop in Exeter (Chudleigh churchwardens' account book).

Turret clocks were just one aspect of Hoppin's trade. A manuscript diary of the household expenses of a fellow parishioner, John Hayne, son of an Exeter serge merchant, has been preserved; it records many transactions with William Hoppin and provides a unique insight into the great variety of work undertaken by the locksmiths of the period. Some examples of payments made to Hoppin by John Hayne are:

> 1635, July 30: More 20d paid Hoppin for making cleane my musket twice, & 6d given his boyes
> 1638, October 9: Paid Mr. Hoppin 18d viz: for an iron plate to mend ye frame of ye greate spruse chest 4d, for mending ye Mousetrappe 2d/for a little iron candlestick to keepe a watch candle in all ye night 12d
> 1639, June 20: More 3s 6d I paid Mr. Hoppin for a Dutch lock and keye for my new presse, and 6d for exchanging a new hamer for an old hamer yt I bought of him formerly, in all 4s.
> 1640, February 5: For a latch (of my Kitchin dore wch he formerly made) . . . with a little invention to open it 1s.

Among other items for which Hayne paid Hoppin were 'a little paire of iron dogges for a chamber chimney 5s', '2 new pans to my fireshovells', '3 Curtins roddes &c to ye grt Bedsteede 4s 8d', numerous metal fittings for furniture, and, surprisingly, '3 doz. of London candles 15s' and '½ a seame of hardwood . . . his carpenters to make pins'. Hayne seems to have acted as a banker to Hoppin, advancing him sums of money in varying amounts from £2 to £14, which Hoppin usually repaid within a month or so. In 1640 the locksmith provided Hayne with a new winding jack at a cost of £1, an old one which was Hayne's father's being taken in part exchange. There was also a payment for 'another chaine for ye jacke of 4 yards ½ longe'. These jacks resembled a weight-driven, vertically framed clock. They were mounted above the fireplace, to one side, and were linked to spits on which joints of meat and poultry were roasted across the front of the fire. Their action ensured that these joints were turned evenly. They were made both by smiths and by clockmakers. An early eighteenth century West Country example has an engraved brass front plate signed by the St Austell clockmaker Thomas Wills; another, by 'Pollard Crediton', can be seen in the kitchen of the Georgian House museum in Bristol.

William Hoppin married Jane Woolcome at St Mary Major parish church in 1626 and re-married after her death in 1639, but then suffered a double bereavement when his second wife Joane and their daughter Agnes died in 1641 and were buried together. Hoppin married for the third time; and his life was caught up in the momentous events leading to the Civil War. In 1638–9 he paid 3s Ship Rate, that most unpopular of taxes, and in 1641–2 contributed 18d towards a rate for new glazing the church windows in St Mary Arches after the old painted glass had been taken down by order of Parliament. He died in 1643, less than two months after Exeter was yielded up to the king's army after a siege, and was buried at St Mary Arches on 2 November.

In his will, Hoppin, whose occupation was given as clockmaker, left 40s to Mr Ferdinando Nicholls, the rector of St Mary Arches, bequests of 20s to the poor of five parishes, £200 to his son William when he was 21, family legacies, and to his wife Alice a tenement and 'all my right in my two houses in Mary Arches Lane and my garden and all my right in the house I now live in and my garden Fryerinhay' and the residue. He also left to Joseph Branscombe, who was a locksmith and one of his apprentices, 'my great buke horne and my vice in the new shop'. The will was dated 24 October 1643 and the uncertainty of the times was reflected by the proviso, 'The several sums of money I have bequeathed to my kinsfolk to be paid in half a year after my death unless great oppression and violence hinder my executor'.

Three of Hoppin's apprentices became freemen of Exeter – Branscombe, Peter Halstaffe the elder, and John Peryam, gunsmith. Later members of the Hoppin family worked as locksmiths in Exeter until well into the eighteenth century.

## JOHN SAVAGE

John Savage (Savidge, Savadge etc) was the first clockmaker so described to be admitted as a freeman of Exeter – on 14 August 1609 on payment of a fee of £1 6s 8d. Like many of the early clockmakers, he seems to have come from a metal-working background and was probably the son or grandson of an Exeter bellfounder named John Savage, who can be traced in Holy Trinity parish in 1555 and was buried there in September 1570. The published Subsidy of 1602 records a John Savidge in St Stephen's and this was almost certainly the clockmaker, for at the time of his death in 1627 he was living in a house in the High Street, by St Stephen's Church and adjoining the New Inn. Savage carried out various repairs to the Exeter Cathedral clock and also installed a set of chimes to play out on the main bells in the south tower. Several decisions referring to this work are recorded in the Act Books of the Dean and Chapter:

> 6 June 1612. They [the dean and chapter] decreed that a chyme be set upp in the said [south] tower by John Savadge, according to an agreement already made betweene Mr Treasurer, Mr Archdeacon Helyar, and Mr Bodly in the behalf of the Chapter and the said John Savadge.
> 24 September 1614. They decreed that John Savadge should amend such things as are amiss about the wheele of the clocke.
> 4 February 1614/15. They decreed that Mr Savadge doe mende the clocke in such reparacions as shalbe necessarye, and that he be payed for his paines therein nowe to be taken, as also at dyvers other tymes heretofore, according as he shall demand in reason.

Five years later he made what would have been a large and expensive turret clock for the Guildhall in Plymouth. Its purchase was recorded in the municipal records: '1620–21. Itm. pd Mr John Savage of Exon for the dyalls and Clocke set up in the Guildhall £32.' The setting-up and materials cost £12 5s 5d more.

On 9 August 1627 when 'weake and sicke in body but of good remembrance', Savage dictated his will and must have died soon afterwards, for it was proved in London the next month. He was a tradesman of substance and held several properties on lease. The value of his estate was put at £326 7s 6d and that of his wife 'which she had of her father' at £55 19s 2d. He is said to have had a son John who was apprenticed in London in 1613 in the Blacksmiths' Company, but this John is not mentioned in the will. The clockmaker, whose surname is given as Savidge on the document itself, bequeathed to his son Thomas, 'my house next the New Inn wherein I now dwell . . . the featherbedd, bolster, pillowe, trucklebedstedd, blankets and covering in the hall chamber on which I nowe lye . . . my third brasse pann and two silver spoones with his owne name'. To his son Peter he bequeathed, 'the use of the shop . . . two houses in the parish of Allhallows on the Walls . . . all my working tools and implements belonging to my trade . . . my best brasse pann and two silver spoones with lyons heads . . . a featherbedd, bolster, pillowe, trucklebedstedd, blankets, and a green rugg in the chamber over the hall'. To his son William he gave, 'my close of land unto Northenhaye . . . a featherbedd, bolster, pillowe, blankets, and a red rugg . . . my second brasse pann . . . one spoone of silver and guilt and one other of silver with his owne name . . . a debt of £8 due to me from George Gill, instrument maker . . . a stone jugg topt with silver'.

In addition, he left legacies of money, furnishings and silver to his daughters Margaret and Mary, £10 to his wife Alice and 'all such goods and householdstuffe which she brought with her when I marryed her', 40s to his sister Avis, 20s to his servant Richard Johns, and 20s to the poor of St Stephen's. His house, which he leased from the Dean and Chapter, consisted of a working room, two well-furnished hall rooms, one above the other, and a cellar. The location next to the ancient and inappropriately named New Inn – a long since vanished hostelry at which many distinguished visitors to Exeter stayed – was in one of the busiest parts of the city and well placed for trade. The accounts of the executors of Savage's will include a payment of £2 5s to a Mr Bridgeman 'for 2 years & qr rent for a forge'. Two of John Savage's sons, Peter and Thomas, became clockmakers, and his daughter Margaret married 'his servant' Richard Johns, who was in fact a clockmaker (see List of Makers).

## PETER SAVAGE

Peter Savage was admitted as a freeman of Exeter by succession on 15 September 1628. In documents he is described both as clockmaker and watchmaker. There are several references to work undertaken by him in the manuscript diary of the household expenses of John Hayne the younger, of St Mary Arches parish, Exeter (Devon Record Office). One of the entries relates to the repair of a country-house clock:

> 1635, November 28: More 5s 6d paid Peter Savidge, viz: for mending and cleansing ye Clock yt stands in ye Hall at Leigh. 3s 6d/ & for 8 yards of corde for ye stringes. 2s/ The whole being 5s 6d.

Leigh House at Winsham in Somerset was the seat of Henry Henly, whose third daughter, Susannah, was Hayne's wife. A further payment shows that the clock was brought by carrier to Exeter for repair:

> 1635, December 2: More 12d given Butcher for bringing ye clocke, and carying it back to Leigh.

Two other payments by John Hayne to Peter Savage are of unusual interest in that they concern the repair of watches. The first refers to a watch belonging to John Henly, of Colmer:

> 1635, March 12: More 4s paid Peter Savage, for a keye to his watch, a new string, & mending it.

The second payment is recorded the next year:

> 1636, December 3: Paid Mr. Savidge for a newe stringe to my watch 1s 6d.

The string, or stringe, must have been more than a mere cord on which to hang a watch. At 1s 6d it was a fairly costly item, compared say with the 12d charged for bringing and returning a clock to Somerset, and was almost certainly the gut-line that was wound between the fusee and barrel of watches of that period.

Peter Savage was a leading parishioner of St Stephen's, serving as churchwarden in 1632. He added to the family's holdings of leasehold property and his two sons, John and Abraham, continued the clockmaking tradition into the third generation. He died in 1657 and his will, proved in London on 27 May that year, is a revealing document. In it, he shows great tenderness towards his wife Susanna and takes pains to ensure that she receives an annuity of £20. He was worried lest his sons should 'turne profuse Careles and Idle, and in a wicked ungodly ranting waye of life . . . consume that idly and loosely which by Gods blessing I have carefully gotten', and should they so do, in the judgment of his overseers or two honest ministers, his bequests to them were to be 'utterly voyd and null'. He left his sons leasehold property in the Exeter parishes of St Lawrence, St Stephen, Holy Trinity and St Mary Major, and at Crediton. There were bequests of furniture, silver, brass and pewter, and John received 'my little vice I did usually worke with wch was my Fathers, and halfe my working tooles of my trade'; and Abraham received 'my bigger vice and half my working tooles'. He desired that there be no discord between them and that they should live 'in the feare of God lovingly and sweetly together as brothers ought to do'. He asked them to be 'dutifull loveing & carefull' towards their mother, and his wish was to be buried near the north door of Exeter Cathedral 'near my Father and mothers graves'.

The Savage family continued as clock and watchmakers in Exeter for another thirty-five years or so after Peter's death. Four of them can be traced in the city around the year 1670: his sons John and Abraham; Thomas, then nearing the end of his career, and his son Thomas. The family worked from premises near the High School in St Lawrence parish and from their shop adjoining St Stephen's Bow in High Street, the lease of which was finally surrendered in 1687 (see List of Makers).

## ROBERT GRINKING

Robert Grinking (Grincking, Grinkinge etc) kept the Exeter Cathedral clock in repair during the Commonwealth period. An order confirming his appointment is to be found in the Act Book of the mayor and chamber of Exeter, under date 19 June 1649: 'Mr Receiver doe see the clock in Peter's Church [the Cathedral] repaired and set in order, and pay Grincking the watchmaker 30s for doing it, and he is to have 10s yearly for keeping it'. This craftsman was resident in Exeter and is not to be confused with the famous Robert Grinkin, of London, who was master of the Clockmakers' Company in 1648 and 1654.

The marriage of the Exeter clockmaker to Elizabeth Toogood took place at Holy Trinity Church on 17 September 1632, and five of their children were baptised there between 1633 and 1644, only one of whom, John, seems to have survived infancy. Robert Grinking himself was very much caught up in the turbulence of the times, getting involved in brawls and uttering oaths, and his name flits through the records of Exeter Quarter Sessions. In 1642 his mother-in-law, Ann, swore the peace against him and he was bound over to appear at the next sessions; he was 'bound' to appear yet again in 1644 when he was accused of 'speaking of traytorous words' touching the king; and on 15 March 1651/2 'Thomas Cheeke of this Citty deposeth that on Friday last hee heard Robert Grincking of the same Citty Clockmaker to sweare eight several oaths by the name of God'. The outcome of these brushes with the law is not clear, but Grinking and his wife can be traced in 1660 in St Lawrence parish (Poll Tax).

Grinking kept Exeter church clocks in repair by the year and in 1650–1 helped to mend the St John's Church clock. In 1655 he and the locksmith Thomas Poling (Fig 1) agreed with the parishioners of Holy Trinity to frame, erect, and 'sett goinge' some chimes on the four bells. The two men also carried out various repairs to the St Mary Steps clock in the 1660s. The St Mary Steps churchwardens' accounts contain this intriguing entry for 1660–1:

> Item payde to Grinking for keeping the Clock this year 6s;
> Item payde to Grinking for Mending the Clock before hee went to London besides his years pay 1s.

Another clockmaker, Peter Halstaffe, is mentioned in the 1661–2 accounts, and Grinking again in 1662–3.

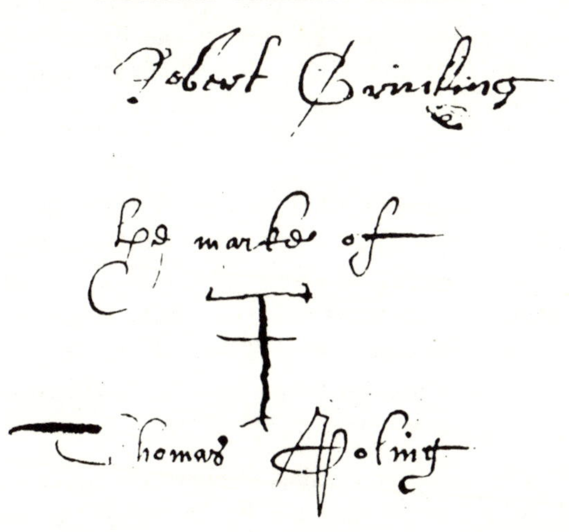

Fig 1 The signature of Robert Grinking and the mark of Thomas Poling in 1655

## AMBROSE HAWKINS

On the evidence of his superb longcase clocks, Ambrose Hawkins deserves to rank among the top handful of English provincial makers. It is not known where he learned his trade, but before coming to Exeter he worked in the cathedral city of Wells in Somerset. A lantern clock with 'Ambrosse Hawkins de Wells ffecit' finely engraved upon the dial is mentioned in *Somerset and Dorset Notes and Queries*, vol 17. The registers of St Cuthbert's parish church, Wells, record that Elizabeth, daughter of Ambrose and Ann Hawkins, was baptised there in August 1690, and they also record the burial of a daughter in 1692. Hawkins was in Exeter in 1695, and in October that year a son, Henry, was baptised at St Martin's Church. Two further sons were baptised there – Thomas in 1697 and Ambrose in 1699. Hawkins was not a freeman of Exeter, but was granted the right to set up shop in the Cathedral precincts by Bishop Sir Jonathan Trelawny. The actual document is among the diocesan records (Devon Record Office).

> Jonathan By Divine Providence
>
> Ld Bishop of Exon
>
> These are to Certifye all whom it may Concern That I doe Hereby Grant and give Liberty and Priviledge to Ambrose Hawkens to Open and Keep Shopp in ye Churchyard of our Cathedrall Church of St. Peters Exon, Wherein to Worke and make Clocks

> Watches and Jacks, And shall protect and defend him therein as Far as the Law and Usage in such Cases will permitt – Wittness my Hand this First day of May 1696.
>
> Jonat: Exon

That autumn Hawkins entered into an agreement with the Exeter Cathedral clerk of works, Peter Passmore, to mend the Cathedral clock for the sum of £17. The Dean and Chapter gave their approval and the progress of the work was recorded in a series of entries, spread over nine weeks, in the fabric accounts. There are some puzzling aspects, however, since not all the money was paid to Hawkins. One wonders if he in fact completed the work.

| | £ | s |
|---|---|---|
| 21 Oct. Pd. Mr. Hawkins ye Clockemaker Five pounds in parte of what hee is to have for mending St. Peters Clocke | £5 | |
| 25 Nov. Pd. Mr. Passmore for Mr. Hawkins ye Clockmaker | £1 | |
| 2 Dec. Pd. Mr. Passmore for Mr. Hawkins ye Clockemaker | | 10s |
| 8 Dec. Pd. Mr. Passmore for Mr. Hawkins ye Clockmaker | | 10s |
| 9 Dec. Pd. Mr Hawkins & his wife by Mr. Ekines order | £2 | |
| 11 Dec. Pd. Mr Richard Smyth by order of Mr. Ekines three pounds in parte of Mr. Hawkins money for mending ye Clocke | £3 | |
| 21 Dec. Pd. Mr. Browne Mr. Ekines servant more in parte of Hawkins money | | 15s |
| 24 Dec. Deducted from Mr. Hawkins money for 100 lb. of leade pd it Mr. Passmore | | 18s |
| pd. Mr. Browne Mr. Ekines servant three pounds seven shillings in Full of ye seaventeene pounds Mr. Hawkins was to have for mending St. Peters Clocke | £3 | 7s |

The final payment is receipted in the margin by Benjamin Browne, a goldsmith who served his apprenticeship in Exeter with William Ekines. It would be instructive to know something of Hawkins's dealings with the goldsmiths; perhaps they embellished clocks and watches for him, carrying out engraving and working on the cases. In 1697 Hawkins sought redress in the Exeter Mayor's Court, claiming £10 from one Amos Needs – possibly a goldsmith – whom it was alleged had lost a clock of his. The clock was valued at £5, a further £5 being claimed as compensation for its loss. Hawkins was involved in further dealings with the law on the morning of 15 January 1697/8 when he seized a man named John Burt, whom he suspected was dealing in stolen goods after he offered to sell him a great brass pan that had been beaten together and put in a bag. Presumably, such metal could be used as raw material for clock parts. Supporting evidence was submitted to the Quarter Sessions that day by two other Exeter clockmakers:

> Walter Trout Clockmaker Precilla his wife W$^{m}$ Hunt Clockmaker Francis Bennam [or Bennum] Brazier all of this City doe depose that about six months since one Jn$^{o}$ Burt of Broadclist Husbandman (who then named himselfe Manly) is the very person that then brought a large brass pann weighing about 20$^{l}$ to ye house of one of these deponents Francis Bennam & afterwards sold the said brass pann to another of these deponents Walter Trout: being feloniously stolen from one Wescott of farrington . . . Further these deponents declare that this 15th day of Jan: this sayd John Burt brought early in y$^{e}$ morning one other great brass pann beaten together & putt into a bagg weighing abt 13$^{l}$ offering to sell the same to Ambrose Hawkins a Clockmaker at 8½d per pound which he supposing might be feloniously stolen seized y$^{e}$ person upon suspicon.

Ambrose Hawkins died in 1705 and was buried in the cloister of Exeter Cathedral. The burial entry copied from the register at the Cathedral is as follows:

> Mr Ambrose Hawkins of this Close was buryed on the Eleventh day of August in the Cloyster by Mr fford 1705.

The register also records the burial of Thomas, son of Mrs Anne Hawkins, on 7 December 1706.

In the Exeter Poor Rate list, dated 2 April 1696, Hawkins was assessed at 1d in St Martin's parish. Dean and Chapter leases and rentals show that he lived and worked in a house behind St Martin's Church and adjoining the now blitz-ruined St Catherine's Almshouses, for which he paid rent in his lifetime, his widow continuing there for several years after his death. In 1711, however, the rentals mention the 'Exors of Mrs Hawkins'. The head lessee of the clockmaker's house was Benjamin Hawkins, a grocer in the adjoining parish of St Stephen, and he provides a link with the Savage family, for his father, also Benjamin, was one of the overseers appointed by the watchmaker Peter Savage in his will of 1656: 'I doe make and ordayne Mr Thomas Chudleigh and my Cousen Biniamin Hawkins, my two loveing and faithful friends to be Overseers of this my will . . . And I give them twenty shillings apeece to buy them rings to weare in my remembrance.'

Hawkins's work as a maker and repairer of church clocks took him over a wide area of the county. He made a large turret clock which has been preserved though no longer in use at St Saviour's Church, Dartmouth. It is signed in capitals along one of the top bars:

IOHN SPARKE MAIOR OF DARTM$^{O}$ 1705
AMBROSE HAWKINS DE EXON FECIT

The four-posted frame of this clock is constructed of wrought iron, and an impressive decorative touch is provided by the 178mm (7in) wide finials on the corner posts, scrolled and dramatically widening in the centre. The frame is just under 1m (3ft) long and the four-spoke wheels are of brass; there is a large outside locking-plate for the striking, and the pendulum measures 1.5m (5ft) approximately. The churchwardens' accounts record that the clockmaker was paid £22 10s for the clock and that at a meeting he was entertained with 'sperett' (spirits). Further outlay concerning the clock, including heavy expenditure on 'painting and gilding ye hands', brought the total bill up to £42 12s 11d – a hefty sum for a Devon clock of that period.

In the same year Hawkins provided an almost identical clock (Plate 7) for the parish church of Modbury in south-west Devon, twelve miles from Plymouth. This, too, has been preserved. An inscription along the top front rail, 'Ambrose Hawkins de Exon fecit 1705', appears to have been etched from Hawkins's own handwriting. The finials on the Modbury clock are 13mm (½in) narrower than those at Dartmouth, but in most other respects the two clocks match part for part, tooth for tooth.

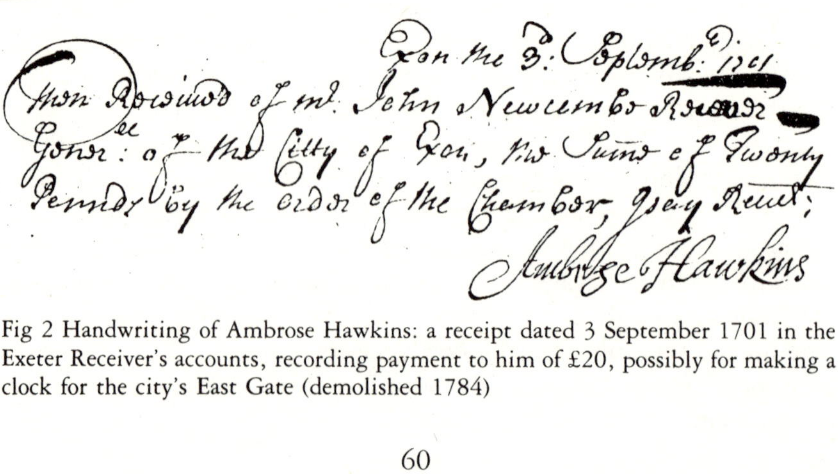
Exon the 3: Septemb: 1701
Then Received of Mr. John Newcombe Receiver Generall of the Citty of Exon, the Sume of Twenty Pounds by the order of the Chamber, I say Receivd;
Ambrose Hawkins

Fig 2 Handwriting of Ambrose Hawkins: a receipt dated 3 September 1701 in the Exeter Receiver's accounts, recording payment to him of £20, possibly for making a clock for the city's East Gate (demolished 1784)

Hawkins carried out repairs to the clock and chimes at South Molton in 1699; and at Crediton, in March 1701/2, he was paid £8 'for righting the Church Clock and making the hand dyall against the Tower' (governors' accounts).

Plate 8 shows the unique dial of a month longcase clock by Hawkins. The hour ring is cut away between the chapters to reveal twelve pastoral scenes. Fishing and shooting are depicted and there are trees, animals, birds, houses, boats and a bridge. Such engraving can sometimes be seen on the dials of seventeenth-century watches and is also reminiscent of designs on Dutch and other porcelain. In the opinion of a former curator of applied art at Exeter Museum, a specialist in antique silver, the engraved scenes are contemporary with the rest of the clock. The dial plate is 300mm ($11\frac{13}{16}$in) square with small apertures for the phases and age of the moon and the day of the month. It is signed across the base 'Ambrose Hawkins de Exon fecit'. A rare feature of the eleven-pillar, three-train movement is that it has split front plates, enabling each train – going, striking, and quarter-striking – to be removed separately. The quarters are struck on six bells and the three brass-cased weights (not original) are each over 305mm (1ft) in length. The clock is housed in a mulberrywood case, the trunk door which has a glass lenticle for viewing the pendulum is inlaid with two eight-pointed stars, and there is a third star on the base.

Another month longcase clock by Hawkins has a walnut case. Its owner on first seeing it thought it must be a Tompion, as the quality was so fine, but was amazed to find on inspection that it was by an 'unknown' Exeter maker who at that time was not even recorded in the standard lists of clockmakers. A third Hawkins longcase clock, this time with grande sonnerie striking, is illustrated and discussed in detail in Tom Robinson's book *The Longcase Clock*.

Several other Hawkins clocks have come to light, including an eight-day marquetry longcase, a thirty-hour longcase, and an attractive single-handed wall clock with tapered arbors. An eight-day clock bought in Dorset some years ago has a 254mm (10in) square dial with spandrel mounts composed of twin cherubs supporting crowns, the chapter ring signed 'Amb. Hawkins Exon'. There are no pillars either side of the hood door; pillars are also absent in a contemporary Exeter longcase clock by Edward Clement. Curiously, a marquetry longcase clock, catalogued as 'by Ann Hawkins, Exon', appeared in a sale at Sotheby's in December 1968 and was sold for £750. The 279mm (11in) dial had crown and cherub spandrels and calendar aperture; the movement had an outside countwheel for the striking; and the walnut case, the waist door with bull's eye, was inlaid with a vase of carnations, tulips, other flowers, and birds perched in branches. The signature on this clock raises two possibilities: either it was sold after Hawkins's death by his widow Ann, or the christian name is not Ann at all but Am, short for Ambrose.

### EDWARD CLEMENT

Among surviving examples of the workmanship of Edward Clement are several longcase clocks, a lantern clock and a silver watch. This maker may be identified with Edward Clement who was apprenticed in London on 28 August 1662 to Thomas Claxton and then 'turned over' to Andrew Prime. He became a freeman of the Clockmakers' Company on 3 April 1671, and can be traced in Exeter in the 1680s. The churchwardens' accounts of the old St John's Church, Exeter, record payments to him 'for work about the clock' from 1684 onwards. In 1690 he repaired the striking, made new wheels for the dial and supplied a new hand to make the clock from '24 howers to twelve'. His bill for this work is in a bundle of parish receipts.

| | |
|---|---|
| Worke Done by Edward Clement<br>To the Parrish clocke of St. Johns in Exon | |
| Item for clensing mending & stopping up<br>severall holes a new spring to the<br>ffly new spring to the hammer & new<br>makeing the hammer worke | £1: 05 |
| ffor makeing new wheeles for ye Dyall<br>& a new hand makeing the clocke from<br>24 howers to twelve | £0: 17: 6 |
| | £2: 2: 6 |

On the back is Clement's receipt dated 6 March 1690/1, which he signs as Ed: Clement.

An early thirty-hour longcase clock, 'Eduardus Clement in Exon fecit', dating perhaps from *c*1690, has an ebonised pine case, a single hand, and an engraved design of flowers in the dial centre. A silver watch signed 'Ed Clement Exon', with a silver champlevé (skeletonised) dial is illustrated in *Transactions of the Devonshire Association*, vol

67, 1935; the owner, a doctor, said that it was still in going order. The case was not hallmarked and had a peculiarly steely appearance which, he suggested, might be due to the presence of a little more copper than in the standard metal of that period. The watch has the appearance of *c*1700, or a little later.

The name Edward Clement recurs in Exeter through several generations: an Edward Clemente, baker, was admitted as an Exeter freeman in 1610–11 and was described as a merchant when succeeded by his son Edward, also a merchant, as a freeman in 1623. The clockmaker Edward Clement is mentioned in the records of Exeter Quarter Sessions when he provided a surety in July 1688. He – or a man of the same name – lived in St Lawrence parish and was churchwarden in 1691–2 and again in 1713. St Lawrence's Church was destroyed in the blitz, but a stone tablet has been preserved recording the consecration of its little churchyard by Bishop Sir Jonathan Trelawny in 1692 when Clement and the pewterer Robert Daw were churchwardens. This tablet is now in St Stephen's Church. Three of Clement's children were baptised at St Lawrence's – Charles in 1684, Joane in 1691 and Jane in 1694; and two were buried – James in 1688 and Suesana in 1693 (in the church). Edward Clement himself was buried there on 28 August 1720. It has been suggested that he was a brother of the anchor-escapement pioneer William Clement, of London, master of the Clockmakers' Company in 1694; but efforts to substantiate this have not so far borne fruit. The family evidently had West Country connections, for a William Clement worked as a clockmaker at Totnes where he was mayor in 1711 (see Chapter 4).

## JACOB LOVELACE

In the first half of the eighteenth century, one Exeter clockmaker stood out above the others – Jacob Lovelace. He lived and worked in St Stephen's parish from about 1712 to 1750 and during that time produced a fine series of walnut and other longcase clocks, the occasional bracket clock and watches. But it is not on these that his fame rests, but upon a very complex calendar, automaton and musical clock that was one of the great showpieces of its age. Known simply as the Exeter Clock, it stood 3m (10ft) high, weighed ·5 tonnes (½ ton), and was remarkable not only for its mechanical ingenuities but for the splendour of its richly decorated case (Plate 9).

This clock, which appears to have been completed in 1739, earned considerable sums for its various owners in exhibition fees. The Liverpool Museum eventually acquired it, but sadly it was largely destroyed after an air raid in 1941 when, in the resulting chaos, it fell victim to vandalism. Some of the parts, however, have since been recovered and carefully restored. These include the dial, a large section of the musical and automata movement, and two of the carved and painted decorative figures from the case. The clock in its glory was a spectacular sight and one who saw it in the nineteenth century described it in these terms:

> The celebrated Exeter clock . . . is not only a timepiece, striking the hours of the day and chiming the quarters, but it is a perpetual almanack, telling the days of the week and month; leap-year when it happens; showing the phases of the moon and its age; moreover, it will be silent if required; and when agreeable, will play a variety of tunes on an organ; Saturn presiding as conductor, and beating time, and Fame and Terpsichore moving to the air. It has also a most musical peal of six bells, with ringers; a moving panorama allegorical of day and night; and a guard of two Roman soldiers, who salute, with their swords, Apollo and Diana as they appear. The soldiers' heads are actually turned when the bells ring – as well they may be.

The traditional account of Lovelace's life as given by T. L. Pridham in his *Devonshire Celebrities* (1869) is largely a fiction, and has led to much confusion among researchers. Pridham stated that Lovelace, a native of Exeter, was born on 15 March 1656 and died aged sixty. He succeeded his father, a clockmaker, in an obscure part of the city and, his habits inclining him to live a very retired life, was seldom seen. The famous clock occupied him thirty-four years and it was to be lamented that he died in the workhouse at Exeter. All this, however, is totally unsupported by fact, and a very different picture emerges from the numerous records of his life that can be extracted from the Exeter archives and from contemporary newspapers.

The clockmaker's earliest years were indeed obscure. Searches of the Exeter parish and nonconformist registers have failed to produce a record of his baptism, and nothing has so far come to light on his apprenticeship. The *Exeter Journal* for 1827, however, describing him as a most ingenious mechanic, states that he was born about 1690 – a date which fits in well with later records of his life. The records of St Stephen's Church, Exeter, show that Jacob Lovelace and Grace Rocket of Exeter were married there on 5 November 1712. His skill with clocks was soon apparent and he was shortly afterwards entrusted with work on the great clock of Exeter Cathedral, as the following entry from the Fabric Accounts (Christmas 1713–14) shows: 'To Mr Lovelace a bill about ye Clocke £21 17s 6d'. Thomas Turtliff, son of a Plymouth cordwainer, was apprenticed to him as a clockmaker in Exeter around the year 1715 and, a few years later, on 21 August 1721, Lovelace was admitted as a freeman of Exeter on payment of £1 1s.

He lived in St Stephen's, in the commercial heart of the city, and, far from being seldom seen, was an active member of the church. His is one of eighteen signatures approving the churchwardens' accounts for 1713 and he was warden himself from 1717 to 1719, having previously had the task of collecting the poor rate in the parish. The churchwardens' accounts for 1719 were prepared by Lovelace as warden and appear to be in his handwriting, a flowing italic, the words 'St Stephens' in the heading being penned in a most flamboyant manner. Lovelace kept the parish clock in repair over a period of about thirty years.

The St Stephen's rate book – the rate was levied 'for the maintenance of the poor in ye Workhouse' – shows that around the year 1718 Lovelace moved house. His new house was among the fifteen highest rated in the parish; but six years later he was back at his former, less highly rated one. *The Postmaster: or The Loyal Exeter Mercury* of 2 October 1724 carried the following advertisement: 'To be let, at a Yearly Rent, THE HOUSE wherein Mr Jacob Lovelace now lives, from Michaelmas next . . . Enquire of John Haddy in Gandy's lane, EXON.' The advertisement is repeated in the following week's issue, but the house is described as 'wherein Mr Jacob Lovelace lately liv'd, adjoining the Printinghouse'. The printing house was that of Andrew Brice, the publisher of the newspaper, and it was opposite St Stephen's Church in High Street. There is now a broad pavement there, and each year countless thousands of people walk over the site of what was once Jacob Lovelace's front shop.

Lovelace worked on the Exeter Cathedral clock again in 1734, when he was paid £2 15s. Two years later, Richard Eastcott, one of the Cathedral choristers, was apprenticed to him as a watchmaker, and in 1739 Lovelace's name was to be found among the subscribers to a book of sermons by John Warren, 'late Prebendary of St Peter's, Exon', and a former rector of St Stephen's. That same year a clock fitting the description of Lovelace's Exeter Clock was put on show in London.

> To be seen at the Leg-Tavern in Fleet Street (Never before exposed) A Beautiful and Magnificent Musical Clock, valuable for its structure and the exact Performance of all its parts; with some other Machines by Clock-work; and the surprising power of the artificial Magnet, made by Jacob Lovelace, of Exeter, who makes and sells all Sorts of Musical and other Clocks, Watches, and artificial Magnets. Note, Constant Attendance will be given for the Space of five or six weeks from the 6th instant, from the hours of Ten in the Morning, till Eight in the Evening, or at any other time if particularly desir'd. Price 1s each (*General Advertiser*, 10 November 1739).

In 1741 Lovelace was paid 10s 6d by the Exeter Receiver for providing lodgings for a judge's servant, and in April 1742 was on a jury at the Quarter Sessions. He and his wife had eight children, but five died in childhood, four being buried in St Stephen's and a fifth in St Lawrence's. The eldest surviving son was John, baptised 1717. Appropriately, he went to Exeter College, Oxford, was ordained, and in 1741 was admitted as vicar of Aylesbeare, a parish some seven miles east of Exeter. He married Anna Maria Locke, daughter of the previous Aylesbeare incumbent, the Reverend Edward Locke, and held the living for twenty-two years, until 1763. A note in the parish register says that Mr Lovelace retired to Essex, probably to Great Waltham. There is no mention of Jacob Lovelace in the Exeter rate assessments for 1752–3 and it is probable that he spent his eventide years at his son's vicarage. He was buried at Aylesbeare in 1755 and the register records: 'Mr Jacob Lovelace (died Nov: 30th) buried Dec. 3'.

The eighteenth-century history of the Exeter Clock is uncertain. According to Pridham it fell into the possession of a Mr Dickenson, of Tiverton, and was then acquired by James Burt who in 1821 exhibited it at a shilling a head at his picture and china rooms on Fore Street Hill, Exeter. Burt was a dealer in antiquities, and in advertisements in

Plate 6 (*above*) Hartland's ancient town clock, made in 1622 by John Morcombe, of Barnstaple. It originally had an overslung foliot (see hole in frame, top left), but was later re-wheeled and fitted with a pendulum

Plate 7 (*below*) Disused church clock at Modbury, signed on top front rail 'Ambrose Hawkins de Exon fecit 1705'. The frame measures 90×67×47cm (35½×26¼×18½in); the finials are 16cm (6½in) wide. The parish paid Hawkins £18 for it

Plate 8 'Ambrose Hawkins de Exon fecit'. Unique skeleton dial with engraved sporting and pastoral scenes between the chapters. It belongs to a mulberrywood longcase clock; the movement is of month duration and has six-bell quarter chimes, hence the third winding square. Hawkins moved to Exeter about 1695, repaired the Cathedral clock the following year, and died in the city in 1705. He was buried in the Exeter Cathedral cloister

1820 and again in the following years offered for sale Old Master paintings and 'rare and curious chime and musical clocks'. The year 1824 was a leap year and great interest was shown in Exeter on whether or not the clock would automatically adjust for it. The triumphal outcome merited a paragraph in *Woolmer's Exeter and Plymouth Gazette*:

> The extraordinary Clock, in the possession of Mr Burt, of this city . . . performed its exclusive movement on the 29th of February, being Leap Year. This is singular proof of the accuracy and ingenious nature of the machinery.

Many years later a correspondent recalled that

> . . . doubts were raised by some, and bets were made that it would not stop at the proper date . . . At a little before twelve o'clock at night parties assembled at Burt's house to watch the result, and when the hand came round and the organ and other instruments began to play, instead of the plate moving on to 30 it unlocked and flew back to 1, much to the chagrin of the losing betting parties.

In 1834, the clock, having been 'put into perfect repair by W. Frost, clock and watchmaker, Paris Street', was removed to London and sold at auction, but was subsequently brought back to Exeter and was owned by Charles Brutton, an attorney. The sale took place at the Exhibition Room, 209 Regent Street, on 11 August, the clock being catalogued as lot 45 – the 'celebrated Exeter Clock, which has been exhibited to almost countless hundreds of admirers, during the present season, and is accounted, at the far-famed city of Exeter, a combination of mechanism so perfect and so wonderful as to be without a parallel'. *The Western Times* carried a graphic account of the sale:

> THE EXETER CLOCK – (formerly the property of Mr. Jas. Burt, of St. David's Hill.) – The celebrated piece of mechanism, formed upon the model of the clock of Strasburgh, and which has been exhibited for several months past, at 209 Regent Street, was placed under the 'superintending care' of that 'Monarch of the Hammer', George Robins, for public competition . . . The room was crowded with connoisseurs in works of art, whose anxiety to possess this rare and splendid – indeed, we may say unique – specimen of mechanical power, was manifested during the biddings. The first bidding was two hundred guineas, and after a lengthened and spirited competition this extraordinary proof of the science, ingenuity, and perseverance of an English artist . . . was knocked

> down for six hundred and eighty guineas . . . It is rumoured that it has been bought for his majesty William IV.

The rumour of royal purchase, however, was, as we have seen, groundless. The clock was subsequently kept at the Judges' Lodgings in Northernhay Place, Exeter, where Brutton lived, and it is said that when the assizes were on, William Frost, the Exeter clockmaker, had to go to put all the different parts of the clock in motion to please the judge and his company. In 1856, Frost's son, a watchmaker of St Sidwell's, Exeter, sued Brutton's executors at Exeter District County Court for £17, the balance of an account for work done by his father. According to a report in the *Exeter Gazette* (17 May):

> It appeared that the late Mr Brutton was the owner of the celebrated clock by Lovelace, and that previous to 1851 he employed the plaintiff's father, who was a clever clockmaker, to clean and repair it. The late Mr Frost was employed two months, with his sons, in repairing it, and then he was sent by Mr Brutton to exhibit the clock in the Great Exhibition in May 1851. Previous to his going to London Mr Brutton paid him £20. After being at the Exhibition six weeks Mr Frost was taken ill, and was obliged to return to his home in this city, where he died in September of the same year; and a short time before his death he instructed his son to make a claim on Mr Brutton in respect of repairs of the clock. The plaintiff, after his father's death, made a claim on Mr Brutton, but it was disputed. The matter ran on until Mr Brutton's death . . . Mr Stogdon, for the defendants, proved that the expenses of carriage of the clock to and from London was paid by Mr Brutton, and contended that the payment of £20 to Mr Frost was made with the understanding that it was to cover all claims, including repairs. His Honour decided that where an opportunity had been offered a plaintiff to bring his action against the defendant during his lifetime he ought to have done so, and not have delayed it until the party had died and was unable to answer it. He therefore gave judgment to the defendants.

The clock was owned briefly by John Stone, an Exeter High Street silversmith, and was then taken to Liverpool where parts of it were cleaned and repaired by J. Condliff in 1857 and 1858. Two lithographs of the clock were published in Exeter – one by Hackett in 1833 (Plate 9) and the other, very similar, by William Spreat. Both show the clock with the cabinet doors open at the base, on which there were paintings of Exeter buildings, including a unique depiction of the ruins of Rougemont Castle. J. J. Hall wrote a paper on Lovelace, which was published in the 1930s. In this he remarks on the comparative smallness of the main movement of the Exeter Clock. Particulars were furnished to him by Mr F. H. Eccles, the then custodian of the Liverpool Museum clocks. It appears from these that the clock had a three-train, fusee spring driven quarter-chiming movement chiming on bells, the dimensions of the frame of this movement – that is, the size of the plates – being 229mm by 229mm (9in by 9in). He goes on: 'There is also a larger movement, also fusee spring driven, which plays tunes on a set of organ pipes, the dimensions of this movement being 18in by 12in by 10in.'

The baroque-style case of the Exeter Clock raises the question of the identity of its maker, for such workmanship was surely beyond the reach of all but the finest of cabinet makers. The answer, however, is not difficult to seek, for there are at Powderham Castle, near Exeter, two magnificent rosewood bookcases which are comparable in design and decoration. They each bear brass plaques inscribed, '17 J. Channon Fecit 40'. The maker, John Channon (1711–83), is associated with the grandest style of English cabinet making, with bold architectural compositions, superbly executed and richly ornamented with engraved brass inlays and gilding.

Channon had served his apprenticeship in Exeter with his elder brother, Otho Channon, who is described variously as a joiner and as a chair and cabinet maker and who also carried out work at Powderham Castle. The family had their origins in St Sidwell's, but by the late 1720s Otho Channon had moved to St Stephen's in the city centre, where he and Lovelace were fellow parishioners. Indeed, from 1736 onwards his name appears on the rating list next to that of Lovelace, which suggests that they were next-door neighbours. John Channon can be traced in London from 1737; his premises were on the west side of St Martin's Lane, near Charing Cross. Assuming that the case of the Exeter Clock was made in his workshops, it is significant that it was in London and not Exeter that the clock 'Never before exposed' was first exhibited in 1739. Otho Channon was churchwarden of St Stephen's in 1735–6, and his sons were given the names of Roman emperors – Otho, Titus Vespasian, Tiberius and Caligula. He died in 1756.

Several watches signed 'Jacob Lovelace Exon' have been noted, including one bearing the date 1718. This is a pair-case calendar watch with a silver champlevé dial; it carries the casemaker's initials 'IH'. Exeter Museum has a Lovelace watch movement, unnumbered and undated. It is known that he made at least one 'table' clock.

Longcase clocks are more plentiful, the earlier ones having square brass dials. The author has examined several examples, all of which have lofty cases, suggesting that they were made to go into large houses. One in a black Japanned case (Plates 10–12) has hammers either side of the bell, the larger one for the hours and the other for the half hours. The later Lovelace longcase clocks have in most instances well-engraved arched dials. The cases are often of finely veneered walnut, and the movements sometimes have engraved moonwork and tidal indication. One clock shows 'High Water at Topsham Bar'. Plate 13 shows a typical walnut-cased Lovelace with a border around the edge of the arched dial and rings around the winding squares. Exeter Museum has a similar eight-day specimen with a pair of engraved birds in flight flanking the seconds dial.

Another Lovelace longcase clock, seen by the late J. J. Hall, had an extremely light escape wheel and the striking work was of 'peculiar construction'. Hall drew an excellent diagram, showing the wheel diameters and numbers (Fig 3). In place of a locking-plate with notches on the strike main wheel, there is a countwheel with raised teeth spaced to allow for the correct sequence of blows. There is only one detent and the system works in the reverse way to the locking-plate in that the detent, instead of falling into a notch, rises up the slope at the completion of striking and there remains (at the top of the tooth) until the next warn takes place. It is a sophisticated system and calls for very accurate spacing of the teeth. A correspondent in Bristol who has a handsome, square-dial eight-day longcase clock by Lovelace, reports that it too has this method of strike control. It is also to be found, in several different forms, in Devon clocks by William Stumbels of Aveton Gifford and Totnes, and has been noted on a thirty-hour clock by William Bradford of East Anstey.

### RICHARD EASTCOTT (EASTCOT)

Lovelace's apprentice, Richard Eastcott, was closely associated with Exeter Cathedral and its music. He sang in the choir as a boy and the Dean and Chapter insisted that he should continue to do so during the time of his apprenticeship:

> 18 December 1736. Eastcot: They [the Dean and Chapter] Consented that Eastcot one of the Choristers should bind himself an Apprentice to Mr Lovelace ye Watchmaker and Keep his place in the Church he giving dayly attendance as other Choristers do (Act Book).

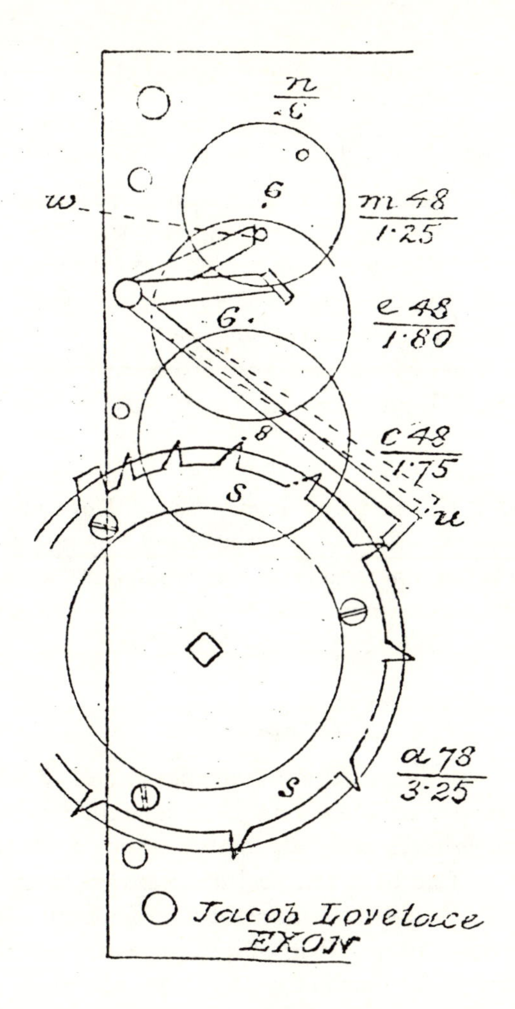


Fig 3 Modified countwheel striking in a longcase clock by Jacob Lovelace, of Exeter. Diagram by John James Hall

In 1739 Eastcott was nominated as a secondary of the Cathedral, an office that entailed helping with the singing, and in 1744 it was agreed to pay him 5s quarterly for looking after the Cathedral clock. He married Ann Carthew of Exeter that same year, and their son Richard was baptised at the Cathedral on 30 April 1745; another son, Sanford, was baptised there in 1749. Their house is mentioned in a notice dated 18 October 1770 in the *Exeter Flying Post*:

> To be Sold: A convenient dwelling house situated in Catherine's-lane, within the Precinct of the Cathedral Church in Exeter, fit for the

> immediate reception of a genteel family . . . also a small tenement adjoining the above, now in the possession of Mr Eastcott, watchmaker.

Eastcott, who was also a lay vicar, resigned as a Cathedral secondary in 1771, and made a present of six anthem books for the use of the canons' ladies. He lived on into his seventies, his burial being recorded at St Paul's, Exeter, on 10 October 1795. A mahogany longcase clock, 'Richard Eastcott Exon', was seen for sale many years ago in Exeter, price £13. It was an eight-day striker with domed hood and 'brass face with fancy corners' (Tapley-Soper Index, Westcountry Studies Library, Exeter).

The watchmaker's elder son, the Reverend Richard Eastcott, was one-time rector of St Edmund's, Exeter, and of Ringmore; and was for many years a priest-vicar of the Cathedral, dying in 1828, aged eighty-three. He was an organist and composer and author of a book, *Sketches of the Origin, Progress and Effects of Music, with an Account of the Ancient Bards and Minstrels*, published in 1793.

### EDWARD UPJOHN

Edward Upjohn was originally a stonemason and self-taught clockmaker at Shaftesbury in Dorset. In 1723 he emigrated with his wife Mary and their children to Philadelphia, moving on from there to Charleston, South Carolina. His health suffered, and in 1726 after an arduous voyage he and his family arrived at Topsham, at that time a small but important port and shipbuilding centre at the head of the Exe estuary a few miles south of Exeter. Edward put his sons to work in the business and at one time five of them were all busily employed by him in making clocks and watches (Brian Loomes, *The Amazing Life of James Upjohn*, see Bibliography).

After remaining at Topsham for twelve years or so, Edward Upjohn moved to Exeter. He lived for a time in Holy Trinity parish, but in the city's rate books for the years 1752 to 1756 is listed in Southgate Street in the parish of St Mary Major. In 1745 he was among a large number of Exeter citizens who declared their loyalty to King George II, at the time of Bonnie Prince Charlie and the Jacobite rising, by forming an association at the Guildhall. Their names were printed on a broadsheet, headed by the mayor, lord bishop, aldermen and magistrates, and they expressed their 'utmost abhorrence and detestation of the present wicked and unnatural rebellion begun in

Scotland, and threatened to be backed by an invasion from abroad'. Edward Upjohn, like many of Exeter's leading craftsmen, served regularly on juries at the city's Quarter Sessions. According to information published by Brian Loomes, he died in 1764, aged over eighty. His widow died in January the following year.

Tom Tribe, co-author of *Dorset Clocks and Clockmakers*, reports the signature 'Edward Upjohn Shaston' on a 305mm (12in) square brass dial of early eighteenth-century pattern. 'Shaston' is an old term for Shaftesbury. Longcase clocks inscribed 'Edward Upjohn Topsham' have also been noted, including an arched-dial example in a mellowed walnut case, which when seen by the author was marking the hours in the dining-room of an old house in the town where it was made. Exeter examples of his work include a longcase clock with a Tudor rose engraved in the dial centre and a gold watch in a repoussé case (recorded by Baillie, see Bibliography). The church clock at East Budleigh was made by Edward Upjohn, and in the Devon Record Office there is a receipt for 5s paid to him in 1749 for supplying a new pendulum spring to Exminster parish church clock.

The St Mary Major registers record the burial on 17 January 1741 of Edward Upjohn, son of Edward. He engraved watches for his father, but died aged about twenty-one (Loomes). Five other sons outlived their father and established watch and clockmaking businesses of their own: William and Richard in Exeter, James in London, Peter in Bideford, and Nathaniel in Plymouth (see List of Makers).

## RICHARD UPJOHN

Some fine clocks and watches by Richard Upjohn, of Exeter, are known. He was Edward and Mary's youngest son and was baptised at Topsham on 7 February 1727/8. He was regarded as being a very lucky child – the seventh son of a seventh son? – and many people came to be stroked by him for the King's Evil and sundry other disorders (Loomes). He worked in the Exeter Cathedral Close and was married in the Cathedral in 1754: 'Richard Upjohn of the Close, Watch-maker, & Eliz: Herridge of the same Spinster were married the 11th day of ffeb: by Virtue of Banns' (Cathedral register).

Three children of the marriage were baptised at the Cathedral and seven at St Martin's nearby. Richard Upjohn's name appears several times in Quarter Sessions jury lists; in October 1761 he is described as of the Close, and in January 1764 as of St Martin's. He branched out from clock and watchmaking into selling jewellery and plate, as the following advertisement in the *Sherborne Mercury* of 17 August 1767 shows:

RICHARD UPJOHN,
CLOCK AND WATCH-MAKER,
At the Corner of Broad-Gate, St. Peter's Church-Yard, Exeter, begs Leave to acquaint Ladies and Gentlemen, and his Friends in general, that he has opened a Shop in the Plate and Jewelery Way with choice of
Gold Chas'd and Plain Gold Watches;
Likewise a Variety of Enamell'd and Metal Watches, for Travelling, quite in the newest Taste; and is determined to sell every Thing as cheap as in London.
N.B. All Dealers in the Watch Way may be supplied with Watches and Materials on the same Terms as in London.

Richard Upjohn was often ill and was much in debt to his brother James, of London (Loomes). He died in May 1778 and was buried in St Martin's on the 26th. Three days after the funeral it was announced in the *Exeter Flying Post* that the business was being continued for the benefit of the widow and family by Richard Bayly, watchmaker and goldsmith from London, 'till the expiration of his engagement made prior to the death of Mr Upjohn, after which time he intends to succeed him on his own account'. This arrangement did not work out. Just a month later, in an announcement dated 29 June the widow, Elizabeth, begged leave to inform that

> . . . the business is carried on in the same house as usual, by her brother, Mr James Upjohn, of London, for the benefit of her and her family: that the said Mr Upjohn has fixed Mr Brown, of London, as his foreman in the shop, for the better carrying on of the trade, and likewise one of his sons, who served his time in London to a working goldsmith.

In a notice dated 5 August 1778 the widow further announced that

> . . . she is remov'd from the corner of Broadgate to the house next door, and has opened a genteel shop where the public may be supplied with all kinds of watches, clocks, plate, plated goods, and jewellery . . . gold enamel'd trinkets and toys . . . and everything that may be wanted will be

> sent with the greatest expedition from Mr James Upjohn, of Red Lion Street, Clerkenwell, London.

Then, shortly afterwards, she announced that the house next to Broadgate was fitted up and was to let. The property consisted of

> . . . a very good underground kitchen, with a good cellar, coal hole, and other conveniences; a good shop with a chimney in it, and a very good and pleasant room over; a hall, parlour, and dining room, with a large closet for a bed; a lodging room even with it, and two upper rooms, all with convenient closets.

Elizabeth Upjohn died in January 1781. One of her sons, Richard, was apprenticed in the watchmaking trade and another son, James, was apprenticed to a tailor in Gandy's Lane, although it was not much to his liking, for in 1787 he ran away (*Exeter Flying Post*, 19 April).

### WILLIAM UPJOHN

William Upjohn and his successors were watch and clockmakers in Exeter from the late 1730s to the mid 1870s, for much of that time at the same house on Bell Hill, Southgate Street (now South Street), three doors above Bear Lane. He and Anna Reeve were married at Old Trinity Church, Exeter, on 11 September 1736. They lived in the adjoining parish of St Mary Major and their thirteen children were baptised there in the parish church, which stood just in front of Exeter Cathedral. Sadly, at least four died in childhood. The clockmaker's father, Edward Upjohn, was in business in the same street and the two were not on the best of terms (Loomes).

William Upjohn was a leading member of St Mary Major Church, serving as one of the parish's three poor rate collectors in 1749–50 and as churchwarden from 1759 to 1761. Like his father, he served on juries at Quarter Sessions – in 1741, 1752 and 1761 – being described as of St Mary Major, watchmaker. He died in 1768 and was buried in the parish on 14 December, the entry in the register giving him the distinction of 'Mr', a prefix sparingly used in those days. The burial of his widow, Anna, was recorded at St Olave's on 13 April 1784.

The Honiton watchmaker Francis Pile jun, who died in 1763, directed in his will dated 22 June that year that his executors should sell his clocks, watches, materials and working tools, 'being first appraised by William Upjohn of ye city of Exon, watchmaker, or some other proper person'.

Longcase clocks signed 'Wm Upjohn Exon' include a thirty-hour example in a burr-walnut case with a fine quality brass dial with a silvered chapter ring and date aperture; F. J. Britten in *Old Clocks and Watches and Their Makers* mentions a watch similarly signed, with silver dial and raised figures, hallmarked 1741.

### THOMAS UPJOHN

Thomas, eldest son of William and Anna, was the maker of some of the finest Exeter clocks of the second half of the eighteenth century. He was baptised at St Mary Major Church on 6 August 1737 and, after serving his apprenticeship with his father, went to London. In 1760 he announced he had opened a shop in Launceston, Cornwall, 'where he makes and sells all sorts of clocks and watches, repeating, horizontal, gold or silver, Pinchbeck, shagreen, tortoiseshell, &c, as good and as reasonable as in London' (*Sherborne Mercury*, 15 September). While in Cornwall he married Charlotte Rouse at Altarnun on 10 April 1765. In the year following there is a newspaper reference to a trickster obtaining goods from his shop by false pretences (*Exeter Flying Post*, 17 October 1766). On the death of his father, Thomas returned to Exeter, and in an announcement in the *Sherborne Mercury* of 26 December 1768 stated that he was continuing the business:

> WILLIAM UPJOHN, Clock and Watchmaker, being dead, the Trade is now carried on, at his late Dwelling-House, in Southgate Street, in Exeter by THOMAS UPJOHN, his Son; who is above Thirty Years of Age, and served his Apprenticeship with his Father: – Where Gentlemen, Ladies, and others, may be supplied with all Sorts of CLOCKS, and Gold and Silver WATCHES, on the most reasonable Terms, and have their old ones mended; and may be assured that they will be served with like Care, Civility, and Integrity, as in his Father's Life Time.
> N.B. Likewise all Sorts of Silver-Smith's Work sold on the most reasonable Terms.

Thomas and Charlotte had a family of at least eight children. Among those baptised at St Mary Magdalen, Launceston, was their eldest son Thomas (1766–1843). He studied at Pembroke College, Oxford, and was later a Devon country parson, being rector of Honeychurch and then of High Bray. Of their five children baptised at St Mary Major, Exeter, John (baptised 1771) and James (baptised 1783) became clock and watchmakers.

The clockmaker Thomas Upjohn was an active member of St Mary Major Church, serving as rate collector in 1768–9 and regularly attending vestry meetings. He died in August 1783 and was buried in the parish on the 24th. His will is dated 20 August 1783, so it must have been dictated almost on the last day of his life. It begins: 'In the name of God amen I Thomas Upjohn of the city of Exeter watchmaker being somewhat indisposed but of disposing mind and memory . . .' The first bequest is £300 to his wife 'as by her marriage settlement', and he goes on to bequeath £30 to each of his children, 'including the child my wife is now pregnant with', when they reach the age of twenty-one, plus a joint share of the residue of the estate. He makes provision for up to £50 – including the £30 legacies – to be spent on each child's education or to bind them as apprentices. He names as his trustees and executors his wife, his mother, his brother William, and his brother-in-law the Reverend William Rouse (Public Record Office: Prob 11/1113 22279).

Thomas Upjohn's apprentices included his brother William and Peter Charlton Owen, who in 1783 was in business opposite the Angel Inn, Crediton.

Many horologists will be familiar with Thomas Upjohn's work through an illustration of an imposing green-lacquer bracket clock by him in Cescinsky and Webster's *English Domestic Clocks.* Other bracket clocks include one in an ebonised case, signed 'Thos. Upjohn Exon', the backplate beautifully engraved with flowers, butterflies and bees. Plate 14 shows a splendid and unusual cartel clock with a 229mm (9in) silvered dial bearing the name 'Thos. Upjohn Exeter' in a sketchily engraved cartouche. It is a timepiece only and is contained in a lovely Chippendale frame of carved and gilded wood, lavish with Chippendale 'Cs' and surmounted by a most aggressive-looking eagle. The frame measures 736mm by 483mm (29in by 19in) approximately. A longcase clock made by Thomas Upjohn for an Exeter Quaker named George Dymond has an unusually plain dial with a raised chapter ring but no 'frivolous' decorations such as spandrel mounts (corner pieces). The case is of walnut with an attractive figured grain on the door. Another Thomas Upjohn longcase clock indicates 'High Water at Topsham Bar'; its case is of mahogany.

Within a few days of Thomas Upjohn's funeral his widow, Charlotte, placed a notice in the Exeter newspapers, stating she intended to carry on the business and for that purpose would engage men properly qualified. However, in the following spring the Bell Hill shop was taken over by a watchmaker from Plymouth called Robert Bake who described himself as 'successor to Mr Thomas Upjohn'. This provoked a response from William Upjohn (ii), Thomas's former apprentice, who placed a notice in the 25 March and 1 April 1784 issues of the *Exeter Flying Post*:

> W. UPJOHN, Clock, Watch-Maker, and Silversmith, opposite the Corn-Market, in the Fore-street, (Brother to the late Mr Thomas Upjohn, on Bell-hill, and the only one of the Name and Trade in Exeter), begs leave to inform the public that his Brother's Widow having declined the Business, he solicits their Favours, having during her Continuance in it, assisted in repairing Repeating, Horizontal [ie cylinder escapement], and other difficult Watches.

Another watchmaker, John Tucker, later succeeded Robert Bake on Bell Hill (South Street); but in 1800 William Upjohn (ii), having traded previously at various premises in Exeter, took over the old family shop where as a lad he had served his apprenticeship. William was probably the maker of an eight-day longcase clock with a six-sided dial signed William Upjohn, Exon, which was found in a saleroom in a back street in Calcutta in 1934 (Fig 4). In 1803 William gave up the business to John Upjohn, his nephew, a notice appearing in the press to the effect that 'John Upjohn, son of the late Thomas Upjohn, clock and watchmaker, succeeds Mr William Upjohn, after having had twelve years practice in London, on the most improved scapements, repeaters and plain watches'.

William Upjohn then went to London where he died in 1812. His son, William John Upjohn (1774–1821), ran a small watch manufacturing business in Clerkenwell and maintained trade outlets in Devon and other parts of the West Country, travelling around the countryside in a gig (see Chapter 5). In the meantime the Exeter business was continued on by John Upjohn through the later years of King George III and subsequent reigns of George IV and William IV into Victorian times. Two of his sons were watchmakers with him in Exeter and two others were in business in Chandos Street, London (see List of Makers). He supplied a new clock for Holy Trinity Church, Exeter, in 1820; and kept the Exeter Cathedral clock in repair. He also had a hand in the restoration of Lovelace's Exeter Clock, it being recorded in *The Western Antiquary* of November 1885 that when James Burt of Exeter acquired it in 1821 he

**EXPRESS AND ECHO, FEBRUARY 3, 1934**

"Exonian" writes me from Calcutta, India: "Mention of old clocks in your 'City Talk' of December 9th prompts me to tell you of an experience here recently. I was wandering around an auction room, in one of the back streets of Calcutta, when I noticed a grandfather clock, somewhat the worse for wear. Imagine my surprise when, on closer examination, I read on the dial: 'William Upjohn, Exon.' The clock has a quaint, six-sided, brass dial, 14in. by 12in., and an arrangement for the date. Inside the case is a printed note giving the method of setting the date, and a name, C. E. Reed. The clock, an eight-day one, has at some time been repaired, and the following appears under the board on which the works are mounted: 'W. S. Payne, silversmith and jeweller, 46, Mary Arches-street, Exeter.'

"I wonder," continues "Exonian," "what is the history of this old clock, and by whom it was brought to India, so far from its home City? It was not in working order when I bought it, for a trifle, but it soon will be, and I hope that some day it may again give Exeter time. All that the auctioneer could tell me about the clock was that it had been lying in an outhouse for some while, and as the people to whom it belonged were going away, they sent it in to be disposed of."

Fig 4 A Devon-made clock is found in Calcutta. Extract from Cit's column in the Exeter *Express & Echo*, 3 February 1934

. . . got it home and found he could do nothing with it. He placed it in the hands of Mr Upjohn, watch and clockmaker, South Street, Exeter, where the clock was taken to pieces and cleaned. The works, as they were taken out, were laid in order on the floor of the room . . . and were at length put together, and it was set going and exhibited on Fore Street Hill, at a shilling a head.

After John Upjohn's death in 1848, aged seventy-eight, the business was continued at first by his sons Robert and James and then by Robert alone, who in 1855 moved to 39 High Street, which was his residence at the time of his death from consumption in July 1866. He was fifty-nine. For a few years the business was continued as Upjohn & Michie. By 1875 it had become Upjohn & Company; finally, in 1877, it was taken over by Samuel Aviolet.

Figure 5 shows a bill submitted a few months after Robert Upjohn's death for a year's maintenance of the Exeter Cathedral clock. It is for eight guineas, the same sum John Upjohn received sixty years earlier for performing the same task.

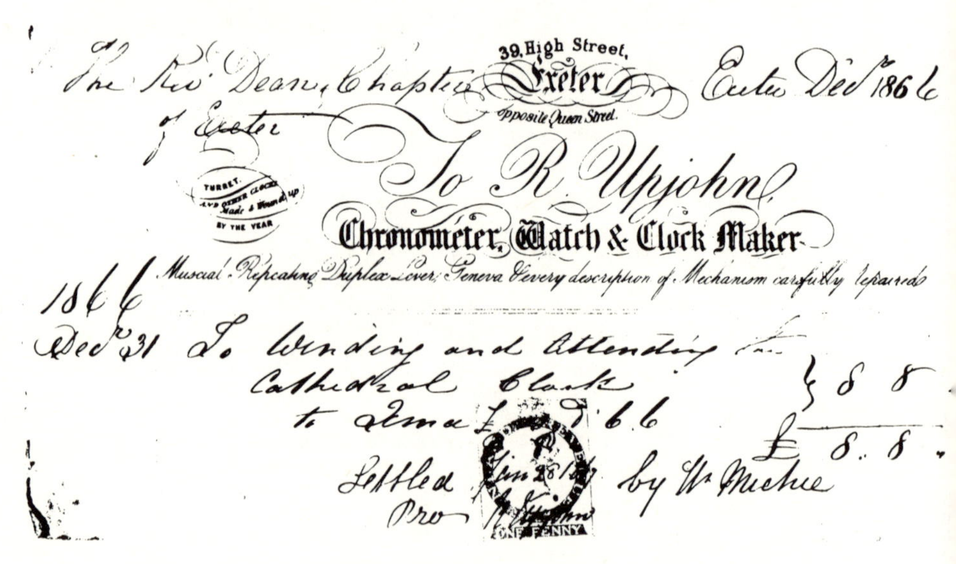

The Revd Dean & Chapter of Exeter

39, High Street, Exeter, Opposite Queen Street. Exeter Decr 1866

To R. Upjohn,

Turret and other clocks made & wound up by the year

Chronometer, Watch & Clock Maker

Musical, Repeating, Duplex, Lever, Geneva & every description of Mechanism carefully repaired

1866 Decr 31 To Winding and Attending the Cathedral Clock to Xmas 1866 — £8 8

£8. 8.

Settled Jan 28 1867 by W Michie Pro R Upjohn

ONE PENNY

Fig 5 Bill for winding and attending the Exeter Cathedral clock

Some rare and fascinating material on the early history of the Upjohn family is contained in a manuscript volume of some 160 pages entitled 'A short account of the life and travels of James Upjohn of Red Lion Street, Clerkenwell, clock and watchmaker and Goldsmith by Company – copied into this book by A. Upjohn, 1784', recently acquired by the Clockmakers' Company, but not, at the time of writing, available for inspection pending repairs and possible publication.

## GEORGE SANDERSON

A walnut longcase clock by George Sanderson, Exeter, was seen by the author some years ago in a saleroom in that city. The case was in a distressed condition and the movement was in a box in pieces; it nevertheless had the makings of a handsome clock of the mid-eighteenth century and was well worth restoring.

Like Thomas Mudge (qv), Sanderson was a man of invention and his name appears in horological reference books as a patentee. In 1761 he was granted the sole use and benefit of invention for fourteen years of some 'tools and engines for preparing, stamping, fixing, turning, cutting and finishing divers parts of a watch'; and the next year he obtained a similar patent for 'a lunar and kalendar watch key, adapted to every watch hitherto made, acting thereon by means of a screw and wheels of curious construction; is capable of showing in the most accurate manner the age of the moon, the day of the month, the revolutions of the tides, &c' (*Calendar of Home Office Papers of the Reign of George III*). In Exeter records Sanderson was usually referred to either as a jeweller or a watchmaker, and he served on Quarter Sessions juries at various times between October 1749 and May 1763. He was married at Exeter Cathedral in February 1746/7; the register entry reads: 'George Sanderson of the Holy Trinity in the City of Exeter, Jeweller, and Margaret Bray of St Petrocks in the said City, Spinster, were married the 13th day of February 1746.'

A son George was baptised at St Mary Major Church in 1749 and the couple had several children who died in infancy. In the Exeter poor rate assessment book for 1752–3 George Sanderson was listed in St Mary Major parish, his house being assessed at 5½d a week and his personal goods at 4d. Another watchmaker in the same parish, William Upjohn, was assessed at 4d and 2d. Sanderson was in St Martin's for the 1754–5 assessment, but moved back to St Mary Major where, in July 1761, he bought the lease of a property in St Mary's churchyard 'over the church porch'. This church – now demolished – was rebuilt in 1875–6, but old prints of the original structure show a gabled building adjoining where the north porch would have been.

Almost in the last year of his life, Sanderson left Exeter for London, where he was in partnership with John Mallett. The *St James's Chronicle* recorded his death which occurred in Clerkenwell on 10 October 1764, and in the same year carried a lengthy advertisement stating that Sanderson & Co's patent lunar and calendar watch keys were obtainable, price 7s 6d, from the shops of a number of London watchmakers and also from Mr Wickstead in the Grove at Bath. The watch key contained 'two figured plates, which though smaller than quarter-guineas, have more teeth than are in all the wheels of a watch, and by only the usual winding of it up, one of them regularly shows the day of the month, and the other the age of the moon'.

Preserved in the Devon Record Office (ref 51/1/10/7) is the assignment of the lease of Sanderson's house in St Mary Major. Dated 4 March 1765, it is between John Mallett of London, merchant, administrator of the goods, chattels and credits of George Sanderson (formerly of Exeter, jeweller, but since of St James, Clerkenwell, Middlesex, deceased, intestate) and Archibald Douglas of Exeter, gentleman, who acquired the lease for £80. The document states that the premises were occupied by a tenant, George Flashman. He was a clock and watchmaker and shortly afterwards moved to the High Street.

George Sanderson's patent watch key is discussed in a detailed article by Rita Shenton in the *Horological Journal*, December 1975, which includes a full specification and diagrams. The key was marketed after Sanderson's death by John Mallett, and an illustration on a trade card shows that the calendar part of the key was similar to the perpetual calendars made by eighteenth-century silversmiths. Sanderson's inventions in watch machinery are mentioned by Thomas Hatton in his *Introduction to Clock and Watchwork* (London, 1773), p383.

## JOHN TUCKER

John Tucker, of Exeter, was the son of the Tiverton clock and watchmaker John Tucker (1730–1807) and grandson of a maker of weavers' sleas (slays) and shuttles in the nearby village of Thorverton, in the Exe valley. He was born about 1760 and when young went to London, later setting up in business in Exeter. In a notice dated 29 October 1789 in the *Exeter Flying Post* he announced he had

> . . . removed from North Street to a convenient shop on Bell Hill, Southgate Street, (late Mr Thomas Upjohn's) opposite Mr Chown's, grocer, where . . . orders for new watches, plain, repeaters, &c. and clocks of all kinds will be executed at the shortest notice, and on the most reasonable terms. Likewise watches and clocks cleaned and repaired.

Ten years later he moved premises to a house 'adjoining the Conduit, in the High Street', and announced that he had the exclusive sale in Exeter of

> . . . the patent polar watches, which, combining elegance with utility, are constructed to admit a mariner's compass . . . the needles are made and suspended with uncommon accuracy and care, and the watches to which they are united, of a superior quality. The great utility of those watches to seafaring men and surveyors, need not be enlarged on, and even gentlemen who travel, will find a watch of this description a serviceable and agreeable companion (*Exeter Flying Post*, 5 December 1799).

A watch fitting the above description can be seen in the Clockmakers' Company Museum at Guildhall Library, London. The white-enamel dial is signed 'Benj$^{n}$. Webb, London, by the King's Patent', and the silver pair-cases are hallmarked 1799. The compass is neatly incorporated in the dial. An American correspondent, in *Antiquarian Horology* (September 1964, pp 251–2), gives a detailed description with measurements of a similar watch by the same maker. The patent referred to was probably that granted in 1798 (No 2,280) to Henry Peckham, of Bermondsey, for the application of a compass to a watch.

John Tucker also offered an early form of hire purchase, informing

> . . . all respectable young men, housekeepers, and others, who may wish to be accommodated with new or secondhand watches and clocks, or other articles, and to pay for them by certain instalments, that he should be happy to supply them on such terms, flattering himself they will see the peculiar advantage of being concerned with a fair dealer on this plan to the pernicious custom of clubs adopted by many.

He later changed his views on the 'pernicious custom of clubs' and himself ran clock and watch clubs at Bampton and Thorverton.

The Exeter Militia List of 1803 records him in St Petrock's parish, aged forty-three, married with three children, and 'willing to serve'. His shop was at 188 High Street, two doors along from the North Street junction. In 1805 he left Exeter to take over his father's business, and in a newspaper notice dated 26 June thanked his friends for their custom and informed them 'that being about to change the place of his abode, by removing to Tiverton, he has declined the business in favour of Mr John Elliott Pye, whom he recommends to their notice as a deserving young man'. At Tiverton the business ran into difficulties and in 1810 both he and his sister, who had a milliner's shop, were declared bankrupt. Tucker's goods were advertised to be sold by auction; they included new and secondhand clocks and watches and several valuable paintings and books.

Tucker subsequently lived at Thorverton and then returned to Exeter, living in Prospect Place, Preston Street. The churchwardens' accounts of the old St John's Church which stood nearby include several payments to him for work on the parish clock between 1817 and 1828. He died on 15 August 1829 while visiting his daughter Betsy in London. One of his sons, John Walter Tothill Tucker, was a watchmaker and silversmith in Fore Street, Tiverton, making the fine clock of 1830 that surmounts Tiverton pannier market and also becoming mayor of that town. Another son, Walter James Kelland Tucker, was a watchmaker in High Street, Exeter. Further particulars on the family will be found in Chapter 5. Exeter Museum has an eight-day mahogany longcase clock, 'John Tucker Exeter', with a square silvered dial, *c*1800.

### BENJAMIN BOWRING

A multi-million pound international insurance, banking and trading group, C. T. Bowring & Co Ltd of Tower Place, London, owes its origins to the enterprise of one man who at the close of 1815 sold up his Exeter watchmaker's business and took his family to Newfoundland. It is a remarkable story and has been the subject of two books: Arthur C. Wardle's *Benjamin Bowring and his Descendants* and David Keir's *The Bowring Story.* These give the overall picture, but can be supplemented from local sources with information on the watchmaker's early career in Exeter.

Bowring came from an Exeter nonconformist family. He was born in 1778 and was only three years old when his father, Nathaniel, died. His mother was from Moretonhampstead, on the fringe of Dartmoor, and it was there that he spent many of his childhood days, being educated, so it is believed, at an academy attached to the Moretonhampstead Unitarian Chapel. He was then apprenticed as a clock and watchmaker with the elder John Tucker, whose shop was on the north side of Fore Street in the market town of Tiverton. After this he presumably gained experience as a journeyman, as was the custom. In October 1803, at the age of twenty-five, Benjamin Bowring announced in the Exeter newspapers that he had opened a shop as a watchmaker, silversmith, jeweller and engraver at premises nearly opposite St Martin's Lane in the High Street. The announcement appeared in the *Exeter Flying Post* of 6 October; and on 8 October, a Saturday, Bowring and Charlotte Price, daughter of Charles Price a watchmaker at Wiveliscombe, were married in Wellington parish church, Somerset.

Bowring's shop was No 225 High Street and was part of a seventeenth-century merchant's house, the façade of which is now incorporated into C & A's Exeter store. In the *Exeter Flying Post* of 2 October 1806 he advertised for an apprentice and announced that he had moved premises to 199 High Street, four doors below the Guildhall. Henry Ellis, who was later to succeed him in the shop, relates that Bowring had 'a tolerably good clock trade' and that he employed a clockmaker named Charles Cross who had served his apprenticeship with Price in Wiveliscombe. Bowring's clocks were of good quality, judging by a well-proportioned mahogany longcase clock, with painted arch dial and moonwork, which is illustrated in Wardle's book. Another Bowring longcase was the subject of an enquiry to Exeter Museum, and a fine mahogany bracket clock by 'B. Bowring Exeter' is also recorded.

Wardle states that in 1811, leaving the Exeter shop in the care of Mrs Bowring and his assistant, Bowring took passage to Newfoundland in order to establish a similar business there. According to hearsay, he was induced to try his fortune after conversation with a customer at his Exeter shop who, on a visit to England in connection with the sale of codfish, purchased several of Bowring's clocks. By 1814 the clockmaker had established links in St John's, Newfoundland, and in Exeter had built the business up to the position where he was able to advertise: 'The trade supplied with clock-movements, clock brass, steel-work and watch materials'. In the following year Bowring decided to leave England for good. The Exeter business was disposed of, together with his agencies for the County Fire and Provident Life Offices, to Ellis, whose Memoirs (see Chapter 5) include details of the transaction, and outline Bowring's reasons for leaving.

In Newfoundland he at first continued in the clock and watch-making line, but then branched out into the general-shop trade and then into trade generally, supplying fishermen with goods and buying and exporting their produce. By 1823 he was the owner of a wharf and the schooner *Charlotte.* The rest of the story is beyond the scope of this book, but readers wishing to follow it further are recommended to read the two previously mentioned biographies. Benjamin Bowring died in Liverpool on 1 June 1846, aged sixty-eight, having returned from St John's in 1834. A ship called the *Benjamin Bowring* was used by the 1979–82 Transglobe expedition.

# 4

# NOTABLE MAKERS: CREDITON, TOTNES, PLYMOUTH AND REST OF COUNTY

## ABELL COTTEY

Abell Cottey (Cotty) was an early Devon maker of lantern clocks. He worked at Crediton and at Exeter; then, in middle life, emigrated to Philadelphia. One of his forebears may have been a certain Cottie who in 1581 assisted with repairs to the parish clock at Halberton, some three miles east of Tiverton. An American descendant, Edward E. Chandlee, devotes a chapter to Abell Cottey in his book *Six Quaker Clockmakers*. Drawing on American records, he states that Cottey was born in 1655 and that his father, John Cottey, died in 1672 at Tiverton, Devon. His wife's name was Mary and they are known to have had three children: John, born in 1677; Abell, born in 1679; and Sarah.

Plate 15 shows a brass lantern pendulum-clock, given to the Royal Albert Memorial Museum, Exeter, in 1899 by R. H. Rooks Bowden. The dial centre has a Tudor rose and other engraved floral decoration and is clearly inscribed 'Abell Cottey Crediton'. Like other lantern clocks of the period it has but a single hand. Empty holes at the front and sides of the top plate indicate that it once possessed ornamental frets. Several similar clocks by this maker have been noted, including one signed 'Abell Cottey de Crediton fecit'.

Cottey did not have the care of the church clock at Crediton; but in October 1695, while still living and working there as a clockmaker, he was granted from Kensington Palace a fourteen-year patent for his invention of a new engine for cutting and rasping logwood, and other woods used by dyers, 'which may easily be worked by hand or horse labour and by water' (*Calendar of State Papers*, Domestic Series,

William III). Cottey then moved to Exeter, possibly to exploit this invention, and can be traced in St Edmund's parish, near the river and in the quarter of the city that in those times was associated with the cloth industry. His name appears as Abell Catty in the Exeter poor rate books of 1697–8 and 1699, and he had to pay one penny a week towards the relief of the poor (Devon Record Office).

His son John had meanwhile trained as a clockmaker, but in 1698 was jointly accused at Exeter Quarter Sessions of stealing from him:

> Forasmuch as it appeareth unto this Court by the oaths of severall Witnesses that John Cotty of this Citty Clockmaker and Thomazine his Wife did lately enter into the dwelling house of Abell Cottey Father of ye sd John Cottey within ye County of this Citty & take and carry away from thence severall Deeds and Writeings relateing to an Estate of the vallew of Fower hundred pounds and they haveing not sufficient suretys for their severall personall appearances at the next Generall Quarter Sessions of ye Peace to bee held for this Citty & County & then & there to answear the sd offence This Court doth therefore for want of such suretys as aforesd Committ the sd John Cottey & Thomazine his Wife to his Majesties Prison at Southgate there to remaine untill they find suretys as aforesd or beefrom thence delivered by due order of Lawe [Book of Recognizances, 3 October 1698, DRO].

The outcome of this case is uncertain, but it is significant that Abell, in his will made nearly twelve years later, left his son a mere 5s.

Chandlee and other writers, unaware of these later Devon records, state that Abell Cottey arrived in Philadelphia in 1682. His name appears in an early list of Philadelphia craftsmen, some of whose names are marked with a cross. The document states: 'The Names that is Crost came In ye furst Ship that Came from England 1681 ye year before ye Propriator In a larg ship ye 11 dy 10 mo landed att Chester the Capts Name was Roger Drew ye Ships Name Bristoll-Facktor.' Cottey is the only clockmaker listed, but his name is not crossed. If it is true that he went out in 1682, then he must have returned to Devon, for it can be clearly established that he left Exeter for Bristol at the beginning of April 1700, and from there sailed to America. His departure, however, led to a flurry of correspondence. The Exeter Friends (Quakers) claimed he had not settled his debts before leaving and, in a letter dated 23 April 1700, R. Lowbridge and John Ganniclife wrote from Exeter to fellow Quakers of the Bristol Men's Meeting informing them that

Abell Cotty of this Citty about two weekes agoe, in y$^e$ night tyme, without Friends knowlidge, Sould some of his goods, and takes y$^e$ rest, with his Family, and goes off, for Bristoll, a great way [in] Debt (, though itts Concluded he hath Sufficient to pay all men,) there hee, with his Effects, are Conseal'd, untill pasage offers for Pensillvania, as we are Informed.

They also complained about another Quaker called John Hurford, but he, on arrival in Bristol, was able to produce a certificate from the Cullompton Monthly Meeting which stated that he left there free of debt. This led the Bristol Friends to send back a sharp reply. However, Exeter then sent certificates to demonstrate that debts due from John Hurford were still outstanding.

You blame us [they wrote on 26 June 1700] for being forward but if we had been more forward perhapes we might a Stopt Abill Cottys going off at Bristoll untill he had payd his just dues, but now hee is gone and left a Stinke behind him, which we offten heare of to y$^e$ dishonour of Truth and reproach of Friends, we doe still thinke . . . some care one way or other ought to be taken, to prevent for ye futer any going off, in such a dishonest way as those have done [Bristol Archives Office SF/A7/4, pp 161–71].

Having arrived in the New World and established himself in Philadelphia, Pennsylvania, Cottey numbered among his customers William Penn, proprietor and governor-in-chief of the province. Edward Chandlee quotes two extracts from Penn's cashbook, including an entry in 1704 recording a payment of 15s to 'A Cotty for scouring etc the clock'. Chandlee also gives details of Cottey's dealings in city property and plantation land, and mentions his own twenty-year search for an American example of Cottey's clockmaking skill. Only one was found, an eight-day longcase with an 280mm (11in) square dial. This was signed on the chapter ring 'Abel Cottey Philadelphia'. The dial corners, in the absence of spandrel mounts, were beautifully engraved and the clock had a seconds dial and a calendar aperture. The front brass plate of the movement had a roughly cut inscription, 'B C 1709 m9 X20 Clock'. The initials B C were those of Cottey's apprentice Benjamin Chandlee, who married his daughter Sarah. 'The date, 1709, is the earliest so far discovered on any American-made clock', wrote Edward Chandlee.

Cottey was described as a watchmaker when he was buried 'on 6 month [August] 30, 1711'. (Before the English calendar change in

1752 the first month was March; hence August was the sixth month.) In his will, made the previous year, he bequeathed 5s as already mentioned to his son John, 'if ever he returns to this Citty of Philadelphia', and to his grandson, Abell Cottey, 'one of my lotts of Land at Nottingham that Lyeth nigh Andrew Jobs Lott when he comes of age or that my executrix can spare it'. The remainder of his estate was given 'unto my dear & Loving wife Mary Cottey'. His furniture and household effects included a best bed, a walnut oval table, pewter dishes, brass candlesticks, sundry old books, a wheel engine and tools, and, valued together at £30, 'Three Clocks unfinished & watch Clock & 2 watches'. A separate inventory of his workshop effects was taken on 8 October 1711 by the clockmaker Peter Stretch and by Caleb Jacob. It is remarkably detailed and shows that Cottey was fully equipped to work on both clocks and watches; it also excellently illustrates the tools and materials needed by an eighteenth-century clockmaker (see Appendix).

Cottey's widow in her will proved in 1714 left property and land to her daughter and son-in-law, Sarah and Benjamin Chandlee, and to her grandsons Abel Cottey and Cottey Chandlee, also 'If my son John Cotty shall come into these parts again I give unto him tenn pounds'. An Abel Cotty was buried at Topsham, Devon, on 7 December 1731 (parish register).

## ARTHUR DAVIS

Arthur Davis was another early Devon Quaker clock and watchmaker. He was closely connected with the Cullompton Monthly Meeting and his name occurs in the previously mentioned correspondence of 1700 concerning the clockmaker Abell Cottey and John Hurford (Bristol archives). Davis seems to have moved from place to place. A bracket clock of *c*1700, in a later mahogany case, is signed 'Arthur Davis Tiverton'; it was seen in a Sussex saleroom in 1979 and had half-quarter marks and very small minute figures on the chapter ring. A lantern clock in the old apartments at Greenwich Observatory is apparently signed 'Arthur Davis de Westleigh'. This Westleigh is probably the small place of that name near Devon's border with Somerset. Davis, described as a watchmaker of Cullompton, was a 'ruler' of the will of Thomas Fry the elder, of Cullompton, proved by affirmation on 22 May 1712. He was also described as a Cullompton watchmaker on the original lease of the Quaker burial ground there, dated 1708; but on another lease relating to it, in 1723, was described as a watchmaker of Kentisbeare (Devon Record Office).

## LEWIS PRIDHAM

Lewis Pridham flourished for many years as a country clockmaker at Sandford, a few miles north of Crediton. Little is known about him, except that his workplace was probably in the hamlet of New Buildings and that he was married three times between 1687 and the time of his death at an advanced age in April 1749. Curiously, at one stage of his life he went by the alias of Greenslade.

Plate 16 shows the square dial of a thirty-hour longcase clock signed on the chapter ring 'Lewis Pridham Sandford fecit' and dating from about 1735. The dial centre and sides are ornamented with a flowing pattern of engraved leaves. There is a lantern clock by Pridham in the Royal Ontario Museum in Toronto, Canada, and he also made some eight-day clocks. Prebendary J. G. M. Scott, who has made a special study of this maker's work, has provided a description of a tall clock dating from Queen Anne's reign which was made by Pridham, apparently for the room in which until recently it stood in a country house near Exeter.

The case is about 3.4m (11ft) high, veneered in walnut on a pine carcase, the trunk well proportioned and slender, and the hood surmounted by gilded finials. The dial is brass with a silvered hour ring, inscribed at the bottom 'Lewis Pridham Sandford fecit'. The mounts in the spandrels show putti supporting crowns. The movement is, most unusually, a chain-drive thirty-hour one, with a separate musical train. The going and striking trains are mounted one behind the other between top and bottom plates which are supported by corner pillars, square in section and moulded. Three pins on the locking-plate operate a lifting-piece to set off the six-bell musical train at 9, 12 and 5 o'clock – traditionally the meal times of the household when the clock was made. The tune is repeated three times, and is timed to follow immediately after the last stroke of the hour.

Pridham made at least one other thirty-hour musical clock; but it is his turret clocks that rank him as a maker of importance. Indeed, it is probably true to say that he made more of them than any other Devon maker of the period, and a high proportion are still doing service including the church clocks at Dulverton and Newton St Cyres, both of which carry his brass nameplate. But three of his earlier clocks – those made for Colebrooke in 1698 (cost £14), Cheriton Fitzpaine, and Alphington – have been replaced, as have the clock and chimes he made for Bampton at a cost of £22 6s in 1728.

His Alphington clock seems to have been his least successful and he was called back in 1713 'to Alter the Clock now set up in the Tower and to make it goe well Thirty howers at each winding up and to make the Index hour hand to Answare the Clock, and the Quarter Clock to be heard over the Church as it is Now and the hower Clock to strike much harder than it does now'. At Alphington also, in 1716, he supplied 42kg (92lb) of new ironwork for the great bell and second bell in the tower. Pridham also made locks. There is a fine wooden-cased example on the door in the church tower at Colebrooke; it bears his initials 'LP' and the date '1730' (Plate 17). The churchwarden's accounts record its purchase from him for 10s. This maker enjoyed a reputation over a wide area of the county and he was employed as a repairer as far afield as Wiveliscombe, Somerset, where in 1714–15 he re-set and wired the church chimes.

Pridham's extant turret clocks are neat and well made; their iron frames exhibit excellent blacksmith's work, the pivot bars are of a distinctive design and the corner posts terminate in attractively wrought finials. At Dulverton in Somerset, Pridham's nameplate is on the top front bar of the church clock. It is inscribed:

IVNE-28th – 1708
HVMPHERY
SAYDENHAM-ESqr
GEORGE-PEPEN-GENT
CHVRCH-WARDENS

LEWES-PRIDHAM SANDFORD
FECIT

The Pridham clock at Newton St Cyres (Plate 18) replaced an earlier one. The churchwarden's accounts record that a horse was hired at a cost to the parish of one shilling 'to go to new building [New Buildings] to fetch home ye Clock & Dyall Table'. Pridham charged £16 and received a further £2 for new placing the clock and dial and making a new hole through the tower. The brass plate on the top rail is inscribed: 'WILLEM LANGWORTHY EDWARD BRAY CHVRCH-WA[R]DENS OCTOBER Y$^{E}$ 8$^{TH}$ 1711 Lewes Pridham Sandfford ffecit'. Forgotten for over sixty years, the clock was restored by a parishioner, John Durrant, and set going again on Palm Sunday 1971. It is now auto-wound, but only the escape wheel is modern. There are scroll finials on each of the four corner posts, the pendulum cock is formed by a double cranked extension of the back pivot bar, and the pallets are retained by a disc-ended key (Prebendary J. G. M. Scott). Another clock dating from 1711 is on display in St Anne's Museum, Barnstaple. It was made originally for the town's Quay Hall and carried the motto 'About Your Business'. The movement has lost its nameplate, but it is undoubtedly a Pridham.

Also in north Devon, in the grounds of a house formerly belonging to the Rolle family, is a charming, brick-built clock pavilion (Plate 19) which retains its one-handed dial bearing the date 1711 and its original Pridham movement. This clock has a longer running period than its maker's church clocks, and is fitted with four-wheel trains. A brass plate on the top front rail reads: 'FEBR[U]ARY 1 1711 LEWES PRIDHAM SAND FORD FECIT'. Bearing in mind that in England prior to 1752 the year did not change until 25 March, ie Lady Day, this clock was made after the Newton St Cyres one, dated 8 October 1711.

Two church clocks attributed to Pridham are Jacobstowe and Colyton. The clocks at North Tawton and High Bickington are also in his style, but could be the work of the Barnstaple maker John Cole whose turret clocks were not dissimilar.

In an agreement concerning the church clock at Alphington, Pridham signed his christian name as Lewes and, as has been shown, this spelling also occurs on some of his clocks. The parish registers of Sandford record that he and Thomasin Parsons were married there on 7 August 1687; she bore him two sons, Noah and John, but died within a few weeks of the latter's birth and was buried on 14 March 1692/3. The clockmaker then married Mary Philip, on 3 May 1693, and there were several more children including a son called Roger. It would seem that Mary died in 1728. Pridham (alias Greenslade) then married Mary Oldden of Coldridge, a widow, at St Stephen's Church, Exeter, on 3 December 1730. He was buried at Sandford on 11 April 1749, probably aged over eighty.

## WILLIAM CLEMENT

William Clement, of Totnes, was a younger contemporary of the Exeter clock and watchmaker Edward Clement. He started in business in the town some time before 1701, was churchwarden of St Mary's in 1705, and served as mayor in 1711. It is tempting to suggest that he was the son of the anchor-escapement pioneer William Clement, of London, master of the Clockmakers' Company in 1694. If so, he was apprenticed to Daniel Steevens in November 1684 and then 'turned over' to his father for the remainder of his seven-year term. Evidence of his arrival in Totnes is provided by a notice published in the *London Gazette* for Monday 23 June to Thursday 26 June, 1701:

> Lost upon the road between Salisbury & Stockbridge the 20th instant, a Silver Pendulum Minuit Watch in a plain Silver Case, with a Bob Ballance, & Glass in the half-cock; the Name William Clement, Totnes, & a sky colour Ribon for the Key. Whoever brings it to Mr Foulks at the Three Lions in Salisbury, or at Mr Halsted's at the Crown & Dial in Fleet Street, London, shall have 20s Reward.

Jacob Good was apprenticed to Clement at Totnes for a premium paid in 1711 of £16 2s 6d (Inland Revenue Apprenticeship Books). The parish registers record the burials of Mary, wife of Mr William Clements, on 27 February 1732 and of Mr William Clement on 18 June 1736.

Outstanding among his surviving clocks is a little square-dialled bracket clock with carrying handle, signed on the chapter ring and backplate of the movement 'Wm Clement Totnes Fecit' (Plate 20). It dates from about 1700 and was bought at auction in 1956 by a leading dealer 'because of its obvious association with the best London work of the period'.

Also illustrated (Plate 21) is the dial of an eight-day lacquer longcase, 'William Clement Totness', dating from about 1720–5. This clock was in a very neglected state when photographed and had lost its second hand and pendulum. Its most remarkable feature, a large casting in the dial arch surrounding a plaque with the maker's name and town (Plate 22), consists of two pairs of cherubs and a crown. The clock has ringed winding squares and a matching calendar aperture, the latter surrounded by an engraving of a pair of birds and a basket of fruit; the movement has an inside countwheel. The arch of another Clement longcase clock has dolphin-shaped castings, in a slightly later fashion. Mr Tapley-Soper recorded a thirty-hour oak longcase with a single hand and a square dial signed 'W. Clement Totnes'. Another thirty-hour clock, reported from Canada, has a round calendar aperture and bird-and-fruit engraving similar to that just described.

## WILLIAM STUMBELS

The tradition of fine clockmaking established at Totnes by Clement, was continued there by William Stumbels, a craftsman of great artistic ability who had previously worked at Aveton Gifford near Kingsbridge. Stumbels catered – though not exclusively – for the top end of the trade and made chiming longcase and other clocks for the owners of some of south Devon's principal country houses. Among his patrons were the Courtenay, Champernowne, Carew and Ilbert families. The late J. K. Bellchambers, in his book *Devonshire Clockmakers*, declared that Stumbels was probably the most brilliant maker of his time outside London.

It is not known where or with whom William Stumbels served his apprenticeship, but he did have a connection with the clockmaker John Belling (i) of Bodmin, whose dates are given by Canon H. Miles Brown in *Cornish Clocks and Clockmakers* as 1685–*c*1761. Possibly Belling was Stumbels's master, or perhaps they worked together as journeymen. They might even have been fellow apprentices. At any rate, Stumbels provided a surety when Belling took out a licence to marry in June 1723.

Stumbels, having learned his trade, established himself in business at Aveton Gifford. He and Sarah Phillips were married there at the parish church of St Andrew on 12 November 1716 and they had a family of at least seven children, only two of whom reached maturity. Stumbels acquired property in the parish, but after two of his children died within weeks of each other in 1729, he moved with the remainder of his family to Totnes, on the River Dart. The name Stumbels has not been noted in the eighteenth-century Totnes rate books, so it must be assumed he worked in rented premises.

The Inland Revenue archives record that in 1751, when trading as William Stumbels and Company, he received a premium of £21 with John Richards, an apprentice in the business. His other apprentices included George Paddon, with whom in 1735 he received a £10 premium, and John Roucklieffe, with whom in 1740 he received £20. Parish accounts provide a sidelight on his work as a repairer. At Modbury in 1730, for example, he was paid £2 10s 'for hanging the second bell'; he repaired Ashburton Church clock several times in the 1740s and, in 1748, according to the accounts of the Wolborough feoffees, supplied two bell brasses and wheelwork for a new dial at St Leonard's tower, Newton Abbot. This last job took him a long time to complete and the feoffees, who were the administrators of parish property, noted that the scaffold was kept up for eight months 'because Mr Stumbells kept ye wheel work of Dial so long before he finish'd it'.

In 1767 his wife Sarah died, just two years before he himself died after a long clockmaking career spanning more than fifty years. He was buried at Totnes on Christmas Day, 1769. There is a copy of his will in the Devon Record Office, dated 18 January 1768 and made by Stumbels when 'weak of body, but of sound disposing mind, memory and understanding'. To his son Bezaleel, clock and watchmaker, of London, he bequeathed 'the sum of one guinea of lawfull British money'; the rest of his estate, including his house, outbuildings and garden at Aveton Gifford, 'which I formerly purchased of one Thomas Martin', and his goods, chattels and tools, went to his daughter Elizabeth, whom he appointed as his sole executrix.

Four Stumbels turret clocks have so far come to light. The earliest of them has a brass plate, inscribed in flowing italic letters:

William Stumbels
Aveton Gifford fecit;
1723

The movement has a four-post wrought-iron frame. It was made for a country house and was followed seventeen years later by a similar but larger clock for St Mary's Church, Brixham. This movement is now displayed in the church, but its arch dial can still be seen on the tower outside. The nameplate from the clock is also arch shaped and has brass supporting scrolls. Upon it is the inscription: '1740 Mr N: Browse Mr R: Lang CHURCH: WARDENS W: Stumbels TOTNES; Fecit'. Another Stumbels turret clock is in Totnes Museum. This movement was housed originally in the porch tower at Dartington Hall, where it operated a single-handed dial and struck on a bell dated

1737. The fourth clock, the latest, is at Forde House, Newton Abbot.

The frames of the three earlier clocks are similar, with cross-pieces set in from the ends, and the octagonally shaped corner-posts curved into scroll finials. The Forde House clock shows Stumbels at his most elegant and original (Plate 23). The whole effect is decorative; the ball and spike finials are gilt and the trains were carefully designed to give symmetry to their neatly turned brass bushes. The nameplate is framed in Stumbels's most exuberant ironwork. Two features reminiscent of longcase clocks are an internal fly-vane at the top of the strike train, and winding ratchets on the barrels. The clock bell in the central cupola at Forde House is dated 1751. At that time the house belonged to the Courtenay family, ancestors of the present Earl of Devon.

Stumbels's wheelwork and design of pivot bar are distinctive. In his striking mechanisms he favoured what has sometimes been described as the modified countwheel system. This takes several forms. The method employed in the Dartington Hall clock, in which a countwheel with raised teeth is affixed to the inner side of the strike main wheel, is similar to that described and illustrated earlier in the section on Jacob Lovelace, of Exeter. The strike in the Forde House clock is controlled by means of pins spaced around the outside of the main wheel. This latter method is also found in an early square-dial eight-day longcase clock made by Stumbels when he was at Aveton Gifford (Plate 24), and in a similar clock made a few years later when he was at Totnes.

This maker is, of course, best known for his longcase clocks, the humblest of which have thirty-hour movements, well-finished dials and simple cases – usually of oak and with an inset glass panel on each side of the hood, to enable the movement to be seen. Delightfully shaped nameplates, as in Plate 25, are a feature.

An eight-day longcase clock, made by Stumbels when at Aveton Gifford, has a revolving moon sphere in the dial arch and a six-bell quarter chime; the chime fly is outside the backplate and there is strike/silent work for the hours and quarters. Interestingly, the movement has three barrels – one for each train – but only two winding squares. A similar arrangement has been observed in a Totnes-made Stumbels which has two sets of chimes, either of which can be chosen to play out after the hours have been struck at 3, 6, 9 and 12. The chimes can be repeated by pulling a string at any time during the two hours following these times. One end of the key of this clock is designed to raise two of the weights, whilst the other end is used to raise the third. On the inside of the case door are 'Rules to be observed in keeping this clock', followed by very minute instructions.

One of the most remarkable Stumbels clocks was described and illustrated by J. K. Bellchambers, who called it the Kingsbridge Stumbels and suggested a date of *c*1735. This clock was seen by the present writer some years ago; it stands in the great hall of a country seat and has a tall case of banded walnut veneer, with gilt enrichments, including a figure of Time standing on a domed hood. On either side of the hood door are detached pillars with Corinthian capitals, and the base is supported by gilt feet. To quote Bellchambers:

> This longcase clock plays quarters or chimes on nine bells, at the whim of the owner. In the arch-dial there is a beautiful tidal dial, with a rotating moon set in a blue sky showing the stars. Underneath this top dial, on either side of an aperture showing the month of the year, are the words 'Stumbels Totnes' [on separate plates]. To top left and right of the main dial there are two further dials, showing the date and the time of rising and setting of the sun. The main dial has only two winding holes for the three trains. In the centre circle [within the seconds dial and marked 'A Table of Equation'] is an aperture showing the difference between the sun time and time of the clock. It is not in fact a true equation clock, the difference being indicated by a disc geared into the annual calendar.

The equation disc is designed to change every Sunday morning. In the bottom half of the dial centre there is a vase-shaped opening showing the day of the week together with an engraved deity and a planetary sign. The gearing of this clock gives compensation for leap-year and, as an added bonus, the clockmaker's original instructions have been preserved, written in a stylish copperplate hand on a paper pasted to the inside of the trunk door (see page 103).

J. K. Bellchambers went on to describe another equally extraordinary clock by Stumbels – his 'thirteen-foot high masterpiece which must now be known to the many visitors to the home of the Earl and Countess of Devon at Powderham Castle near Exeter. It is basically the same as the Kingsbridge clock, but in addition it has maintaining power on the going train and the layout of the dial is somewhat different; instead of having the equation disc, there is the full equation kidney and mechanism, and the case is slightly bulbous

Plate 9 Lovelace's celebrated Exeter clock. It was first exhibited in 1739. The magnificent case was probably made by John Channon, of London. Lithograph by Hackett, Exeter, 1833. The initials top right are those of the clockmaker Benjamin Lewis Vulliamy, onetime owner of this particular print

Jacob Lovelace
EXON

Rules

To be observ'd in case this Clock shou'd be Stop't or let down

If you are mind to stop the quarters shift the Piece on the right hand and it will alter them into Chimes every four Hours at 8, 12, & 4 but you must not shift it, but at 1, 5, & 9. And if the Piece on the Left-Hand is to Silent you cannot shift it at all, for Strike-Silent on the Left-hand stops all except the going Part without any detriment. The day of the Month by being turn'd till the Month of the Year is right, sets Sun-Rising and Sun-Setting right. The Equation Circle moveth every Sunday Morning so that if you observe the Months as under it will alway be right.

| Jan$^{y}$ | 1$^{st}$ | 8 Min | 59 Sec$^{ds}$ | | July | 1$^{st}$ | 4 Min | 43 Sec$^{ds}$ | |
|---|---|---|---|---|---|---|---|---|---|
| Feb$^{y}$ | 4$^{th}$ | 14 | 4 | faster | Aug$^{st}$ | 5$^{th}$ | 3 | 49 | |
| Ma$^{r}$ | 4$^{th}$ | 7 | 10 | Ditto | 7:$^{ber}$ | 2$^{d}$ | 4 | 9 | Slower |
| April | 1$^{st}$ | 0 | 49 | Ditto | 8:$^{ber}$ | 7$^{th}$ | 14 | 31 | |
| May | 6$^{th}$ | 4 | 11 | Slower | 9:$^{ber}$ | 4$^{th}$ | 14 | 55 | |
| June | 3$^{rd}$ | 0 | 40 | Ditto | X:$^{ber}$ | 2$^{d}$ | 5 | 13 | |

in the Dutch style'. The clock has been at Powderham since the mid-eighteenth century when some panelling in the hall was cut to fit it. On the inside of the trunk door are 'Mr Stumbels's Directions to the Person who has the Care of this Clock' (Fig 6). The complex dial is perhaps the finest Stumbels ever made (Plate 26); the layout is masterly and the finish luxuriant. The clock's splendidly gilded and carved case – the hood is shown in Plate 27 – was probably made by the Channon family of cabinet makers (see page 71), who were employed extensively at the castle. The bulbous base appears to be made of rosewood and could be a replacement, as the rest of the case is walnut. There was incidentally a connection through marriage between the family which commissioned the Kingsbridge clock –

Plates 10 & 11 (*left and top right*) 'Jacob Lovelace Exon'. Black Japanned longcase clock with gilt and brown decoration, *c*1715, about 2.4m (8ft) high excluding top finial. The eight-day movement has two hammers, one on either side of the bell, the smaller one for sounding the half-hours. The 30cm (12in) square dial is signed both in the centre and on the chapter-ring. The lever marked NSN above the XII is for silencing either the hour or half-hour strike

Plate 12 (*bottom right*) Hood of black Japanned longcase clock by Jacob Lovelace, of Exeter, with turned wood finials and a caddy top, *c*1715. The silvering had worn off the brass chapter-ring when this photograph was taken, hence the dull appearance

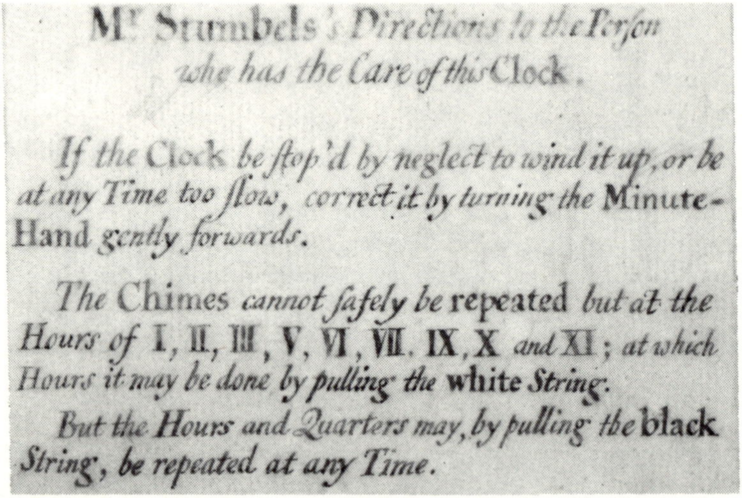

Mr Stumbels's Directions to the Person who has the Care of this Clock.

If the Clock be ſtop'd by neglect to wind it up, or be at any Time too ſlow, correct it by turning the Minute-Hand gently forwards.

The Chimes cannot ſafely be repeated but at the Hours of I, II, III, V, VI, VII, IX, X and XI; at which Hours it may be done by pulling the white String.

But the Hours and Quarters may, by pulling the black String, be repeated at any Time.

Fig 6 Directions on the inside of the case door of the William Stumbels clock at Powderham Castle

which also has a case in the Channon style – and the Courtenays of Powderham. The castle clock has an eleven-bell musical train fitted with double hammers; the pin barrel at the back of the movement is supported by scroll-shaped metal brackets – a typical Stumbels touch. The letter S (for Stumbels) is fashioned on the tail of the hand within the small Sun Rising/Sun Setting dial. An S-shaped winding key from another of his longcase clocks is illustrated (Plate 28).

Other notable longcase clocks by Stumbels include a 'regulator' with a superb dial and mahogany case (Plates 29 and 32); a very tall one in a walnut case with a universal tidal dial in the arch (Plate 30); a month clock with moonwork; a 2.7m (9ft) high eight-day clock in a green lacquer case with gilded brackets immediately below the hood; another similar, but not so tall (Plates 33 and 45); and an oak-cased movement showing 'High Water at Totnes'.

Two Stumbels bracket clocks have been noted, both with ebonised bell-top cases. One of these is illustrated (Plate 31). The dial has silver spandrel mounts, mock pendulum and calendar apertures, a particularly elegant nameplate and, in the arch, a dial which appears to be for the seconds but is, in fact, for regulating the pendulum. The backplate is engraved with a cherub, birds and scrollwork, and the movement can be made to repeat the quarters. The second clock has an all-over silvered dial and is, strictly speaking, a spring timepiece with alarm. A Stumbels watch, number 733, hallmarked London 1754, is in the Ilbert collection at the British Museum; a second specimen is in Glasgow, and notes have been taken of two more which were sold at auction. Both were verge movements with enamel dials in silver pair-cases and were numbered 1200 and 2807 respectively. Two newspaper advertisements were quoted in *Devonshire Clockmakers*; the first was from *Aris's Birmingham Gazette*, dated Monday, 29 September 1777:

> Early on Saturday morning the 27th instant, the Bristol Diligence was robbed within half a mile of Birmingham, by two men on horseback who, beside cash, and other valuables, took from me a silver watch, Maker's Name Wm. Stumbles, Totness, No. 3300 and some odd number forgot, with an enamelled Dial Plate, with a plan in it. If the said watch is offered to sale or pawn pray stop it and give notice to me at Teignmouth, in Devon, or to Mr. Ed. Ruston, in Birmingham, and they shall be well rewarded by me. W. MORTIMOR.

The second, from the *Public Advertiser*, London, October 1779, referred to 'a plain double cased metal watch, name William Stumbels, Totnes', which was in the Public Office at Bow Street, 'stopped on Thursday night supposed to have been stolen'. The owner was desired to apply immediately to the office.

It was perhaps characteristic of William Stumbels that he gave his son Bezaleel the name of an Old Testament master craftsman who was just such an artist as he was himself:

> And the Lord spake unto Moses, saying, See, I have called by name Bezaleel the son of Uri, the son of Hur, of the tribe of Judah: and I have filled him with the spirit of God, in wisdom, and in understanding, and in knowledge, and in all manner of workmanship, to devise cunning works, to work in gold, and in silver, and in brass, and in cutting of stones, to set *them*, and in carving of timber, to work in all manner of workmanship (Exodus 31: 1–5).

Bezaleel Stumbels, clock and watchmaker of London, was born at Aveton Gifford on 30 March 1728 and baptised there a month later. He married Sophia Rafugeau at St Paul's Church, Covent Garden, on 26 June 1760; and the baptisms of at least seven of their children have been noted in the registers of several London churches between 1761 and 1776. Recorded examples of Bezaleel's work include a watch and a longcase clock.

Plate 34 shows the arch dial of a longcase clock made at

Kingsbridge by William Stumbels's former apprentice George Paddon. The lettering on the chapter-ring is similar to that on the thirty-hour Stumbels clock shown in Plate 25.

The name of William Stumbels has been linked with Adrian's Clock in the Victoria and Albert Museum, London. The mechanism of this imposing piece of clockwork incorporates a carillon and the whole is majestically housed in a walnut-veneered, two-stage cabinet (Plate 35) upon the top of which, under an ornate hood, is an arched dial such as might be found on a longcase clock. The hood, in turn, is surmounted by a crown-like structure terminating in a pineapple finial. Four other gilt pineapples ornament the corners of the main cabinet top.

The clock, dated by the museum as *c*1725, was made under the direction of Maine Swete, an amateur metalworker and bell enthusiast, as a gift for his brother Adrian who lived at a house known as Traine, at Modbury in south-west Devon. Adrian Swete was high sheriff of Devon in 1725, and such was the family eccentricity that when he died in 1733 his executor asked that his funeral might take place in Modbury Church at five in the morning! The vicar, Archdeacon Baker, not unnaturally refused, so it took place at Ermington instead. The house was inherited by Maine Swete who had earlier lived in London and on an estate in Antigua in the West Indies. In 1810 Prebendary John Swete, of Oxton near Exeter, the then owner of the clock, wrote of him as follows:

> Mr Maine Swete appears to have been a man of much ingenuity, which led him to be somewhat peculiar in his choice of amusements. He learned to work as a blacksmith and had a forge in the Traine orchard adjoining the highway, where a friend of his who died some years since at Modbury often assisted him with the sledge. He delighted much in bell-ringing and bell-music and once (in the disguise of a blacksmith) headed the ringers of Modbury in winning a prize at Totnes. An elegant musical clock was made by his direction at the expense of near £200 and is now in the hall of the present Mr Swete [ie of the writer].

Maine Swete died in 1735, but his widow – his second wife Esther – survived him by forty-six years. Known as Adrian's Clock after its original owner, the clock was acquired by the museum from the actor E. Lyall Swete, in 1925. Notes furnished by the museum state that inside the top section of the cabinet is a bell to strike the hours, and twenty-seven other bells tuned in semi-tones to play three octaves by means of a small keyboard or by a cylinder set with pins in the manner of a musical box. The latter can be made to play one of three tunes every three hours – 'Hark! the cock crows', 'The Beau's Delight', and a rigadoon – and the cylinder can be replaced on Sundays to play a psalm. The cabinet is said to have been made by a cabinet maker in Plymouth; the actual maker of the clockwork is not recorded, although the clock may once have carried his name, for there is a blank silvered disc in the shallow arch of the dial. The clock is weight driven and has quarter chimes; the chapter ring is bevel edged and the drive is taken to the hands by lead-off rods, as in a turret clock.

Stumbels was one of the few south Devon clockmakers capable of working on such a clock and it is perhaps significant that at the time it was made he was living at Aveton Gifford, just three and a half miles from Modbury. This clock is discussed in some detail in Tom Robinson's book *The Longcase Clock* (pp 240–2).

## SAMUEL NORTHCOTE

The first Samuel Northcote, watchmaker, was the father of the artist James Northcote RA, and is chiefly remembered for his reluctance to allow his son to embark on his chosen career. Before being permitted to pick up his brushes in earnest, James was obliged to serve a full seven-year apprenticeship with his father as a clock and watchmaker, and continued to work in the business until he was nearly twenty-five. As James wrote in his autobiography:

> I was born at Plymouth, in the county of Devon, on the 22nd October, 1746. My father, whose name was Samuel, was by trade a watchmaker, valued by all who knew him for his great integrity, abilities, and general knowledge. He also was born and dwelt in the town of Plymouth, and was a pious, studious, humble and ingenious man, one better calculated to make a good use of money when got than to get it; and although an ardent admirer of the art of painting, he could never be persuaded to give his consent to my choice of it as a profession. One reason for this disapprobation was that his father had been an unsuccessful painter, which had taught him to consider it as an art difficult of attainment and uncertain of success, therefore not proper to be the profession of one whose sole dependence was to rest on his own industry. For, as I before observed, being more honest than wealthy, it was not in his power to give anything but his good example to his children (three of whom, out of seven, arrived to years of maturity).

The artist's father was born in the early part of the year 1708/9 and was described as a watchmaker, of St Andrew's, Plymouth, when he took out a licence on 18 December 1740 to marry Mary Dennis of the adjoining parish of Charles. The three children who survived were Samuel (*c*1742–1813), James (1746–1831), and a daughter Mary who kept house for James when he finally settled in London.

Sam Northcote's shop, said one who knew him, 'had two windows. In one he exhibited watches under repair, and in the other his wife exposed thread, tape, sleeve buttons; and such small matters for sale. He was a quiet and not ill-informed man, and some of the better sort of inhabitants would resort to his shop and chat for an hour or two on the ordinary topics of the day'. Among these callers was a Mr Lyne (or Line) Brett who, in a poetical letter to a friend, penned the following:

> The morning past and dinner done,
> At Sam's I spend the afternoon
> And while he works we two discourse
> Of you, Bulls, earthquakes, motion, force,
> Of gunpowder, and civil wars,
> Volcanoes, riddles, coxcombs, stars,
> Of parsons, pendulums, and tides,
> And twenty other things besides.

Brett and Northcote were founder members in 1740 of the Otter Club. This was set up by a group of young men who used to meet of a morning to bathe in the sea and also met once a fortnight for supper at a Plymouth tavern. The club had twelve members, among them Dr John Mudge, brother of the watchmaker Thomas Mudge, and each member had a silver medal which was worn on the evening of the meetings. Sam Northcote was Lyne Brett's most valued friend and when Brett died, aged twenty-eight, of consumption in January 1741, he was left £100 together with a good part of Brett's library.

Northcote himself had a long working career. He outlived his wife by thirteen years but, when old, suffered 'repeated paralytic strokes' and was in a good deal of pain from 'gravel' and other disorders. His elder son Samuel wrote:

> After the last paralytic stroke I believe he never went into the shop any more though he talked every day that he would have a fire made there the next day as he had a quadrant which he must do, but whenever he got up from his chair and attempted to stagger out Jenny and I prevented him, telling him it was too cold today for him and that he should stay till tomorrow: and then he answered, 'Well, if you think so, I am easily

persuaded, for I must confess I am very loath to quit the fire.' And so we kept on till the third evening before his death.

The *Exeter Evening Post* in November 1791 reported the death at Plymouth of 'Mr Northcote, a very worthy man, many years an eminent optician and watchmaker'. A mural tablet with an urn, salvaged after the blitz in 1941, can be seen in the north aisle of St Andrew's Church, Plymouth. It has an inscription:

To the memory of Samuel Northcote, of Plymouth, who died on the 13th of Nov. 1791, in the 83rd year of his age: and of Mary, his wife, who died on the 3rd of Sept. 1778, in the 67th year of her age. In this church also are interred, the remains of all those of his family, who have deceased since its settlement at Plymouth, about the year 1630.

An eight-day black and gilt lacquer longcase clock, signed 'Samuel Northcote Plymouth' on a disc in the dial arch, was presented to Plymouth Museum by members of the Harmsworth family in 1937, and still proclaims the hours.

A quarter-repeating bracket clock by this maker has a red japanned case. Another bracket clock has a plate in the dial arch which rises and falls to indicate the state of the tide (Plate 36). Signed 'Samuel Northcote Plymouth', it was made to a design by the astronomer James Ferguson, who visited Plymouth and stayed for several months as a guest of Northcote's friend, Dr John Mudge. The original clock of this kind was contrived by Ferguson in 1764 for Captain Hutchinson, dockmaster at Liverpool, but it was from Plymouth in May 1766 that Ferguson sent a detailed description of it to the Royal Society, together with a large drawing of the dial and its mechanism. To quote this description, it was 'a table clock, showing the hours and minutes, the day of the month, the moon's phases, age, and time of coming to the meridian, with the time of high water every day, and the state of the tide at any time of the day or night, by inspection'. Northcote's clock has rack striking, and dates from 1766 or a year or so later.

Samuel Northcote's apprentices, in addition to his son James, included John Hutton with whom in 1743 he received a premium of £40. The business was carried on after his death by his other son, Samuel.

### SAMUEL NORTHCOTE (ii)

In 1812 the authors of the *Picture of Plymouth* wrote

> In enumerating men of splendid talents, we must not omit Mr Samuel Northcote, a native of this town, his father having lived here as a watchmaker, which business has been carried on by his son. He is the elder brother of Mr James Northcote, RA. We cannot mention this gentleman without expressing deep regret that his talents should have remained so unproductive. Intimately acquainted with all the branches of natural philosophy, and possessing a mind that might have applied them to various purposes, they have remained dormant, and have been confined to a small circle of friends, where his society has always been highly esteemed from the vast fund of information his mind is stored with, and the happy mode he has of communicating it to others.

When the watchmaker was in his late twenties he went with his brother to London. James Northcote was anxious to further his career as a painter and asked Dr John Mudge and another friend in Plymouth, a senior alderman called Henry Tolcher, for letters of introduction to Sir Joshua Reynolds, the president of the Royal Academy, who was himself from Plympton, Devon. In his autobiography James noted: 'It was on Whit Sunday in the month of May, 1771, at five o'clock in the morning of a beautiful day, that I left Plymouth, armed with my ten guineas, to seek my fortune in the world, and began my journey accompanied by my brother.'

Most of the journey was accomplished on foot. The brothers enjoyed a good night's rest at a small alehouse, but when they reached the Woodyates Inn, a few miles from Salisbury, the landlady refused to serve them as they were foot-travellers, and they slept in the hayloft with the grooms and postboys – 'We performed the last part of our journey on the top of the stage-coach, and entered London, I believe, on the fifth day of our journey.' James called on Sir Joshua the next morning, took lodgings, and decided to stay on; but his brother, 'soon growing tired of London', parted from him about a week after their arrival and returned to Plymouth. The following notice in the *Exeter Evening Post* shows that he moved from there to Plymouth Dock, then a separate town; its name was changed in 1824 to Devonport:

> LOST on Friday, the 31st of March 1775, on Dean-Clapper Hill, between Ashburton and Ivybridge, A LADY'S GOLD WATCH, in a green Shagreen case, Maker's name Richardson, No. forgotten. The Watch, when lost, had an old fashioned Gold Chain to it, the impression a Caesar's head. Whoever will bring the same to Mr Samuel Northcote, Jun, Watchmaker at Plymouth Dock; or to the Prince George Inn, in Plymouth, shall receive FIVE GUINEAS Reward.

He later went back to Plymouth, presumably to take over his father's shop, and can be traced in 1796 with premises 'at the corner of Colmers Lane, near the Guildhall' (*Exeter Flying Post*). This lane is marked on a map engraved in 1820 by John Cooke.

In 1790, Northcote and Dr John Mudge were present at the Pope's Head Tavern in Plymouth when the Otter Club celebrated its half-century. Two years later Mudge sent his Otter Club medal to Northcote with a message, 'Will you have the goodness to return my medal to the Club with my best wishes for its happiness and permanence . . . I have worn it full half a century near my heart!' In 1793 the watchmaker was one of seven pall-bearers at the doctor's funeral.

Allan Cunningham in his *Lives of Eminent British Painters* says, in passing, that Northcote made excellent watches and improved the barometer. His death on 5 May 1813 was recorded in *Flindell's Western Mercury* and, subsequently, in the *Gentleman's Magazine*: 'Devon – At Plymouth, aged 71, Mr S. Northcote, a man of genius and science.' Northcote, who, like his painter brother, was unmarried, had left Plymouth about a year before to visit him in London, and had only been back in Plymouth a few weeks when he died. A contemporary noted that his remains were interred in the family vault in St Andrew's Church, and the service was 'very mournfully attended by a large number of the most respectable persons in the town and neighbourhood'.

As might be expected, there are a number of extant portraits of various members of the Northcote family. An early example shows Samuel Northcote sen at the age of four painted by William Gandy who died in 1729. 'He came to Plymouth about the year 1714,' wrote James Northcote, 'and was then a man advanced in years. My grandfather was a great friend to him; but Gandy quitted Plymouth much in his debt, departing secretly and leaving only a few old books and prints behind him.' James regarded the portrait of his father when a child of four as excellent.

> One of my father's mother is likewise extremely fine [he wrote], although Gandy from his ill nature was quarrelling with her the whole time he was painting it. The drapery of this picture is painted in a slovenly manner from a print after Kneller, but there is a hand in it very freely executed . . . He [Gandy] resided a long time in the house of my grandfather, who admired his talents, and esteemed him the greatest artist of his time.

James Northcote painted a portrait of his father which was engraved in mezzotint by Samuel William Reynolds, and did a series of self-portraits. He also painted some dashing portraits of his brother; one of these, now in the Plymouth Art Gallery, is reproduced as Plate 37; another shows him with a hawk. The Royal Albert Memorial Museum in Exeter has a portrait group entitled 'The Northcote family'. A monument by the sculptor Sir Francis Chantrey depicting James Northcote, seated and holding brush and palette with the manuscript of his *Life of Sir Joshua Reynolds* underfoot, can be seen in the north transept of Exeter Cathedral – by the clock, appropriately enough.

A tiny precision clock with a regulator-type dial, signed simply 'Northcote Plymouth', was sold at Sotheby's in April 1968 for £1,700. It was catalogued as 'a fine small bracket clock with pivoted detent escapement and a large balance running on anti-friction discs, the silvered dial with subsidiary seconds and hours and regulation scale in the arch of the dial; contained in an ebonised pearwood case outlined in brass mouldings, with a loop carrying handle and bracket feet, 9½in high'. A longcase regulator with an eight-day movement, wooden pendulum, maintaining power, adjustable pallets and with endless chain winding, was illustrated in *Antiquarian Horology* in December 1961 (p250). The influence of Thomas Mudge has been detected in these clocks. Northcote watches include an example with a silver champlevé dial and some with plain enamel dials. A barometer signed 'Samuel Northcote Plymouth fecit' was noted many years ago in the hall at Widey Court, Plymouth (Tapley-Soper Index, Westcountry Studies Library, Exeter).

Because their careers overlapped, it is not always possible to tell the work of the elder Samuel Northcote from that of his son.

## THOMAS MUDGE

Thomas Mudge (1715–94), pioneer of the lever escapement and one of the most brilliant watchmakers the world has seen, was a member of a remarkably talented Devonshire family. His father, the Reverend Prebendary Zachariah (Zachary) Mudge, was described by James Boswell as 'that very eminent divine . . . who was idolised in the west, both for his excellence as a preacher and the uniform perfect propriety of his private conduct'. One brother, Dr John Mudge, was a Fellow of the Royal Society and a winner, in 1777, of its prestigious Copley gold

Thomas Mudge's family tree

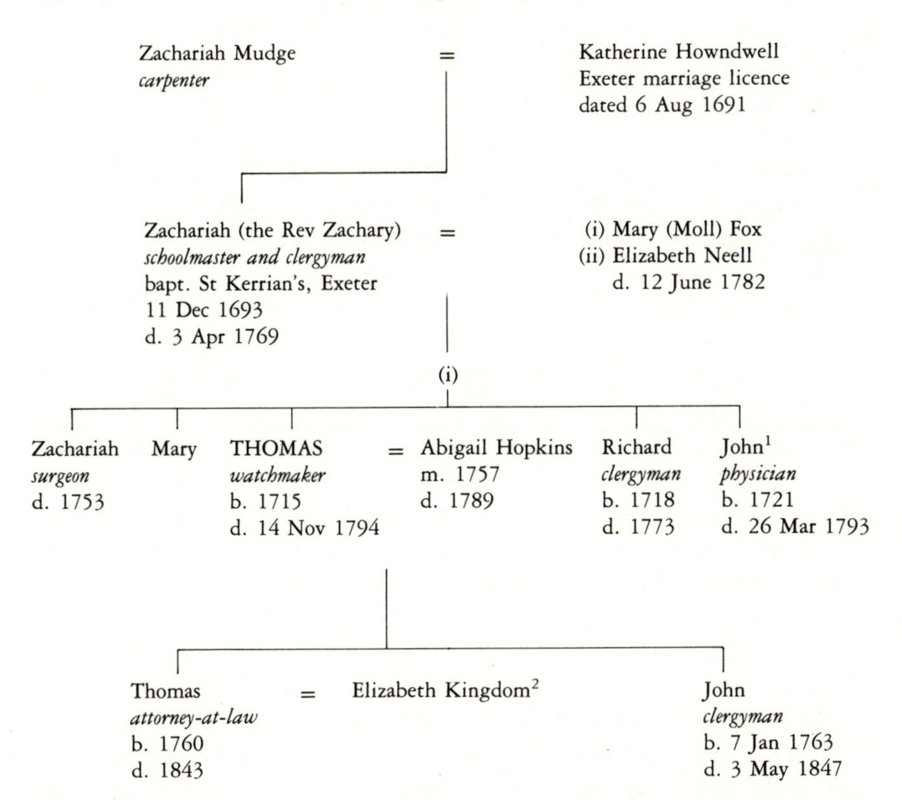


[1] Dr John Mudge was thrice married and had twenty children. One of his sons became a major-general, and another an admiral. A daughter Jane (Jenny) married Richard Rosdew, of Beechwood near Plympton, in 1783. In his notebook Dr Mudge wrote: 'Jenny was this day married, dined at Beechwood, and gave Jenny the watch'.

[2] She was a daughter of William Kingdom, naval agent and mercer of Plymouth. Her sister Sophia married Sir Marc Isambard Brunel and was mother of the engineer Isambard Kingdom Brunel. Another sister, Sarah, married Thomas Dutton, son of William Dutton, Thomas Mudge's partner in the Fleet Street clock and watchmaking business; a brother, William, married William Dutton's daughter Hephzibah (*Antiquarian Horology*, Winter 1980, pp 439–40).

medal for his work on reflecting telescopes; another brother, the Reverend Richard Mudge, was a composer whose playing on the harpsichord impressed even the great Handel, who declared that he was second only to himself.

Thomas was sent to London when young and worked there for forty years, but then returned in 1771 to live in semi-retirement at

Plymouth, where he toiled for another twenty years to try to perfect a timekeeper of exceptional accuracy for use in determining the longitude at sea. His original lever escapement watch, made in 1770 for Queen Charlotte, wife of King George III, is at Windsor Castle; his first marine timekeeper is in the British Museum in London, and his other two, known as 'Blue' and 'Green' because of the colour of their cases, are in overseas collections – the former in Dresden and the latter, having been sold at auction in Geneva in November 1976 for Swiss F250,000 (£62,500), in the Time Museum at Rockford, Illinois. A biography entitled *Memoirs of the Life and Mechanical Labours of the late Mr Thomas Mudge* was published in the *Universal Magazine* for July 1795. This formed the basis of the Thomas Mudge section in Stamford Raffles Flint's subsequent memoir of the family (1883); but Flint chose to alter some of the dates concerning the clockmaker and his, at times unreliable, material was followed by the compiler of the Thomas Mudge entry in the *Dictionary of National Biography*.

Efforts to find a record of Mudge's birth or baptism in parish and nonconformist registers appear to have been unsuccessful; but Zachary, his father, was baptised at the old St Kerrian's Church in North Street, Exeter, on 11 December 1693 and received his early education at the city's free grammar school, proceeding from there to Hallet's academy in Exeter to train for the Presbyterian ministry. John Fox, who met him during those early years, later recollected that Zachary's 'father, as I have heard, was by trade a carpenter but died while he was very young. His mother maintained herself by attending lying-in women and had the character of a careful, sober woman'. He received an annual grant, but was criticised after taking a journey to London for wearing a fine wig and not going as one 'maintain'd by charity' (*Minutes of the Exeter Assembly*). He married Mary (Moll) Fox about 1713 and, needing an income sufficient to support a family, took a post as usher (second master) at the free grammar school, where the headmaster was John Reynolds, a half brother of Samuel Reynolds, the father of the famous artist Sir Joshua Reynolds who was later to paint three portraits of Zachary.

Zachary Mudge taught at the school for three years and then left to become master of Bideford Grammar School, Reynolds stating that he believed him 'to be perfectly well qualify'd thereto by an extraordinary knowledge of ye Greek and Latin tongues, & a person of great application & diligence in ye teaching of boys'. He severed his links with the Presbyterians, later repaying the grants he had received, and at St Paul's Church, Exeter, on Sunday 31 March 1717, 'receiv'd the Sacrament according to the usage of the Church of England'. He was ordained in 1729 and in December that year was instituted as vicar of Abbotsham in north Devon. In August 1732 he obtained the valuable living of St Andrew's, Plymouth, and was later made a prebendary of Exeter which, wrote Fox, 'entitled him to a scarf and enabled him once a year to make himself known at the Cathedral'.

His wide circle of acquaintances included John Smeaton (1724–92), the civil engineer and instrument maker who built the third Eddystone lighthouse on the notorious reef fourteen miles seaward of Plymouth Hoe. In 1759 after 'Laus Deo' had been cut upon the last stone set over the door of the lantern, Smeaton conducted Mudge to the summit of his 'tower of the winds'. There in the lantern, upon Mudge's lead, the pair 'raised their voices in praise to God, and joined together in singing the grand Old One Hundredth Psalm, as a thanksgiving for the successful conclusion of this arduous undertaking'. The lighthouse, built of interlocking stone and shaped so as to resemble the trunk of an oak tree, was replaced in 1881 and now stands on Plymouth Hoe. F. J. Britten in his *Old Clocks and Watches and Their Makers* stated:

> At Trinity House is a clock bearing the name of Jno. Smeaton, the celebrated engineer, the initials 'JS' being on the hour socket. There is also preserved at Trinity House a longcase clock, on a brass plate affixed to the door of which is inscribed, 'This time-keeper was placed on the old Eddystone lighthouse by John Smeaton, CE, FRS, on the 8th October 1759'.

Smeaton in his *Narrative* on the building of the lighthouse recorded that on Monday, 8 October, 'a time-piece I had provided was set in motion'. In a footnote he added: 'A dial for the regulating of the clock, would naturally arise from observing when the sun-beams of the south windows, directly tended to the opposite ones; as this would point out the time of noon. This time-piece, by a simple contrivance, being made to strike a single blow every half-hour, would thereby warn the keepers to snuff the candles'. An inquiry about the clocks to Trinity House in London in October 1983 produced a reply that it was feared, but could not be confirmed, that they were lost in the fire there during air raids on 29 December 1940. Smeaton designed an escapement for the turret clock at the Royal Naval College, Greenwich. This is still in use and is described and illustrated in *Rees's Clocks, Watches and Chronometers* (pp 202, 215). S. R. Flint, the Mudge family historian, wrote that when Smeaton came to Devon about the lighthouse, Zachary's youngest son, Dr John Mudge, 'who always associated himself with all matters of scientific interest', at once sought him out, 'and the engineer became a guest in his house during his residence in Plymouth. This was the commencement of a friendship which lasted through their lives'.

Zachary Mudge died on 3 April 1769 and a tribute, written by Dr Samuel Johnson who had met him during a visit to Plymouth, was published shortly afterwards in the *London Chronicle.* His portrait by Lucy Deebonison, a copy of an original by Sir Joshua, hangs in the Clockmakers' Company Museum at Guildhall Library, London; but of his portrait bust in St Andrew's Church, Plymouth, only the head remains following war damage.

Thomas was Zachary and Mary's second son. According to the *Universal Magazine* biography, he was born in Exeter in September 1715 and, following his parents' move to north Devon, was educated in Bideford at his father's school. His mechanical genius became evident at an early period of his life and, while a schoolboy, he could with ease take to pieces a watch and put it together again. At the age of fourteen he was apprenticed in London with the most eminent watchmaker in the country – George Graham, the successor of Thomas Tompion. The premium his father paid was £30 and the Clockmakers' Company *Register of Apprentices* records that the term was seven years from 4 May 1730, Mudge becoming a freeman of the company on 15 January 1738. It is said that he soon attracted the particular attention of his master, who so highly estimated his mechanical powers that upon all occasions he assigned to him the nicest and most difficult work. Once, when there was a complaint from a customer, Graham answered that he was certain it could not be well founded as the work had been executed 'by his apprentice, Thomas'. And so it proved, for on examination it was found the work had been executed in a masterly manner and the supposed defect had arisen entirely from the unskilful management of the owner.

On completing his apprenticeship, Mudge – who did not marry until his forties – took lodgings and continued to work privately in London for some years. Then, in 1750, he was joined by William

Dutton who had also been an apprentice of George Graham, and they took a house in Fleet Street at the Sign of the Dial and One Crown (later No 151), opposite the Bolt and Ton. The clocks and watches made and sold there were at first signed with the name of Thomas Mudge, but in the mid 1760s the style was changed to Mudge & Dutton and continued so until Mudge's death.

It has been frequently stated that Mudge succeeded Graham at his shop in Fleet Street. This was not so. Their shops were a few doors apart, and when Graham died in November 1751 his business was continued by Thomas Colley and Samuel Barkley. Mudge, however, sought a share of the custom and placed a series of newspaper advertisements, stating he was 'late apprentice to Mr Graham deceased' and 'carries on business in the same manner Mr Graham did . . . and employs (as he always has done) the same workmen Mr Graham did'. The circumstance which first rescued Mudge from obscurity, as it were, is related in the *Universal Magazine*:

> Mr [John] Ellicot[t], who was one of the most distinguished watchmakers of his time, and who had been often employed by Ferdinand VI, king of Spain, was desired by that prince to make an equation watch. Mr Ellicot[t] not being able to accomplish the undertaking, applied to Mr Shovel, an ingenious workman, to assist him; but he also being unequal to the task, mentioned it to Mr Mudge, with whom he was very intimate, and who readily undertook to make such a watch. He not only succeeded to his own satisfaction, but to the admiration of all who had the opportunity of inspecting it. This watch having been made for Mr Ellicot[t], his name was affixed to it (as is always customary in such cases) and he assumed the whole merit of its construction. An unfortunate accident, however, did justice to the real inventor, and stripped Mr Ellicot[t] of his borrowed plumes. Being engaged, one day, in explaining his watch to some men of science, it happened to receive an injury, by which its action was entirely destroyed; and he had the mortification to find, moreover, upon inspecting the watch, that he himself could not repair the mischief. This compelled him to acknowledge that Mr Mudge was the real inventor of the watch, and that to him it must be sent to be repaired.
>
> This transaction having, some way or other, come to the knowledge of his catholic majesty, who was passionately fond of all mechanical productions, and particularly of watches, that monarch immediately employed his agents in England to engage Mr Mudge to work for him; and such was his approbation of his new artist's performances, that he

honoured him with an unlimited commission to make for him, at his own price, whatever he might judge most worthy of attention.

Among the several productions of Mr Mudge's genius, which thus became the property of the king of Spain, was an equation watch, which not only showed the sun's time, and mean time, but was also a striking watch and a repeater; and what was very singular, and had hitherto been unattempted, it struck and repeated by solar, or apparent time. As a repeater, moreover, it struck the hours, quarters, and minutes. From a whim of the king's, this watch was made in the crutch-end of a cane, in the sides of which were glasses covered with sliders, on the removal of which the work might be seen, at any time; and his majesty being very fond of observing the motion of the wheels at the time the watch struck, it was his practice, as he walked, to stop for that purpose. Those who have seen him on these occasions have observed, that he ever showed signs of the most lively satisfaction. The price of this watch was 480 guineas, which, from the expensive materials and nature of the work, afforded Mr Mudge but a moderate profit for his ingenuity, as he was strongly urged by several of his friends to charge 500 guineas for it, which the king would have readily paid. To this Mr Mudge answered, that 'as 480 guineas gave him the profit to which he was fairly entitled, as an honest man, he could not think of increasing it, and he saw no reason why a king should be charged more than a private gentleman'. Indeed, the king of Spain had such a high opinion of his integrity, that he not only used to speak of him as by far the most ingenious artist, as a watchmaker, he had ever employed, but excelling also in his sense of honour and justice.

Two original letters, written to Mudge from Madrid by the Spanish court clockmaker Michael Smith, and supplementing this account, were published by A. J. Turner and A. C. H. Crisford in *Antiquarian Horology*, Winter 1977 (pp 580–2). In the earlier of these, dated 3 July 1752, Smith stated that he called several times at Mudge's shop while on a visit to London that year. Mudge showed him a piece of watchwork or part of a movement that was intended to show the 'equation of the sun' and was the same as the one made for Ellicott which the king had. On his return to Spain, Smith recommended Mudge to the Court in preference to Ellicott as 'the most capable person now Mr Graham is dead'. He also told 'his majesty of the Eaquation watches you made which Goes in Ellicotts name w$^{th}$ maney Circumstances in your favour for which Reason the Prime minester and severall of the nobility hath taken down your name'. Smith anticipated that this would lead to 'Great orders from this Corte' and hoped

Mudge would make him some allowance. In the second, dated 26 December 1757, Smith ordered on behalf of the king a large gold, quarter-striking, minute-repeating watch and, for the Duke of Alba, an equation month regulator in a mahogany case not exceeding 1.8m (6ft) in height. Also mentioned in the letter are the cane-head watch, another in a crook head, and a large quarter-clock watch which Mudge made for the king four years previously and 'which ye King wares & allows to be the best watch for performance he ever had'.

Writing in 1792, Thomas Mudge, jun, stated: 'The abilities and integrity of my father were so well known to the King of Spain . . . that he had, from the death of Mr Graham, until the King himself died, an unlimited commission from his Majesty to exercise his mechanical talents in the production of anything that he thought would be deserving of his majesty's attention' (Preface to *A Reply. . .*).

Between May and the end of October 1755, Mudge was employed on making a marine timekeeper for the Swiss astronomer Johann Jakob Huber, who designed for it a free or constant-force escapement which was not dissimilar to that later used by Mudge himself in his marine timekeepers. A man called Howel made the case for Huber's timekeeper and the mechanism was tested in front of the fire in Mudge's shop and then in the cold; the making of a satisfactory balance spring caused the watchmaker a great deal of trouble. Huber recorded that on 9 June that year he was shown a drawing of a new deadbeat escapement which Mudge had invented, and which operated with less friction than Graham's (T. P. C. Cuss, see Bibliography, page 348).

In a document dated 25 June 1763 Mudge set down his *Thoughts on the Means of improving Watches; and more particularly those for the use of the Sea*, and in 1765 was a member of a committee which examined John Harrison's masterpiece – his H4 watch which kept such accurate time on voyages to the West Indies that it won one of the great prizes offered by Parliament for a practical method of finding the longitude. In 1766 Mudge was called to the livery of the Clockmakers' Company.

For King George III, Mudge succeeded where others had failed in mending a large and complicated watch. The two often discussed mechanical subjects and in 1776, on the death of George Lindsey, the king, who was himself an horologist, appointed Mudge as his watchmaker. Queen Charlotte also held Mudge in high esteem and on one occasion presented him with the then large sum of 50 guineas for merely cleaning a watch; and it was through her recommendation that his second son, the Reverend John Mudge, obtained the living of Brampford Speke in Devon.

Among Mudge's friends in London was James Ferguson the astronomer, who lived nearby at 4 Bolt Court, Fleet Street. Ferguson noted in his Commonplace Book that he observed some sun spots from the watchmaker's house in November 1769. Two years later Mudge made a present to him of a complex, German astronomical clock dated AD 1560.

Thomas Mudge's apprentices included William Hull, son of a London saddler John Hull, who was apprenticed on 3 September 1753 with a premium of £52 10s; Lawson Hudleston, son of Prebendary William Hudleston of Wells, Somerset, who was apprenticed on 21 December 1759 with a £50 premium; and Matthew Dutton, son of his partner William Dutton, who was briefly his apprentice in 1771 before his move to Plymouth.

Mudge's reasons for quitting London – and for leaving the shop with William Dutton – were that his father had lately died and he wished to be near his brother, Dr John Mudge. 'Besides, Sir,' he wrote to his patron Count Bruhl, 'when I left business I gave up a clear £500 per annum to be freed from the trouble that attended it, and to be at liberty to indulge my own inclinations.' He lived at his father's old house in Plymouth. Samuel Northcote (ii), who knew the family well, related that:

> After his father's death he [Dr John Mudge] inspected the fitting up of the house for his brother Tom and when they came to new paint the fore parlour he observed the window seat, where his father used to sit and study, to have the old paint quite worn away to the board at that spot. He said, 'God! it went to my heart to paint over the place'.

In a letter to Count Bruhl, dated 19 February 1773, Mudge described how he set up his regulator clock in the house. He wanted it to be as solidly fixed as possible and used

> . . . a piece of wainscot, fourteen inches square, and four inches thick, which, when screwed against the wall, lies exactly even with the surface of the wainscot of the room . . . I will just observe, that when I had fitted the wainscot to the wall, I took it off, and plaistered the wall thick with mortar of a somewhat thicker consistence than usual, and then screwed it up firmly against it, so that it became, in effect, part of the wall itself. The

> back of the case was screwed firmly to this wainscot, with four large wood screws, two just above the holdfasts near the point of suspension, and the others as near as possible to the setting board under it.

Some seven years later he altered the escapement and mounted the pendulum on the wall itself; this led to 'an amazing increase in the arc it vibrates'.

The saga of Mudge's marine timekeepers has been told many times. Briefly, he completed the first of them in 1774 and, with the encouragement of £500 from the Board of Longitude, constructed two more. All three were rigorously tested, notably by the Astronomer Royal, the Reverend Nevil Maskelyne, at Greenwich Observatory; but while performing at times with incredible accuracy – No 1 especially on two voyages to Newfoundland – they suffered from a defect in the compensation for heat and cold and the board eventually decided that Mudge's efforts did not qualify for further reward. A great controversy ensued, centred mainly on the Astronomer Royal's methods of testing. Mudge's son, Thomas jun, plunged into print on his father's behalf and Maskelyne was forced to reply. A committee of Parliament then investigated the matter and in 1793 recognised the watchmaker's great skill and ability with an award of £2,500.

Mudge's will, made after the death of his wife, is dated 31 May 1789. It gives his address as Dyers Buildings in the City of London and was witnessed by Robt. Bayly, John Kingdom and Sarah Kingdom. He left his property to his sons Thomas and John and mentions an interest in an estate in Worcestershire 'called or known by the name of Cakebold' [Cakebole].

His death was briefly recorded in a 1794 obituary list in the *Gentleman's Magazine*: '14 November. In an advanced age, at his son's house in Walworth, the ingenious Mr Mudge, late watchmaker in Fleet Street'. There was a suggestion that he should be buried in Westminster Abbey, but in accordance with his wishes his funeral took place on 21 November at St Dunstan in the West and he was buried in its Fleet Street churchyard (register).

A striking clock (Plates 38 and 39), described as the last of Mudge's performances and the only work of his remaining with him at his death, was afterwards given by his sons, as a memorial and as a token for many kindnesses, to Mr Richard Rosdew of Beechwood near Plympton, Devon. A small oil portrait of Mudge, painted about 1770 by Nathaniel Dance, can be seen in the Clockmakers' Company

Museum in London. Copies of it were engraved by Charles Townley, L. Schiavonetti, and J. Baker.

Plate 40 shows a watch known as the 'Flint' Mudge which came to light through J. K. Bellchambers's researches. It is an unsigned copy, hallmarked 1795, of Mudge's lever escapement watch and has an inscription around the dial: 'The original was invented and made by Thomas Mudge for her Britannic Majesty AD 1770'. Unfortunately, it does not retain its original escapement.

An ambitious attempt by Thomas Mudge jun, to manufacture marine timekeepers on his father's plan, ended in financial failure. Several highly skilled workmen were employed, including William Howells, Robert Pennington and Richard Pendleton; but the complexities involved were great, the costs were high, and the instruments – about twenty-seven were produced – were not as accurate as the originals.

In addition to Mudge's famous first marine timekeeper, the collection at the British Museum includes his watch No 574 with perpetual calendar mechanism; his 'lever' table clock which once belonged to Brunel and has a lunar train with an error of only 0.2 of a second in 29½ days; and the first of his son's copies of his marine timekeeper, inscribed 'No. I, Howells and Pennington for Thomas Mudge'. A cast-iron framed turret clock, typical of those made by Thwaites of Clerkenwell, but allegedly made by Mudge, was formerly in the tower of St Andrew's Church, Plymouth. It was sent for scrap earlier this century, but an illustration of it appears in *Devonshire Clockmakers* by J. K. Bellchambers.

## PHILIP BROCKEDON

Philip Brockedon, clock and watchmaker, of Totnes, was the father of the artist William Brockedon whose vast canvas, 'The Acquittal of Susannah', hangs in the Crown Court of the Castle at Exeter. The family came from the Kingsbridge area, where they had held land at Dodbrooke from the days of Henry IV.

Some of Philip Brockedon's earlier clocks were signed with the place-name Kingsbridge. His marriage at Totnes parish church on 28 December 1784 was reported in the *Exeter Flying Post* newspaper; the bride, Miss Mary Bastow, being described as 'a very agreeable young lady, possessed of every qualification requisite to render the marriage state happy'. They lived in a small house, near the east gate of the town, which later became a public house called the East Gate Inn. A contributor to the 1877 volume of *Transactions of the Devonshire Association* stated that although only in a small way of business as a watch and clockmaker, Brockedon seems to have been a man of genius. In the latter part of the eighteenth century during the Napoleonic War, the clockmaker was accustomed, when the *Courier* newspaper arrived in Totnes, to meet with some of his fellow townsmen in the Church Walk, the Old Exchange, and read to them the latest intelligence from the seat of war.

In October 1798 Brockedon was commissioned by the corporation to make a new town clock to be placed in the tower of the parish church. It is related that his son William helped him to make this clock, cutting the fly-pinion out of the solid steel. Later, when his father became ill, William carried on the watchmaking business for twelve months. Philip Brockedon died in September 1802, but his widow lived on until 1837.

He made both brass-dial and painted-dial longcase clocks, including a three-train example illustrated in Cescinsky and Webster's *English Domestic Clocks*. Among church clocks he repaired were those of Paignton in 1789–90, and Modbury in 1792, the churchwardens' accounts recording that for the latter job he received £14 14s. His Totnes clock, after being replaced in the tower by the present chiming movement to commemorate Queen Victoria's 1887 jubilee, was sold to the Mayor, Mr H. Symons, who utilised it at his paper mills at Tuckenhay, a few miles further down the Dart. It is still in existence and is a fine clock of the period; it carries an inscription, 'Reconstructed by Sainsbury Bros, Walthamstow, for S. E. Burrow, Totnes'.

The painter William Brockedon (1787–1854) was also a scientist and inventor. He wrote books on Alpine travel and was a friend of the Italian statesman Cavour, whom he met while sketching in the mountains.

## JOHN THOMAS TOWSON

John Thomas Towson, watch and chronometer maker, was born at Fore Street, Devonport, on 8 April 1804 and died in Liverpool on 3 January 1881. He was the son of John Gay Towson (qv) and his wife

Elizabeth Thomas. Around the year 1826 he received a Vulcan medal and £10 from the Society of Arts for a chronometer banking. He was one of the earliest pioneers of photography and also made a notable contribution to the science of navigation. The following obituary notice appeared in *The Athenaeum*, 8 January 1881:

> Mr John Thomas Towson, well known for being the first to direct attention to the advantages of sailing on the great circle, died on Monday, January 3rd, at his residence in Liverpool, in the seventy-seventh year of his age.
>
> Mr Towson was born in Devonport in 1804, where he was educated and trained to his father's trade of a watchmaker. In 1838–39 he associated himself with Mr Robert Hunt, then resident in Devonport, and together these gentlemen devoted considerable attention to the Daguerreotype process, just then introduced. In November 1839, Mr Towson published in the Philosophical Magazine a paper 'On the Proper Focus for the Daguerreotype', in which he demonstrated the important fact that 'the mean chemical influence lies without the limits of the luminous portion of the spectrum, very near the extreme violet ray'. Acting upon this, Dr Draper, of New York, obtained the first photograph from life. Towson's active mind was constantly creating new and often successful combinations, and with his colleague he succeeded in producing very highly sensitive photographic papers, for the sale of which they appointed agents in London and in the provinces.
>
> Mr Towson was induced to give instruction in navigation to a few young men in the Naval Yard. This directed his attention to the subject, and led to the suggestion that the quickest route across the Atlantic would be by sailing on the great circle. He communicated this to Sir John Herschel, who immediately replied that he was astonished that a thing so obvious had been overlooked so long.
>
> Sir John Herschel drew the attention of the Admiralty to Mr Towson's discovery. Towson invented and constructed a table for the reduction of ex-meridian altitudes, a work highly valued in the mercantile marine, and he also arranged tables to facilitate the practice of great circle sailing. Mr Towson gave the copyright of those works to the Admiralty, and they were published by that department. In 1850 Mr Towson was appointed Scientific Examiner of Masters and Mates in Liverpool, which he held until 1873, when he retired, still holding an appointment as Chief Examiner in Compasses. In 1857 the shipowners of Liverpool presented Mr Towson, as a testimonial, with a Dock bond of the value of a thousand pounds, and a handsome additional present.
>
> In 1854, at the recommendation of the British Association, the

> Liverpool Compass Committee was formed, and three reports were subsequently presented to both Houses of Parliament, these being in the main the result of Mr Towson's labours. In 1863 the Board of Trade instructed Mr Towson to prepare a manual on the deviation of the compass, which was subsequently published under the title of 'Practical Information on the Deviation of the Compass, for the Use of Masters and Mates of Ships'; and in 1870 the Board of Trade adopted a syllabus for examinations in compass deviations, prepared by Mr Towson.
>
> It appears that early in December Mr Towson fell and fractured two ribs. He never rallied from the effects of this fall, and ultimately the liver became congested and death ensued.

Towson's method of banking in a chronometer prevented the balance from describing more than a complete turn in either direction; it is discussed in a chapter headed 'Miscellaneous Mechanical Developments (Chiefly of historical interest only)' in Rupert T. Gould's *The Marine Chronometer* (1923).

## JOHN BULKELEY

For John Bulkeley the spectre of those twin Dickensian horrors, the debtors' prison and the workhouse, became grim reality. In 1815 when imprisoned as an insolvent debtor 'in his Majesty's Gaol of St Thomas the Apostle, Exeter', he was described as 'late of the parish of Dawlish, watchmaker' (*Exeter Flying Post*, 6 April 1815). Some years later, in 1838, he was to be found eking out a pauper's existence in Kingsbridge Union Workhouse. At that time he was chargeable to the parish of South Huish, but two magistrates were called in to examine him and it was ordered that he should be removed to Dawlish, 'his last place of legal settlement'. Bulkeley made a statement, as follows:

> I am about 53 years of age. I have heard and believe I was born in wedlock, and that my parents at the time of my birth were residing in Starcross within the parish of Kenton and had a settlement in that parish. When I was about the age of 17 years old I was bound an apprentice to Mr William Welshford, of West Teignmouth, clock and watchmaker, for five years by private indenture. I was then unmarried. My father was present when the bargain was made and approved of it. I lived and served my time with him from the time I was bound [to] within seven months of the end of my apprenticeship when my master, Mr William Welshford, removed to Plymouth, but I did not go with him. I then went to Bristol and worked as a journeyman at my trade for two years. I then went into Wales

> and worked there for nine months. I then came back to Starcross and opened a shop at my business for one year and a half, where in the year 1809 I got married by licence in the parish church of West Teignmouth to Charlotte Taylor, my present wife, by whom I have had six children. I lived with my father while there but rented the shop at £3 per year. I then went to Dawlish to live and opened a business in a house of Mr Nicholas Tripe and gave sixteen pounds per year for it. I lived there one year and a quarter. I then went to Newton Abbot and opened a business there in a house of a Mr Phillips for which I gave £8 per year. I lived there between seven and eight years. I then separated from my wife and family. I then travelled the county as a journeyman and continued to do so up to this time. I am now actually chargeable to the parish of South Huish, not being able to maintain myself. My wife Charlotte is now living at Teignmouth [with] my children, John, aged about 24 years, Elizabeth, aged about 22, James, aged about 18 years, Margaret, aged about 16 years, William, aged about 14 years, [and] Henry, aged about 13 years. I have not maintained my said children since myself and wife separated about 13 years since – my wife maintains most of my children as I believe.

Bulkeley's chequered life would have gone largely unrecorded but for the discovery of his statement among the South Huish parish settlement papers (Devon Record Office). As far as can be gathered, he was listed as a watchmaker in only one trade directory of the period, Pigot & Co's 1823–4 *Royal National and Commercial Directory and Topography*, which recorded him in business in East Street, Newton Abbot.

### GEORGE ROUTLEIGH

The famous and much-quoted epitaph to the watchmaker George Routleigh can still be read on his tomb just outside the south porch of St Petrock's parish church, Lydford.

Here lies in horizontal position
The outside case of
GEORGE ROUTLEIGH, Watchmaker,
Whose abilities in that line were an honour
To his profession.
Integrity was the mainspring,
And prudence the regulator
Of all the actions of his life.

Humane, generous, and liberal
His hand never stopped
Till he had relieved distress.
So nicely regulated were all his motions
That he never went wrong
Except when set agoing
By people
Who did not know
His key.
Even then he was easily
Set right again.
He had the art of disposing his time
So well
That his hours glided away
In one continual round
Of pleasure and delight
Till an unlucky minute put a period to
His existence
He departed this life
Nov. 14 . . 1802
Aged 57
Wound up
In hopes of being taken in hand
By his Maker,
And of being thoroughly cleaned, repaired
And set agoing
In the world to come.

These words, the church guidebook states, were first published in an almanac for the year 1797 by an American negro scholar named Benjamin Banneker. The epitaph in similar but abbreviated form has also been noted at Bolsover in Derbyshire on the grave of Thomas Hinde, clock and watchmaker, who died in 1836.

Routleigh is known to have worked at Launceston, Cornwall, some fourteen miles from Lydford; his work, however, sometimes brought him to Devon. In 1780 he was one of several clockmakers who submitted proposals to the twelve governors of Crediton parish church about erecting a new set of chimes. He and James Pike, of Newton Abbot, were each paid 15s for their trouble, while the local men,

George Newman and James Bucknell, both of Crediton, received 5s each. The job, however, went to Thomas Bilbie of Cullompton, although payments in the governors' accounts indicate he was assisted in the work by Routleigh. In January 1781, Routleigh received a guinea 'ffor altering ye Tune of ye Chimes as by agreement'.

5

# A WATCHMAKER'S AUTOBIOGRAPHY: HENRY ELLIS (1790–1871)

Henry Ellis, with the precision one would expect from a watchmaker, begins his story with the exact time of his birth, at a quarter past ten in the morning on Tuesday, 3 December 1790, in a little cottage in Friars Walk, Exeter. He was the son of Henry and Elizabeth Ellis, his father at that time being a clerk in the Exeter wharfinger's office.

Ellis relates that when his grandfather died in 1800 he left him his silver watch, made by Matthew Sayer of Exeter, a relative of the family; and he then goes on to describe how through a boyhood friendship with John W. T. Tucker (qv) he became a watchmaker. The year was 1804.

> About this time, I formed an intimate acquaintance with a lad of the name of Tucker, a schoolfellow, whom I used frequently to visit. His father [John Tucker] was a watchmaker; and he, having a vacancy for an apprentice, just as I was about leaving school, it was not a very difficult task for young Tucker to persuade me to fill it. It was shortly after agreed that I should (as is the custom) be with Mr Tucker a month on trial prior to my becoming an apprentice.

The month passed agreeably and he was duly bound apprentice on his fourteenth birthday for seven years. His father paid a premium of £30, for which consideration Ellis was to board in John Tucker's house and 'be taught and instructed in the art, trade or mystery of a clock and watchmaker and engraver by the best and most approved means and method'.

It was an emotional occasion when he left home and his mother and sisters observed the time-honoured custom of throwing an old shoe after him for good luck. But things did not turn out well for the young apprentice and a harsher side of life was soon revealed. He notes:

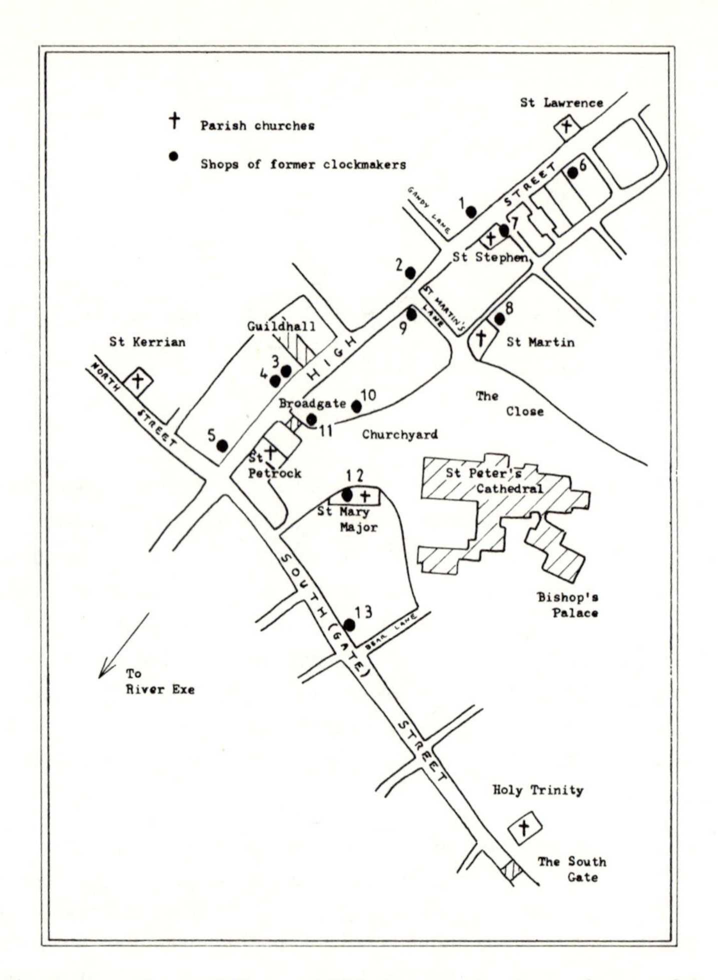


Fig 7 Sketch map of central Exeter *c*1800 showing the shops of some of the city's former clock and watchmakers. Key: 1 Jacob Lovelace, 'adjoining the Printing-house', High Street, *c*1718–24; 2 Benjamin Bowring, 225 High Street, 1803–6; 3 Henry Ellis and successors, 200 High Street, 1828–85; 4 John Balle, 199 High Street, late eighteenth century, Benjamin Bowring 1806–15, Henry Ellis 1816–28; 5 John Tucker, 188 High Street, 1799–1805, John Elliott Pye 1805–*c*1813; 6 John Skinner and sons, 20 High Street, 1813–46; 7 John Savage, Peter Savage and successors, 'next the New Inn', *c*1602–87; 8 Ambrose Hawkins, Catherine Street, *c*1695–1705, Richard Eastcott (here or nearby) *c*1770; 9 R. W. Upjohn and successors, 39 High Street, 1855–77; 10 Henry and William Gard, Cathedral Yard, *c*1767–1805; 11 Richard Upjohn, 'at the corner of Broadgate, St Peter's Churchyard', *c*1764–78; 12 George Sanderson, 'over the porch' of St Mary Major parish church, *c*1761–4, George Flashman 1765; 13 William Upjohn, Bell Hill, South(gate) Street, 'three doors above Bear Lane', *c*1736–68, Thomas Upjohn 1768–83, Robert Bake 1784–8, John Tucker 1789–99, William Upjohn (ii) 1800–03, John Upjohn 1803–*c*1848, R. W. & J. R. Upjohn 1848–56

Mr Tucker was considered a good practical workman, and, possessing a considerable share of mechanical knowledge, was well qualified to instruct, had he been gifted with some other requisites equally essential. It was not long before I discovered that my master, although professing to be very religious, was possessed of an ungovernable temper; he was a bitter sectarian, a radical reformer, almost a revolutionist – in heart a very tyrant, whose word was law, ruling his wife and family with a rod of iron, while he continually exclaimed, 'I will reign su-p-preme in my house!' (He had a great hesitation in his speech, and was a great stammerer, which made the matter still worse, for not being able to deliver his words, it added to his fury).

## 1805

When I had been apprenticed about six months, my master talked of leaving Exeter, to succeed his father in the same line at Tiverton, who having become aged and infirm, wished to relinquish his business. [Whilst Tucker prepared to leave Exeter he placed his son, John Walter Tothill Tucker, and Ellis with an engraver with whom they remained six weeks.]

Ellis left Exeter on 8 September 1805:

On arrival at Tiverton I found things more pleasant than I had anticipated. Mr Tucker's father was an agreeable old gentleman and appeared willing to take pains in instructing his grandson and myself. He was a sound workman, had laboured hard as a mechanic, and I have no doubt that had he moved in a different circle, would have ranked among the eminent men of his day in his profession.

Young Tucker and myself were placed in the clock-shop, while Samuel Kendrick (who was the elder apprentice and who also came from Exeter with my master) worked in the front-shop at watch-work. In this comfortless back shop, frozen with cold in the winter, and scorched with the burning rays of the sun in summer, did I drag on four years of my apprenticeship, the latter part of it quite alone. This I considered too long a time to be kept at clockwork; but remonstrance was in vain; and as Kendrick and John Tucker were in the front shop at watch-work, there could be no opening made for me, until the former had completed the period of his apprenticeship. (Poor Kendrick was the son of an innkeeper at Brixham; he was thoroughly good natured and full of mirth and hilarity. After having endured all the severity of a harsh master through his apprenticeship, he went to Spain during the Peninsular War, followed his trade there, working for the Army, returned to his native place, set up business, married, and died early in life, leaving a widow and five young children).

## 1808

Ellis occupied his leisure hours in the cultivation of poetry and literature. He writes:

> Early in the year 1808, an additional spur was given to my literary pursuits by a young man from London named Joyce, who was placed with my master to learn the business. He was my companion in the clock-shop, and I had to instruct him in that branch as far as I was able. (Joyce left about 12 months after). During the time I was in the back-shop I manufactured many 30-hour and several 8-day clocks, besides the ordinary work of repairing.
>
> Among other things, my master had the looking-after the church clock at Bampton, about seven miles from Tiverton, and I made frequent journeys thither respecting it, and other business. A new dial was put to the clock, and the clock itself removed from one part of the tower to another; I had the making of many parts of the new work, which I found both dirty and laborious. About this time my master established a clock and watch club at this place, and another at Thorverton which required to be attended once a month. I used frequently to be sent there, and spent many an evening in the village alehouse, with a set of bacchanalian rustics, not much to my comfort or edification . . . I cannot, however, help sometimes laughing outright, at the idea of my presiding at these meetings, (boy as I was) with a long pipe in my mouth, in a place 'all redolent of beer and baccy'. Enveloped in a cloud of smoke, hour after hour was I obliged to listen with complacency to the ribald jests of a parcel of country bumpkins. By a little management, however, I did contrive to get through this business, without, I hope, any injury to my morals or reputation. The plan of these clubs was for each member to pay a sum monthly; and the whole number of subscribers thus paying, made up the price of a clock or watch, which was drawn for, and a clock thus sold and paid for every month. But had the trouble and loss of time been taken into consideration, it would have been found that there was little gained by such transactions.

## 1809

Samuel Kendrick, my elder fellow apprentice, having at length completed his term, I was promoted to the front shop, and worked on mechanism of a smaller kind than that to which I had been accustomed; and here while attaining a knowledge of it, I was doomed (being more under the immediate surveillance of my master) to endure no small share of harsh language . . . Amongst other things, he was most severe with regard to allowing his apprentices the least relaxation after the hours of business, and when he could prevent it would never suffer us to take a walk even on a Sunday!

## 1810

In this year Ellis's apprenticeship was interrupted abruptly when his master, who had engaged in 'the fatal but too prevalent system of drawing "accommodation bills" ', was declared bankrupt. The stock in trade, household goods and other effects were sold by auction. Another watchmaker, Samuel White, who at that time was in business in a bow-fronted shop on the Quay at Bideford and who had served his apprenticeship in Tiverton with Tucker's father, stepped in with an offer of help, and it was agreed that John Tucker, the apprentice, should go to Bideford and work there for two months, then Ellis should go. This was a particularly happy arrangement, for it was at Bideford that Ellis met and fell in love with his future wife Mary, who was Samuel White's third daughter. In the meantime the shop at Tiverton was reopened, although exhibiting signs of poverty. Ellis decided to return there to complete his apprenticeship. He writes:

Mr Tucker's business began by degrees to revive a little. He was at this time employed on a curious piece of mechanism, a small-sized eight-day alarm timepiece, which was to repeat the hours, quarters, and minutes. It was rather a complicated piece of work, and in which I assisted him; we both spent a considerable time over it, but I never saw it brought to completion.

## 1812–13

After completing his apprenticeship, Ellis decided to leave Devon to seek work in the watch-manufacturing trade:

Mr White had kindly made enquiries while in London with a view of procuring employment for me; the trade still continued very dull there, but he was promised work for me at the house of Messrs Jefferys and Ham, watch manufacturers, in Salisbury Square on my paying a certain premium (I think ten pounds) for being initiated in the art of finishing.

Ellis set off for London in 1812. The idea of paying 'a certain premium' did not appeal and on arrival he lost no time in looking out for a situation, setting off the next day for Clerkenwell where he called at the premises of William Upjohn, watch manufacturer, at 11 St John's Square. He took with him a letter from John Tucker, speaking in high terms of his (Ellis's) character.

> I found Mr Upjohn, (who was himself from Exeter,) on the eve of setting out on his western journey: he received me very kindly; and after a few inquiries desired me to call again, which I did, agreeable to his appointment, the same afternoon. He at once engaged me to work at the board in the warehouse, desiring me to draw what money I might require during his absence, and on his return (if we suited each other), we could arrange as to terms, and enter into a more permanent arrangement. (This was afterwards done; and I received twenty-six shillings per week during my stay in London, besides taking tea with the family). Mr Upjohn's family consisted of himself, his wife, and one of his sisters afterwards Mrs Mallett; the latter was employed in the business, and with Mr Wittingham, myself and an errand boy formed the establishment at the warehouse, nearly the whole of the work, as is usual in the trade being executed out of doors by workmen [not literally, but by outworkers] in the different branches into which the 'art and mystery' of watch manufacturing is subdivided . . . Mr Wittingham who sat at the bench with me was a good workman as a finisher, and contrary to the generality of journeymen, ready to assist and give me any instruction in his power. I was employed principally on jobs from the country; fitting movements to the cases, being what is technically called 'boxing in', and occasionally finishing.

While in London, in 1813, Henry Ellis wrote the following account of watchmaking:

> Many country watchmakers, or rather menders, form but an inadequate idea of the method of getting-up watches. The manufacturer is one who procures the different parts of the watch from the several makers of them, sees them all rightly adjusted, and put together in a fit state for the wearer. The movement maker sends home the movement in a round tin box: it consists of the frame, barrel, fusee, great, centre, third, contrate, and balance wheels; the verge balance, pottance, &c. To furnish these several parts alone, many hands are employed by the movement maker, some in making the pinions, some one particular wheel, some another. Movement making is generally considered a good business, the masters being often men of property, and having a large number of persons in their

Plate 13 Hood and arch dial of another quality longcase clock by Jacob Lovelace, of Exeter. This one dates from about 1730 and has a walnut-veneered case

Plate 14 Giltwood cartel timepiece by Thomas Upjohn, of Exeter. The frame measures 74 × 48cm (29 × 19in). It dates from the period 1768 to 1783 when Upjohn was in business on Bell Hill, South Street, Exeter

Abell Cottey
Crediton

employ. Movements are charged from four shillings to half-a-guinea, and upwards for superior work. The best are manufactured in Lancashire, and command a higher price than Town-made ones. The frame is next sent to the enameller, who makes the dial; this also goes through several hands. A thin piece of copper is first fitted to the frame with dial feet, the glass is then laid on, and the surface, while yet hot, is smoothed over with an iron tool, the figures are then painted and baked in, or the dial is what is called 'fired'. Common dials are rendered about ninepence each, an extra threepence being added when the edge is gilt; seconds dials, and others requiring more work, are of course charged higher. The next operation is to send the frame and dial to the case-maker, who having first fitted the box to the frame, then sends it to Goldsmiths' Hall to be assayed and marked. On the repeal of the Watch Tax, the duty on the silver and gold used in the manufacture of the cases was taken off. To prevent imposition, however, the cases are still assayed and stamped; in which the Assay-master is very particular, breaking up such as are not found to be of proper standard. The rate of charging is silver at per ounce and the fashion: common pairs of cases are made at six shillings, plain hunters eight shillings and sixpence and the better sort of work in proportion. Gold cases are charged much higher – for no other reason, than I can learn, than that they are gold. The case-makers are generally the most wealthy of those who carry on the different branches in the watch trade; some of them employ between twenty and thirty men, and have been known to retire to an ample independence.

Meanwhile, back in Devon, Samuel White of Bideford had moved from his shop on the Quay to one in Allhalland Street, on the corner of Bridge Street. There he was joined in 1813 by Ellis, who some months later returned to Exeter and engaged premises at 263 High Street to commence business on his own account. The shop was opened in May 1814.

## 1814

Number 263 . . . was a small shop divided off from the one adjoining, and as the front only contained nine panes of glass, I had to make the most of it, to show it off to the best advantage. Small as the front was, some ingenuity was required to extend my small stock over that small surface. I

Plate 15 'Abell Cottey Crediton'. Brass lantern pendulum-clock with single hand, *c*1680. Empty holes at the front and sides of the top plate indicate that it once had ornamental frets. Cottey (1655–1711) emigrated to Philadelphia in 1700 spread my watches about the window by means of arches of a novel description divided into compartments, and I made use of the door to gain, apparently, more frontage, by covering it on the inside with green wire lattice, on which I exhibited my spoons and other small silver articles. A horizontal showcase filled the bottom of the window, which held my gold chains, seals, and keys, comprising nearly all the articles of jewellery I thought of keeping at that time, such being more immediately connected with my own particular branch, of a Watchmaker. Immediately behind was my workboard with my tools arranged in drawers underneath . . . the few clocks I had were placed against the walls, and one fronting the door stood in the passage leading to the room behind.

He writes to Mary White:

My little shop has attracted much attention. Since opening it has not ceased to draw constant crowds around it all day long . . . They all say it is the prettiest little shop in Exeter; while some go so far as to assert that it beats Mr Skinner's large one hollow. Be that as it may, I cannot but feel gratified that it excites so much notice after the remark made by Mr Skinner himself that it would 'not answer' . . . On the first two days alone of my commencement, I have taken in from mere chance custom, half a dozen watches to repair; sold half a dozen silver teaspoons, a pair of sugar tongs, a gold seal and key, besides watch glasses, gilt keys and other minor articles.

Ellis had trade cards and watchpapers printed. 'The latter bore my name, occupation and address, and were placed in the bottom cover, or the outside case of all watches I sold or repaired.' His stock in trade on beginning business amounted to £150 1s.

My capital being so small, it was my intention at first to confine myself wholly to the clock and watch trade, considering this the legitimate business to which I had been brought up, selling also those things more immediately connected with that branch, such as gold chains, seals, keys &c, without launching generally into the jewellery line. But I soon found that one thing led to another, and by little and little I added to my stock the more general articles sold by jewellers. This line I was imperceptibly drawn into by the temptation of travellers showing their goods, backed with the most pressing solicitations to purchase.

I recollect the first wholesale jeweller who called on me (a few days only after my opening) was Mr Carter, of the firm of Jones and Carter, of Birmingham, who prevailed on me to open an account with them, and with whom and their successors I have continued to do business to the present day. Mr Joseph May from the house of Mr W. H. Smith called on me about the same time, and who not long after commenced on his own account. I remember well his pleasant and agreeable manner as a young man, and his pressing me to spend the evening with him and go to the theatre. To a stranger like myself, unused to such attentions, the civility thus shown may be said to have been flattering; but I steadily refused the invitation. These two wholesale dealers were soon followed by a host of others.

The 9th Dragoons were at this time quartered at the higher barracks in this City, and my shop being the nearest, was visited nearly every day by some or other of them. I sold several watches to the non-commissioned officers and privates, and had a great many brought me to repair. This regiment after remaining about six months was exchanged for the Scots Greys, who in like manner, became customers, but were ordered off after a short stay, and joined in the bloody fray at Waterloo.

The demand for clocks being very disproportionate to that of watches, my sale of them was not very extensive. They were supplied to me by Messrs. Wasbrough and Co. of Bristol, and Messrs. Selby of Wareham in Dorsetshire.

Having now brought my narrative to the close of the year 1814, I will here mention some particulars, which may not have been noticed, or else slightly passed over in my correspondence. To my friend Mr Upjohn, my principal creditor, I made remittances whenever I had any cash to spare, thereby establishing my credit with him – what few *London* orders I received out of my regular trade as a watchmaker were done through Mr. Upjohn. Mr. Hicks [Joseph Hicks, of Exeter] was my working silversmith, and who from his general knowledge of the trade, and the description of goods likely to suit my sale, was of great service to me as a young beginner. He was a shrewd, clever man, one who had seen a great deal, and was thoroughly acquainted with the ways of the world. His anecdotes, and relations of his personal adventures, were truly original and highly amusing. Notwithstanding all this, however, he was wholly unacquainted with books; and was often heard to declare, that the only books he ever read were his Ledgers!! But he was an accurate observer

> Much had he *seen* – he studied Men,
> And in th' original perused Mankind.

Though I occasionally employed one or two others, John Sweet principally attended to my jewellery jobs; I found him very useful, and who with his son continued to work for me many years.

Mr Upjohn at whose manufactory I was employed while in London, was the principal person from whom I obtained my supply of watches. I did not at first venture much on gold ones, confining myself principally to silver, but my account with him at first inconsiderable, gradually increased until it became one of the largest, or perhaps the largest in his books. Although not inclined at that time to speculate on gold watches, Mr. U. having left one with me on approbation, I shortly after had the good fortune to sell it to a gentleman and lady of Moreton. (I find in my day-book the following entry: '1814, July 13 Mr Martin, Moreton-hampstead, a new gold Watch H. Ellis Exeter No. 21, £14; Gold Key, 12s.')

This watch being a solitary specimen (Hobson's choice), it was not without some difficulty that I managed to dispose of it. And after all this difficulty was, as I considered, overcome, I nearly missed the sale from another cause. My name not being engraven on it, I mentioned the circumstance, requesting that the watch might be left for the purpose, but the lady was anxious to take her new purchase with her, it being a wedding present from her husband. By dint of much persuasion, however, I at length prevailed on her to leave it a couple of days, and forwarding the name plate in a coach parcel to town immediately backed with half a crown to the guard to ensure a speedy delivery, I succeeded in getting it completed by the time it was called for. Had it been otherwise, there would have been an end of this, to me important matter, for in the mean time the lady had cooled on the bargain, having seen a watch she liked elsewhere, and evidently came prepared to decline the purchase had it not been ready.

Ellis lived at home during this period. He recalls:

My father's charge for my board was very moderate – four shillings a week. This of course was for my dinners only, but after taking into account bread, butter and tea, a moderate allowance of porter, and a glass of wine now and then, the whole of my domestic expenditure was very inconsiderable. I kept a journal of my daily earnings and also a cash account detailing my expenses. This rule in some shape or other I have continued to follow up through life, and the necessity of which all my past experience has fully persuaded me of. How many young tradesmen have I seen ruined from the want of this system! They have never known their profits, and never known their expenditure; while in the full receipts of money they have fancied it all their own, not reflecting that although they had obtained credit, the day of reckoning must surely come!

## 1815

At midsummer 1815, Ellis moved to a larger shop at 246 High Street, Exeter, which had formerly been occupied by a watchmaker named Hercules Rickard. It was directly opposite the premises of Ellis's great trade rival, John Skinner, at that time Exeter's leading watchmaker and jeweller. Ellis's name and profession were raised in solid letters on a frieze on the shop front; this was then a novelty, most being painted flat. Things were moving apace for the young watchmaker. On 9 September that year he and Mary White were married at Bideford parish church, and he took on an apprentice, Edward Piper. Writing of the latter part of 1815, he says:

> An unexpected event occurred, which as it materially affected my future prospects, I shall here fully record; it being no less than an offer to succeed to the business of another watchmaker and silversmith, Mr Benjamin Bowring, residing at No 199 in the same street [High Street].
>
> I had known Mr B's family for many years; his mother lived on the Friars Walk near my father's, and was an acquaintance of my mother's; and Mr B. had served his apprenticeship at Tiverton, in the same house as myself, with my master's father. I had, however, very little knowledge of him personally, having left Exeter shortly after his commencement in business. On my return, (ten years later) I began on my own account, which circumstance in itself was not likely to increase our acquaintance.
>
> Mr B. was carrying on a very fair business at this time, supporting his wife and family in credit and respectability, but he fell anxious to enter into a more extensive line, and may perhaps have had some other reason not stated for wishing to quit Exeter: in fact he had made himself busy in party politics and religion; felt dissatisfied with the government of the country, and had a misunderstanding with the members of the congregation of dissenters to which he belonged, on the occasion of an organ being erected in George's Meeting House, to which he strongly objected. These things combined made him determined on seeking his fortune elsewhere and for the purpose he embarked for St. John's, Newfoundland.
>
> Mr B. on leaving Exeter had not made known his intentions of settling abroad, but understanding that he was absent from home, I offered my assistance to Mrs B. if I could be of any service to her in the business. This offer it appears she considered kind, most other persons in the trade having exhibited a different feeling towards her. In the month of October 1815 Mrs B. received a letter from her husband dated St. John's, in which he informed her that having found a good opening for business in that place, he desired her to dispose of the concern at Exeter with as little delay as possible, and join him with the family. Thus situated Mrs B. at once thought of me, and made a proposition offering me the business.

Ellis needed a little time to make up his mind, as he had just fitted up new premises at some expense. He was also apprehensive of another removal, bearing in mind the maxim that 'three removes are as bad as a fire' and

I never saw an oft'-removed tree,
Nor yet an oft-removed family,
That throve so well as them that settled be.

His wife, however, thought the offer advantageous, and Ellis's creditors, the principal one being his friend W. J. Upjohn of London, agreed to help. He continues:

> I saw Mrs Bowring again. Fearing perhaps from what had passed, that I should not be inclined to treat for the business, she now informed me to my great satisfaction that although she would rather have parted with the stock as well as the business, had I been so disposed, yet she would not make this a material consideration, and that I might select any portion of it I pleased. She also gave me to understand that no 'good will' would be expected, her principal object being to get the house with the fixtures taken off her hands at Christmas. She also said that in addition to the business there was the agency of the County Fire Office and Provident Life Office, which she had no doubt, on making proper application would be continued with Mr Bowring's successor.

The stock purchased by Ellis consisted principally of clocks, clock cases and dials, watch glasses and materials. The rest was sold off at 'prime cost'. Ellis adds: 'Mr Bowring returned to England early in the spring (1816) in order to finish what matters he had left unsettled and take his wife and family to Newfoundland.'

## 1816

The new shop at No 199 was opened on 4 January 1816. The business prospered, and Ellis took over Bowring's insurance agencies. These brought custom to the shop, as did the setting up of a public select library in a room above. He relates that his former master, John Tucker, then living at Thorverton

> . . . had now a man from Coventry with him of the name of [Timothy] Stafford and he also worked for me, until at length my business still increasing I engaged Stafford to work in the house. He was a good workman; but like too many of his class not a good man; occasionally breaking out and getting drunk, returning with a black eye or a bandaged head; this ended at last in his dismissal. He married a decent young woman in Exeter and commenced business on his own account, but as might have been expected from his dissolute habits, spent all her little money, half ruined her father and mother and became a bankrupt.

## 1817

At the commencement of the year 1817, having a great abundance of jobbing, I took into my employ as a watchmaker a young man of the name of Traies, who had just completed his apprenticeship with Mr [John] Upjohn in South Street; he worked piece-work with me for several months, then went to London, where he married and commenced business as a working silversmith, principally in the spoon and fork line, for which articles I became a customer to him, and continued for several years afterwards. Forks were now getting more generally into use, which were not so well made in Exeter as in London.

My predecessor Mr Bowring had a tolerably good clock trade attached to the business, which I was desirous of continuing. He kept in employ for several years a steady workman of the name of Cross, who had served his apprenticeship with Mrs Bowring's father, Mr Price, of Wiveliscombe. He had left Mr B. on his going abroad, to seek another employer; my first object therefore was to seek him out, but to my great mortification I found that he had just engaged himself with Mr Bradford, of Newton Abbot. Mr B. offered Cross's services to make clocks for me at a cheap rate; and in this way I employed him for some time.

Ellis kept in touch with several friends who had shared his youthful interest in literature and poetry. These included John Treadwin, an Exeter bookseller, and Robert Loosemore of Tiverton. He was helped in his business from time to time by his watchmaking friends, John Walter Tothill Tucker of Tiverton, and the younger Samuel White of Bideford, his brother-in-law. Tucker, his schoolboy companion and fellow apprentice, had by this time established himself as a watchmaker and silversmith in Fore Street, Tiverton, and had married. Ellis writes:

Mr and Mrs Tucker came to visit us on Good Friday, returning on the evening of Easter Day by way of Thorverton, where Mr and Mrs Tucker, Sen., still resided. Dear Mary and myself accompanied them as far as Cowley Bridge [on the outskirts of Exeter]. A cold north-easterly wind at this time prevailed, which was so piercing and raised such a cloud of dust, that it caused an inflammation in my eyes, from the effects of which I did not recover for some time. This attack, I remember, gave me much concern, my eyesight being of vital importance to me in my business. I placed myself under Dr Eaton's care, and not feeling altogether satisfied with his treatment, I went to the Eye Infirmary, which had not then been long established, with the intention of having an operation performed for the removal of a speck which had formed on one of the pupils of my eyes.

Not applying, however, at the right hour and the surgeons being absent, I thus escaped the infliction of a painful process, and my eyes from that time began to get better, although it was a long while before the speck wholly disappeared.

On the Easter Monday of this year I first came into office in the parish [St Pancras, Exeter], having been chosen under warden, a nominal office for this year, but with the understanding that I was to be head-warden the year following, and go through all the parish offices.

When Mr and Mrs Tucker paid us their last visit, it was agreed that dear Mary and I should return it; and in the month following, on a fine Sunday morning I drove her to Tiverton, taking the baby with us. We spent a most delightful day there, and set out on our return early in the evening, several of my Tiverton friends accompanying and walking with us some little distance out of the town. But here unfortunately an accident occurred to the gig, from the breaking of one of the springs just as we were getting into it. Instead, however, of calling it unfortunate, I might say perhaps it fortunately happened here, as not being a mile out of the town; I was enabled to get it temporarily repaired, by means of a strong bar of wood firmly fastened with cord, binding the broken parts together. This was a sad delay, and we did not arrive home until after dark, the most unfortunate part of the business being that the gig being borrowed of a commercial traveller in our line, who had offered it for the occasion, I could do no less than pay for a new spring which was rather expensive.

Ellis relates that he usually spent his evenings at home with his wife, but on one occasion he caused her considerable uneasiness by not returning until after midnight:

I had sold an 8-day clock to Sir Freeman Barton, who had just taken, and was about to enter on, a house at Up-exe, a few miles distant. It being necessary for someone to go there to put up the clock, I borrowed a horse for that purpose, and started early in the afternoon, the clock, as I supposed, having already been forwarded by Mr Bennett, who was furnishing the house; on my arrival, however, it had not reached its destination, and hour after hour did I wait for it until my patience was fairly exhausted; when at length, it being nearly 10 o'clock the waggon with a heavy load of furniture and amongst it the clock hove in sight. My own work was soon completed; but not liking to return alone at such a late hour by an unfrequented road, I waited while Mr B. with his assistants unpacked the furniture, and it was not until the Cathedral clock of St Peter had announced the 'small hours' that we found ourselves again within the precincts of the ancient and loyal city. On reaching my home I found my dear and affectionate Mary waiting up for me and in tears, fearing some accident had befallen me.

In the summer of 1817 new friendships were made which widened the circle of clockmakers and jewellers in which Henry and Mary Ellis now moved. Ellis explains:

> On the 7th of July dear Mary left me for the purpose of visiting her parents, taking our babe with her, while her brother Samuel came up to assist me during her absence. She remained at Bideford until the beginning of August, and then returned with Mr Upjohn in his gig direct from Barnstaple, Mr U. happening very opportunely to be on his Western journey at this time. It was here [Barnstaple] that dear Mary got first introduced to Mr and Mrs Mallett, Mr M. [John Mallett, watch and clockmaker and silversmith, 4 High Street, Barnstaple; see List of Makers] having some little time before married Miss Susan Upjohn, Mr U.'s sister, who was in the warehouse at the time I was working for improvement with Mr U. in London. Thus an intimacy was first formed between the two families which has continued through life.

Also in 1817, Ellis notes: 'My old master had continued working for me at Thorverton, but had lately located himself in Exeter, residing in Prospect Place, Preston Street.'

## 1818

At Easter 1818 Ellis was 'fully installed in the office of churchwarden of the great and important parish of St Pancras, so different in extent and population to that of its namesake in London!' He was also appointed the parish's collector of rates and taxes. The tiny parish church of St Pancras is now a central feature of Exeter's Guildhall shopping precinct.

On 25 April 1818 Ellis set out for London, Bowring's late journeyman Cross having finished his engagement at Newton Abbot and agreed to come back again to the old shop, to work for Ellis.

> One principal object which I had in going to Town was . . . to gain information with respect to the manufacture of clocks, on which I was now fully determined; intending to get castings of work for spring clocks, and to select cases of the most approved patterns . . . I purchased at Walker's tool shop several sets of brass and steel work for spring clocks; also inlaying for [Nathaniel] Tucker who was to make the cases for them, one or two of which I purchased as patterns for him to copy.
>
> This clock manufacturing, however, did not turn out much more to my advantage than the other speculations of this period. The ornamental

ormolu French clocks were now being imported more and more every day. Shoals of English were constantly visiting the French shores, and they must needs bring back with them some souvenirs from Paris to talk about, and thus make it known to their friends and acquaintances that they had been there – and what so likely to attract attention as those glittering pendules! Hence in a great measure the lessened demand for more homely looking articles of English manufacture.

## 1819

I had also this year another opponent in business, Mr Westaway. He was a native of Exeter, and had served his apprenticeship with Mr Pollard, clock and watchmaker in St Thomas's . . . He opened his shop which was not large in a dashing sort of way (he had been working as a journeyman in London), although beginning with little or no capital . . . He had lately married and with a wife as gay as himself, was 'fond of visiting and being visited' . . . He was not satisfied with having an establishment in Exeter only, but during the season opened another in Teignmouth where he 'combined business with pleasure', having his visitors from the city, and living away most gaily . . . this state of things could not be expected to last very long; 'the day of reckoning' came and he failed to the tune of some thousands; his stock was brought to the hammer, and he to his original nothingness.

## 1821

I had this year to regret the loss of my much esteemed friend Mr Upjohn, who took a severe cold while on a journey, from which he never recovered. He had always been exceedingly kind and friendly towards me. He had always a kind word to say, and cheered me onwards at my outset in life by observing, 'Don't expect too much at once Mr Ellis, you will do very well. Remember that Rome was not built in a day.' The business was carried on by his widow for a short time after his decease, when it was transferred to Messrs Hoskins and Bird – the former from Bristol (and who succeeded me in the establishment on my leaving Mr U.) and the latter also from Bristol, who had become acquainted with Miss Henrietta Upjohn, Mr U's sister to whom he was not long after married.

I was now busily employed with my workman Cross in the construction of a new regulator (the one I have at present) which occupied a good deal of my time and attention [Plate 42]. I was desirous of adopting all the modern improvements, and rendering it in a great degree equal to the one at the Devon and Exeter Institution [in the Close, Exeter], which had been erected some few years since by Mr Skinner of this city at an expense of 70 guineas. The brass and steel work in the rough I procured from London, where I had the dial engraven; Nathaniel Tucker made the case on the most approved plan, and after the model of regulator cases in London, on which description of work he had been almost wholly engaged while there. The mode of hanging the pendulum, and locking it in the crutch I learnt from a number of the Magazine of Arts & Sciences just then published; the balancing of the hands I obtained from some other quarter, also balancing the wheels after they were mounted with small bags of shot. In short, I read the best publications on the subject, and gathered all the information I could procure before commencing the undertaking. The compensation or mercurial pendulum, which cost me £7 10s, was furnished me by Thwaites of London.

I had to form a niche for the reception of the regulator when completed, by excavating the walls, and which to avoid any interruption on the part of my landlord (who lives next door) I caused to be effected at an early hour in the morning. I had also to prepare and support the floor under the cellar, in order to get a solid bearing for the clock to rest upon. These operations occasioned some delay; for the wall required to be plastered, this was soon done, but not so the drying, for that took many days, during which I kept a charcoal fire in a brazier burning in the recess, the effluvia from whence was anything but agreeable.

The preparations having been completed, one night after the shop was shut, we got the machine fixed up; upon which occasion I gave my workmen and apprentices a little entertainment. They were all very merry, wishing success to me and my new timekeeper, and loud they were in the praises of both. Some of the toasts proposed were ridiculous enough it must be confessed. Among others, Tucker the casemaker gave 'May the clock be well oiled'. But this, I inferred, related more to the oiling of his own tongue, which ran very glibly, betokening that it was time to depart, it being 'long after the witching time of night'.

My regulator when put up attracted no small share of attention among my customers, many of whom were daily visitors to the watch department – coming to regulate their watches and 'compare notes'. Mr Charles Sanders I remember used to be frequent in his calls, and introduced his friends to see this new standard of time, which he observed did me great credit (my old regulator, made by John Tucker on going into my second shop I parted with to Messrs White for seven guineas).

The making of this new regulator led to my having the care of the one at the Institution very much to the annoyance of Mr Skinner who had the supplying it, and who although it had cost such a large sum, asserted that he had not got one farthing by the transaction. Be this as it may, on his first hearing of my appointment, he fell upon Mr Squance the librarian, whom he rated soundly for getting it taken out of his (Skinner's) hands; to which the other replied, 'Why I don't see Mr Skinner how you can charge me with that, since you have never had it in your hands, or touched the hands of it, from your first putting it up until now', alluding to his having paid no attention to the regulating it, since it was first erected.

Having to look after my new regulator as well as that at the Institution, it became necessary to be more particular than heretofore about getting the correct time. I had been indebted to my old master [John Tucker] hitherto for giving me solar time who obtained it by means of a rude meridian line which he had constructed. I now put one of my own in the library [the Public Select Library that he ran at his shop until 1825]; on which I bestowed a vast deal of pains; but could never satisfy myself as to the exact correctness of it, from the impossibility of clearly defining the end of the shadow, and consequently of taking a very correct observation. I remember the Rev Geo. Bl——e calling on me, also visiting the old Mr Tucker, and interesting himself very much about this matter. He thought, he said, that it was a great pity that in such a place as Exeter, possessing such an expensive instrument as they had at the Institution, and with such a timekeeper as my own, that the citizens should be without a better method of obtaining true time.

This timekeeper is still in existence. It was sold by Ellis's successors in January 1893 for £30. The purchaser was Sir Henry Hepburn, owner of the paper mills at Hele, near Bradninch. In a letter written in 1920, Mr John Young, of the Exeter firm of Depree & Young Ltd, stated: 'We have good cause to remember that sale, as our workmen lost the metal parts of the pendulum, on the way to Hele, it having slipped out over the tail board of a furniture van between Exeter and the Paper Mills; but was found later on.'

## 1822

In the month of June Mr Hoskins visited Exeter on behalf of Mrs Upjohn, his late employer's widow, with whom I still continued to do the principal part of my watch trade, buying a few watches only occasionally of Messrs Bidlakes, Sharp & Son, and the Messrs Bracebridges. About this time also Mr Biggs, a jeweller of London, called on me in his threefold capacity of jeweller, watchmaker and silversmith. He carried a large and elegant stock of gold and silver watches belonging to Mr Gibbons; also a stock of silver plate manufactured by his brother, consisting principally of tea services, children's mugs, and mustard-pots from which I made several purchases;

his stock of jewellery was small, but select, and wholly London workmanship.

In the fancy spoon and fork line I confined myself to Mr Traies, who was also a manufacturer of 'Argentine' (German silver) spoons and forks, and from whom I got supplied with those goods – my friend Mr Hicks continuing to furnish me with the plainer articles of the fiddled pattern in silver.

Mr John Stevens, jeweller of London, this year first introduced himself to my notice; he was in high dudgeon with Mr Adams of this city, with whom he had hitherto done considerable business, but who now refused to see him in consequence of an affair that had occurred between S. and Mr Brogden, also a manufacturing jeweller. The latter horse-whipped Stevens for some expression he (B.) supposed S. had made use of affecting his honour, the result of which was an action at law, in which Stevens recovered from Brogden £800 damages with costs! – A pretty expensive morning's amusement! Stevens employed the celebrated Charles Phillips on the occasion, who drew in his flowery style, such a vivid picture of the injuries his client had sustained from an attack which he designated as being 'unprovoked, wanton, malicious and cruel' that the jury to mark their sense of such conduct awarded the above-mentioned damages, which were considered excessive. Nor did it end here, for Stevens had the trial published, sending a copy to everyone in the trade with whom he did business, and it was this which caused Mr A. to take up the cudgels in favour of his friend B. who was represented as a high-minded and honourable man, who had been carried away by the impulse of his feelings at the moment, without stopping to enquire into the real grounds of the alleged offence S. was supposed to have committed.

## 1827

John Sweet, my working jeweller, had for some time past become a spoon maker; he was provided with a press, and had discovered the means of manufacturing spoons with great facility; but this he kept to himself as a great secret. At length he consulted me on the subject; and this led to an increase of his account. I could not, however, do a great deal with him in this branch, without withdrawing in some measure from Mr Hicks, who had always been exceedingly kind, rendering me every assistance in his power in my business.

Dear Mary's father [Samuel White] was with us in July, and when he returned home his daughter Eliza and her friend Miss Bastard accompanied him. This journey, however, was likely to have been attended with serious consequences; for the coachman on turning the corner of the road near the river at the entrance into Bideford, upset the coach! It was fortunately high tide at the time, so that the passengers escaped with little more than a good ducking and a few slight bruises. Mr Hoskins (Hoskins & Bird) who was at Bideford at the time, walked out some little distance to greet the party, and on meeting the coach he was requested to mount. He had dressed himself very sprucely for the occasion; but after the capsize as he emerged from the mud and saltwater, his white ducks [trousers] presented themselves in a most pitiable plight: in short from head to foot he was regularly bedaubed and bedabbled. This accident, however, from which no other unpleasant effects had resulted, afforded when the danger was over, no small degree of merriment, particularly with regard to poor Mr White, who is his encounter with the briny had lost his wig! It was carried away with the tide, passed under the bridge, and would in all probability have gone to sea, had it not been fished up and restored to him shortly after his return home.

At Tiverton that summer the shop of Ellis's friend, John W. T. Tucker, which had been burgled in 1826, was broken into yet again by thieves who stole sixty-two watches and a quantity of plate. Two men, George Champion, aged twenty-one, and John James, twenty-two, were subsequently brought to trial for this second burglary, found guilty, and publicly hanged on 24 August 1827 at the drop in front of the Devon County Gaol in Exeter. These events attracted considerable publicity, as serious crime was virtually unknown in rural Devon at that time. Ellis gives a detailed and graphic account:

In the summer of this year we also went again to Tiverton (Aug. 12, 1827. Journey to Tiverton 21s). I had not visited my friend John Tucker since the death of his wife [Elizabeth, in January 1823]. He had now removed into a new house and shop, which he had caused to be built opposite to the old premises where I served my apprenticeship, and on the site of the splendid mansion formerly occupied by the Dickinson family. On the death of Mr T.'s grandmother (Mrs Tothill) John came into possession of some little property which enabled him to purchase the land and erect three houses thereon. His sister was at this time residing with him and his children, and acted as manager of the family. We took some of our children with us, returning again in the evening in the car engaged for the occasion; leaving, however, at the pressing solicitations of Mr T and his sister, our daughter Elizabeth behind, who did not return until the 10th of September.

About two months previous to this visit (8th of June) my kind friend who had been robbed during the previous year, became the victim of a more extensive depredation, his house having been burglariously entered and his shop in a few hours plundered of the most valuable part of his stock. This was about one of the most daring robberies on record, and such as none but practised thieves could have accomplished. I remember being perfectly astounded on receiving the first intimation of it through John's father, as it seemed beyond belief, so short a time having elapsed since the last offender had been brought to justice. Here was a house in the centre of the principal street of a populous market town, inhabited by a tradesman, his sister, a family of children, a servant, and one or more apprentices; the doors and windows being all properly secured, and a dog kept by night in the shop. Notwithstanding all this, the villains (for there were two concerned), effected their purpose undisturbed. Having previously reconnoitred the premises and planned their mode of operation, they got into the first-floor window, first breaking a pane of glass by the process known to housebreakers, unfastening the spring-bolt, and lifting the sash. The field thus entered upon, they proceeded cautiously and quietly down stairs; picked the lock of the shop door; quieted the dog by some means understood by the members of this fraternity, and thus in possession, deliberately selected every thing that was portable and valuable, separating the gold and silver goods from such articles as were only gilt and plated; and with their booty, after unfastening and letting themselves out by the side door, they decamped without the least molestation, or any alarm being created!

This wholesale robbery on being discovered the next morning, not only overwhelmed my unfortunate friend with dismay, but excited quite a consternation in the town; calling forth the kind assistance of the authorities, as well as that of many friends, who interested themselves in the affair. Mr T. having obtained some scent of the track of these merciless plunderers, started, himself, in pursuit of them; and instead of sitting down under his losses, and crying 'God help!' as is too often the case with many, he seemed, after the first shock, to be armed with unwonted energy and resolution, and set out with a fixed determination, at whatever risk, to recover his property, and bring the offenders to punishment. With a brace of loaded pistols about his person, he visited the rendezvous of thieves and pickpockets in all the dens of infamy in the city of Bristol and its neighbourhood, but without success. A circumstance in itself trivial, led, however, fortunately at last to a discovery.

In looking over their spoil at a cottage (one of their haunts) some miles on the road to Taunton, the wretches had accidentally thrown away outside the door some small pieces of brown paper, which formed the wrappers of certain penknives, and on these envelopes most fortunately appeared Mr Tucker's private marks. Here at once was the clue – some of the inmates of the cottage were immediately apprehended, remanded, and after repeated examinations apart from each other, under the terror of being implicated in the robbery, one of them turned King's evidence. In the meantime the real burglars had been so closely pursued that in consulting their own safety, they had no opportunity of making away with their ill-gotten treasure; and by one of the parties in their confidence, the very spot where the goods were deposited was pointed out:– the principal part of them being buried in a hole in a hedge, on, or near Blackdown where, in a confused mass, watches and jewellery were found huddled together. The two thieves on the evidence of an accomplice were shortly after detected, fully committed, and tried at the ensuing Exeter Assizes, where they were found guilty and both of them executed. They belonged it appeared to an extensive gang of burglars; and although young in years, had been concerned in many other robberies of the same nature.

After sentence of death had been pronounced on these wretched men, my friend tried all in his power to save their lives; he visited them in their cells, heard the tale of their guilt from their own lips, and waited on the Judge to intercede for them. – It was however of no avail:– the Judge received him kindly; but said, that, much as he approved of his humane exertions on behalf of the unhappy culprits, yet such was the daring nature of the offence, that the law must take its course, and such desperate criminals be made an example of for the purpose of deterring others.

The year following the principal planner of the robbery, with his daughter were tried at our Assizes, as accomplices; when the former 'Old Dally' – the Fagin in this affair – was found guilty and sentenced to 14 years transportation, while the latter was acquitted. Dally was convicted principally on the evidence of a notorious horse stealer, named Watts, with a number of aliases beside, who was then a prisoner, and who had been himself an inmate of nearly half the gaols in the kingdom.

This robbery had the effect of arousing me to a sense of my own

Plate 16 (*opposite top*) 'Lewis Pridham Sandford fecit'. Dial of thirty-hour oak longcase clock *c*1735. The engraved leaf pattern extends to the sides of the dial, with the sun peeping out beneath the XII. The hands are not original

Plate 17 (*below*) Wooden-cased lock by Lewis Pridham on a tower door at Colebrooke parish church. It carries his initials and the date 1730. The churchwarden's accounts confirm that it was purchased from him in that year for 10s

Plate 18 (*overleaf top*) Newton St Cyres Church clock, made by Lewis Pridham and carrying the date 'October ye 8th 1711'. It is now auto-wound; the escape wheel and perspex setting dial are modern

Plate 19 (*below*) This charming little building in north Devon houses a 1711 turret clock by Lewis Pridham, of Sandford, with a one-handed external dial

Plate 20 Bracket clock in an ebonised case with carrying handle, signed on the chaptering and on the backplate of the movement, 'Wm Clement Totnes fecit'. It dates from about 1700 and has an obvious association with the best London work of the period

Plate 21 (*below left*) Arch dial of unrestored William Clement eight-day longcase clock, *c*1725, in lacquer case. The regal ornament filling the arch is a rare feature; the hands are not original. Clement was buried at Totnes in 1736

Plate 22 (*below right*) Detail showing the inscription 'William Clement Totness' in the dial arch of the longcase clock shown in Plate 21.

Plate 23 Wm Stumbels, of Totnes. Exceptionally decorative turret clock movement at Forde House, Newton Abbot. The bell upon which it struck is dated 1751. The strike (left) is controlled by pins spaced around the back of the main-wheel. Frame dimensions: 50 × 50 × 30cm ($19\frac{3}{4} \times 19\frac{3}{4} \times 12$in). Other Stumbels turret clocks have similarly shaped pivot bars and octagonal corner posts

security. Ere this I had fancied myself safe, having iron bars to my shutters, which were also lined inside with pieces of iron hoop placed diagonally. But now I began to think that these precautions were not in themselves a sufficient safeguard. I had therefore some loose sheets of iron put up at night between the shutters and sash frames, and bells attached to the doors; and reflecting on the nature of the late burglary, considered it necessary to have shutters made for the windows of the first floor. Hitherto I had kept a dog in the shop, but my faith had now been shaken in this mode of giving alarm [the watchdog kept in J. W. T. Tucker's shop remained silent throughout the events of the great Tiverton burglary].

In a footnote about dogs, Ellis adds:

We found they were a perfect nuisance in a small confined house such as ours. I kept them, however, from time to time, of all sorts and sizes from a yelping cur to a long-bodied mongrel. They were, however, of little service; for they became familiar with strangers, from being accustomed to the sight of so many coming to the house, so that they would scarcely ever bark at them. When we had a new dog, however, for the first few nights, being in a strange place, they were so troublesome, that they kept barking and howling all night long to the no small annoyance of ourselves and neighbours. I fastened up one of these newcomers at the back door, which was considered a vulnerable point. The brute yelped incessantly and I recollect Mr Smith, our next door neighbour, springing out of bed one night in a great rage (for we were also kept awake by the dog's music, we heard the whole performance) throwing up the sash in a fury, and making a yell almost as unmusical as that of the dog 'Yah! Ya-h! Y-a-a-h!' But all to no purpose; he might as well as whistled to the winds. He then slammed down the sash again, as if he would have broken every pane of glass in it, when the poor disconsolate beast groaned and howled more hideously than ever. Had any robber at that time made an attempt, he would most assuredly have been scared and frightened away.

Mr Horton coming here shortly after when the subject [the burglary] was the common talk of persons in general, and of the whole trade in particular, strongly advised that a person should sleep in the shop, furnished with a brace of loaded pistols. Then arose the difficulty, as to where a person could be found, sufficiently vigilant as well as trustworthy. He suggested that no young person should be chosen as he would in all probability sleep too soundly. It then occurred to me that my working jeweller John Sweet, on whom I could place the utmost reliance would be just the sort of person required, and he very good naturedly volunteered his services without waiting to be asked. A folding bedstead had now to be provided, such as might be readily removed, as well as bed and bedding. I had the recess fitted up with doors under the new glass case at the bottom of the shop, wherein it was deposited every morning. I commissioned Mr Horton to send me from Birmingham a pair of horse pistols (the same which are still in the shop) and which with a bullet mould and powder flask soon arrived. So many accidents happening from firearms being incautiously often left loaded with ball; after a short time, I caused them to be charged with powder only, laying them at night close to the bed of my trusty guardian together with the leaded pills to be administered when required, for which, happily for me, no candidate ever made his appearance.

To these precautions, however, and the keeping a light constantly burning both in the shop and front bedroom, I no doubt owe my having escaped a visit from these nocturnal depredators. From the many robberies which have come to my knowledge in my own line of business since my commencement, I am the more fully convinced that too much vigilance cannot be observed. There is scarcely a jeweller's shop in Exeter (and the remark holds good with respect to other places) but what at some time or other has been robbed. Even my own with all my precaution, has been several times attempted.

## 1828

The watchmaking, jeweller's and silversmith's side of the business continued to expand during the 1820s, but the manufacture of clocks did not prove remunerative and was discontinued. Charles Cross, however, continued to work for Ellis on clock jobbing at his own house in Guinea Street. In the meantime, Ellis's family was expanding, too – Emily, their eighth child, being born in 1828. That same year the business was moved to much larger premises next door at 200 High Street. The new shop was opened on 24 November and a special display was mounted in the window. At night the whole was brilliantly lit up with gas.

## 1829

In the month of July this year, Messrs Hoskins and Bird, with whom I still did the principal part of my watch business, and who had hitherto packed my weekly parcels from London, became (to the great surprise of most persons) Bankrupts.

In the month of August . . . my old master Mr John Tucker departed this life. He was at the time on a visit to his daughter in London, who had been married some time previous to Mr Rossiter, a relation by marriage to her brother John's wife, and who was carrying on his business as a watch and clockmaker in the Borough. The poor old gentleman was ill but a short time and his remains were interred in the burying ground of the parish in which Mr R. resided. The following notice of his death appeared in the Exeter papers: 'Died Aug. 15: On Sunday last, in London, Mr John Tucker, many years a respectable watchmaker in this city – a very exemplary and pious character.'

## 1832

Ellis employed at various times a succession of journeymen watchmakers, working silversmiths, jewellers and shop assistants. Among the longest serving was his first apprentice, Edward Piper, who remained with him sixteen or seventeen years. 'Edward Piper's father,' he writes, 'kept the Three Cranes public house in the Corn Market, which was pretty much resorted to by farmers on market days. He had formerly lived as a groom in a gentleman's family in the neighbourhood.' The premium for Edward's apprenticeship was to have been £50, of which £30 was paid initially, but Ellis had great difficulty in obtaining the remainder and in fact only received a further £10. He goes on: 'His father had turned out to be a very bad character; having decamped from an inn at Cowley Bridge, deserting his wife and family, and taking with him every thing which he could convert into money, thereby cheating his creditors. Edward had behaved tolerably well on the whole, considering the way in which he had been brought up.' Things deteriorated, however, and in February 1832 Piper was dismissed. Ellis writes:

> Having known him so long, I could not but feel sorry at parting; but his habits had become so intemperate that notwithstanding he had now a wife and family, I was obliged to give him notice to quit. To this course I was driven, after bearing with him until his conduct could be tolerated no longer. He would sometimes absent himself for whole days together; and one evening in particular, after being wanting all day in one of his drunken frolics, he came rambling through the shop to the great peril of my French Clocks on the counter, shaping his zig-zag course through the narrow opening into the counting-house, where I was busily engaged; and while there began talking so incoherently, that I was much afraid one of his old fits was about to return. At length I succeeded with some difficulty, in getting him home, and with the assistance of his wife undressed and put him to bed. I did not think this a proper time to upbraid him with his folly; but he became self convicted, exclaiming, 'Oh! I am ashamed, Sir, that you should see me in this state! You who

have been so kind to me, and always conducted yourself so differently.' After this confession I was greatly in hopes it would have had the effect of reforming his behaviour; but no; bad habits once formed are not so easily eradicated.

On his leaving me, he opened a small shop near the Cornmarket, whilst his wife carried on a millinery business in a house higher up the street. The expense of two shops, and his being removed from his wife's control by being out of her sight, did not tend to improve his condition, besides he had now lost his banker, having been in the habit of frequently borrowing of me sums of money for the avowed purpose of carrying on his wife's business.

That same year Piper's father, having returned to Exeter, was 'carried off' in the great cholera epidemic. Piper then foolishly opened a shop at 201 High Street, adjoining Ellis's, but shortly afterwards his effects were sold off for the benefit of his creditors. After this he removed with his wife, Matilda, and family to London, where he worked as a journeyman watchmaker. Ellis continues:

I had now to look out for another workman to supply Piper's place, and finding the watch part of my business so very important, I was determined to procure if possible a first-rate hand in that department. This I at length effected through the agency of Mr Bird. The new workman's name was Bristow; he had been used to first class work in London, having been in the employ of Barraud, Earnshaw, Desbois and Wheeler, and several other men of eminence in their line. It was to use a common expression, rather coming down a peg, for a man who had been thus engaged to take a situation in the country at the low salary of 30s a week, which he had consented to do, on the promise of an advance if I found him to suit me. Bristow like too many others of his grade, who consider genius and vice must be allied, had, I afterwards discovered, led a very dissolute life in London, bringing himself, his wife and family to beggary. He intended in all probability when the storm was blown over to return again to Town; or he may possibly have felt a desire to reform his conduct, an opportunity being thus afforded him of leaving his old haunts and associates. He commenced on a very wise plan, that of only drawing 15s per week of his wages for his own maintenance, and allowing his wife, who was still in London, the remainder, which she received every week through Mr Bird. At first B. was remarkably steady and attentive, and I was not long in discovering that he was really a first-rate workman.

The knowledge of this soon became evident to my customers, among whom were many who carried watches of the very best description. Mr Chas Sanders and Mr Neave were among my patrons, and used to talk with Bristow about their watches . . . Finding after some time, that his superior knowledge in watchwork was becoming a source of considerable profit, he applied for an advance of wages, urging at the same time the reasonable desire of bringing down his wife and children. He now candidly admitted that when he left London, it was never his intention to remain in the country . . . Still he was sure that with my connections I could make it worth his while to remain with me; that my establishment was and must continue to be the head quarters in Devonshire for that part of the business at any rate, standing as it did at what he termed 'No. 1 A'. I felt too fully the value of B.'s services to part with him on a mere matter of wages, and therefore at once agreed to give him two guineas a week, with a further advance of half a guinea at the end of twelve months. On these terms he agreed to send for his workboard and the remainder of his tools.

In the meantime I was still employing Cross out of the house, who received constantly 20s per week, while the remainder, he being engaged on piece work, he allowed to run up in my hands. My sale of clocks, not being very extensive, I caused them to be manufactured at Bideford by John Williams, a person in the employ of Mr White. My watch business at this time continued to increase to such an extent that with my present force of workmen it was almost unmanageable, and I could not ensure any job brought in being completed in much less than a month, as there were generally 70 or 80 watches in the house untouched. Under these circumstances, I found it necessary to engage another workman to be employed on some of the common jobs; he came from Plymouth and was called James; to him I gave 25s a week; he proved, however, but an indifferent hand, and remained with me only a few months.

## 1835

Towards the close of the last year it was proposed by Dr Shapter (one of the St. John's Charity Trustees) that the old clock at the Blue School [at St John's Hospital, High Street] should be put in repair and fitted up with an illuminated dial. Having obtained the sanction of the other trustees, the doctor applied to me for plans and estimates which I lost no time in preparing . . . I set about the job in right earnest; and on the 28th of February the first illuminated dial in this city was open to the public and gave great satisfaction [though not to Ellis, for it was many months before he was paid].

The following notice appeared in the *Exeter Gazette*:

The new dial recently fixed in front of St. John's Hospital was illuminated for the first time on Saturday evening, and the novelty drew a large concourse of persons to view it. It answers exceedingly well, producing a fine light by which the time of the night may be seen very distinctly. It has also a very good appearance by day, and is not only of great utility, but is a most creditable and handsome ornament to the city, indeed one of the best improvements recently effected, for which not only the citizens of Exeter, but strangers will be much indebted to Dr Shapter who suggested the alterations and used his exertions to procure subscriptions for carrying into effect this most desirable object. The dial was executed and the clock put in complete repair by Mr Ellis who will superintend it.

## 1836

In 1836 Ellis noted that French clocks with rocking ships were then in fashion.

Bristow, his most skilled watch repairer, contracted a fatal illness and died on 17 December. Ellis purchased from Mrs Bristow the mahogany workboard, drawers etc that her husband had used in the watchmaker's shop.

## 1837

Ellis declined to tender for a new clock for St Sidwell's parish church, Exeter:

> My past experience with regard to a similar job (St. John's clock) was . . . fresh in my recollection; when I came to the conclusion that 'the public' collectively is but an indifferent employer – ever ready with her thousand tongues to grumble and find fault, but not so ready to settle the bill when called upon to do so.

His father-in-law, Samuel White, the Torrington and Bideford watchmaker, died on 12 October 1837, aged eighty-one. He was the second son of a Torrington saddler and had married a Miss Mary Green, by whom he had three sons and six daughters. The business was continued by his son, Samuel. The father had served his apprenticeship in Tiverton with the elder John Tucker, and Ellis preserved a letter that Tucker wrote as a reference for Samuel White, referring to the time of his apprenticeship between 1771 and 1777.

> Tiverton
>
> Dear Sam!
>
> Mr John Warren hath been here and informs us that he saw you at Exeter in your Way to Bath & that you do desire me to send you your

character – Being willing to do you all y$^{e}$ good I can, I can freely say that you have been with me as an apprentice for six years from y$^{e}$ 11$^{th}$ March 71 to y$^{e}$ 11$^{th}$ d$^{o}$ 77 And for a considerable part of that time I intrusted you with all the business of My shop & found you faithfull & honest & do belive that any man may safely intrust you with untold gold.

Wishing you helth and prosperity I remain y$^{r}$; friend &c.

John Tucker

## 1843

In this year the younger Samuel White died, aged forty-nine. Ellis selected several items from his effects including 'a table of the equation of time by George Lindsay, published nearly a century ago, and by which dear Mary's father had corrected his clock or regulator ever since commencing business'.

Another horological relic that came into Ellis's possession about that time was a handbill of John Darke, clockmaker, of the late 1760s (Fig 8).

*JOHN DARKE,*

Clock and Watch-Maker, from *London,*

IS now ſettled at HOLSWORTHY; where he makes and ſells all Sorts of CLOCKS and WATCHES at the moſt reaſonable Rates; and begs Leave to acquaint the Publick, they may depend on having foul Clocks and Watches well clean'd, and when out of order carefully mended after the neateſt Manner, and at the loweſt Prices, by

*Their humble Servant,*

JOHN DARKE.

*N. B.* He gives the beſt Price for SECOND-HAND WATCHES, OLD GOLD and SILVER, and all Kinds of METALS.

Fig 8 Holsworthy clockmaker's handbill

Ellis remarked that 'John Darke subsequently removed to Barnstaple, where himself and sons after him carried on a thriving trade for many years. After the decease of the sons my friend M. [John Mallett, of Barnstaple] purchased the stock when this paper came into my possession'.

## 1844

On a page heavily edged in black, Henry Ellis recorded the sudden death of his wife, Mary, on 1 April 1844, aged fifty-two. Apparently she was engaged in knitting one afternoon in the company of her daughters when she was taken alarmingly ill and died before medical aid could reach her. 'In all relations of life', said the *Exeter Gazette*, 'she was most exemplary, and her sweetness and amiability of temper endeared her to all who knew her.' It was a terrible blow to the watchmaker:

> As shortly after the funeral as time and circumstances would permit I caused a tomb to be erected in the cemetery over my loved ones remains, and in the following autumn planted the willow which now so sadly but gracefully bends over it, and added the other funeral shrubs. And there also in the following year I planted the sweet clematis.

## 1847–8

In the years following the death of his wife, Ellis ceased to take so active a part in the business. His son, Henry Samuel Ellis, had been apprenticed to him as a watchmaker in 1840, and in 1847 registered a silver safety shawl brooch, an example of which is preserved in the Royal Albert Memorial Museum, Exeter. He writes:

> The safety brooch was made in a variety of forms and had a great run for a considerable time, until the novelty became no longer a novelty. In conjunction with the new brooches, (which were principally manufactured in silver), the silver mine at Combe Martin in North Devon, was then in operation, and the new brooches and other articles were made of the silver therefrom to a large amount, and the goods were advertised under the head of 'Devonshire silver', attracting no small share of attention.

Ellis adds: 'When the Royal Squadron anchored in Dartmouth Harbour on her majesty's visit in 1847 the Queen was seen sketching the castles [which flank the mouth of the Dart]; and this circumstance suggested to the inventor of the brooch the choice of the old fortresses for his design.' Queen Victoria later bought a brooch depicting Kingswear Castle joined by a chain to Dartmouth Castle, and the nearby church of St Petrox, and four others of different patterns, all made of Devonshire silver. 'At the same time a piece of smelted silver, and a piece of the ore from which the silver is extracted was presented to Prince Albert.' An additional purchase was reported in *The Times*:

> Devonshire silver – On Thursday Mr Ellis jnr. of Exeter had the honour of submitting for the approval of the Queen at Osborne House a basket of silver net-work, made of the silver extracted from the lead mines in Devonshire, and manufactured by the firm with which he is connected, by Order of Her Majesty.

In Ellis's business accounts the following entry appeared: '1848, April 12: Her Majesty the Queen. A Devonshire silver wire gauze basket £36 5s.'

Both Henry Ellis and Henry Samuel Ellis were subsequently appointed silversmiths in ordinary to the Queen and the Royal Arms were set up over their shop. This was done quietly at daybreak and the Arms were 'in position and properly secured before the busy world was astir'.

In 1848 the younger John Mallett, of Barnstaple, returned from London

> . . . where he had been improving himself under a first rate artisan and was now assisting his father at Barnstaple . . . He married and succeeded his father who retired with his two daughters to a pretty cottage at Instow . . . Mr M. added to his retail business that of a watch manufacturer.

It was decided to send Ellis's third son and ninth child, William Horton Ellis, on trial to John Mallett. On arrival he wrote to his father:

> I work in the front shop; they are rather late people here, and the shop is not opened till half-past eight o'clock at which time I begin work, leaving off at half-past eight o'clock in the evening or more frequently nine o'clock.

He was duly indentured as an apprentice on 12 June 1848 for a premium of £150.

In a later letter to his father, he writes:

> Our old clockmaker is about to leave us next month – the clock-work having so much fallen off, there is not sufficient for him to do. Mr Mallett has engaged a man, who not only does clocks, but repairs jewellery, engraves, and can do a little watch-work if required. Mr M. is to pay him 15s a week – a much more useful man it would appear than our —— and at a less price. There are plenty of such men in the trade to be had it seems – they only require a little looking up.

## 1850

On 15 August 1850, Ellis married for the second time at Ringmore, Shaldon. His bride was Miss Mary Cooke Snell; he refers to her in his Memoirs as 'Mary secundus'.

Henry Samuel Ellis, in Paris, writes:

> I have bought a few French clocks from Raingo in the Louis Quatorze style (one of which may be of your taste for your own use) and a few others from Rollin, who manufactures for Berens – the latter is a very desirable account for us. Mr Hall, of Manchester, to whom I recommended him has bought a large number. Mr Banks has kindly given me his opinion in each instance. The value of those I have purchased are from 75s to £7 each.

## 1852

Ellis recalls that, when he was on a visit to London, Henry Samuel

> . . . wrote a few lines, desiring me to call on Shepherd to enquire about the electric clock he had ordered (which was now become his favourite hobby). This new motor, electricity, thus applied, was a novelty recently introduced, but it was going a round about and expensive way to produce a result accomplished by the simple means of a line, pulley, and weight! And as we already had an excellent regulator, on which I prided myself, from its having been designed by myself and made under my own eye in the workshop, and my standard of time for so many years past, I considered this innovation superfluous and unnecessary.

The Shepherd referred to was Charles Shepherd of 53 Leadenhall Street. In 1849 he patented (No 12567) an application of electricity for impelling clocks, and afterwards supplied an electric clock for driving the large dial just outside the gateway of Greenwich Observatory (F. J. Britten: *Old Clocks and Watches and Their Makers*).

## 1854–6

In 1854, Ellis, having moved to a house called Grovelands in St Leonard's, Exeter, gave up the business to his sons Henry and William and it was continued under the name of Ellis Brothers. He writes that from the window of his sitting room, as the autumn denuded the trees, he could see the massive square towers and upper portions of the venerable cathedral, while a great part of the city 'which a quarter of an hour's walk would take us into the midst of, lay stretched before us in repose, its busy hum unheard'.

In 1856 he records that his old friend at Tiverton, John W. T. Tucker, was suffering from an extreme defect of vision which ultimately terminated in total blindness.

## 1871

Ellis died on 18 July 1871 in his eighty-first year. The *Western Times* remarked

> . . . he had seen most of the friends of his youth and early manhood go before him. In truth the old class of Exeter tradesman have nearly all disappeared . . . The deceased was the founder of the firm known in the city as Ellis Brothers and was the architect of his own fortune. In early life he occupied his leisure hours in the cultivation of poetry and general literature. In 1811 he published a poem on the state of Europe . . . so popular in its day that it rapidly ran through several editions . . . A more amiable or more conscientious man than the deceased has rarely discharged the duties of every relation of life; a high sense of honour actuated him in all his transactions. His clear intellect, kindly feeling, delicate appreciation of humour, admiration of the poets, and love of culture remained with him to the close.

### *Note on the Memoirs and the family*

Henry Ellis began to compile his private Memoirs – the source of this autobiography – in the late 1840s, and was still at work on them in 1866. They consist of eight strongly-bound, handwritten volumes, some of them 400 pages long, and have been preserved by his descendants, together with a further volume of correspondence. Included among them are the notebook he kept as an apprentice, examples of his watchpapers and trade cards, and some delightful illustrations drawn and coloured by his friend Charles Townsend. They conclude with the events of the year 1859.

Both Henry Samuel Ellis and William Horton Ellis (see List of Makers) served Exeter as mayor. After the former's death in 1878 the business became Ellis, Depree & Tucker and was moved to imposing premises built in 1881 on the corner of High Street and Bedford Street, Exeter, where, from about 1885, it was continued by F. Templer Depree.

6

# LIST OF MAKERS IN ALPHABETICAL ORDER

## ABBREVIATIONS

| | |
|---|---|
| Baillie | G. H. Baillie. *Watchmakers and Clockmakers of the World*, second edition, 1947. |
| Bellchambers | J. K. Bellchambers. *Devonshire Clockmakers*, 1962; revised second edition in manuscript, notes and letters. |
| Britten | F. J. Britten. *Old Clocks and Watches and their Makers*, second edition, 1904. |
| Buckley | Francis Buckley and Dr George Buckley. Manuscript list of provincial clockmakers, Guildhall Library, London (MS 3355). |
| DNQ | *Devon and Cornwall Notes and Queries.* |
| DRO | The Devon Record Office, Exeter. |
| EFP | *Trewman's Exeter Flying Post* newspaper. |
| EPJ | The *Exeter Pocket Journal.* |
| D | Trade directories and gazetteers. |
| Loomes | Notes supplied by Brian Loomes, compiler of volume 2 of *Watchmakers and Clockmakers of the World.* |
| Miles Brown | H. Miles Brown. *Cornish Clocks and Clockmakers*, third edition, 1980. |
| Moger | Miss O. M. Moger. Abstracts of Devon wills, Westcountry Studies Library, Exeter. |
| Scott | Information supplied by Prebendary J. G. M. Scott, Exeter diocesan adviser on church towers, bells and clocks. |
| Tapley-Soper | H. Tapley-Soper. Card index of makers compiled when he was Exeter city librarian, Westcountry Studies Library, Exeter. |

*after an entry denotes that the main text should be consulted for fuller biographical or other information.

Note: Details of apprenticeship premiums are taken from abstracts of the

Inland Revenue Apprenticeship Books (IR 1) in the Public Record Office, Kew.

ABRAHAM, ABRAHAM: Plymouth, 13 Lower Batter St, 1830 (D).

ABRAHAMS, AARON: Plymouth. Watchmaker, described in *West Briton*, 16 Mar 1821, as 'late of Lostwithiel, removed to 23 Nut St, Plymouth' (Miles Brown).

ABRAHAMS, M: Plymouth Dock. Watchmaker, North Corner St, 1814 (D).

ADAM, JOHN RICHARD: Teignmouth, 29 Brunswick St, 1897–1902 (D).

ADAMS, ANN: Exeter. Widow of John Adams whose business she continued after his death in 1806, later joined by her son Edward and trading as Ann Adams & Son until 31 Oct 1814. 'To be sold, a very well-toned grand piano. Price Twenty-five Guineas. Apply to Mrs Adams, jeweller, High St, Exeter' (EFP, 12 Apr 1810).

ADAMS, CHARLES: Plymouth. Union St, East Stonehouse, 1873–8 (D).

ADAMS, CHARLES RICHARD STRIPLING: Torquay. Fore St, St Marychurch, 1883–9 (D).

ADAMS, EDWARD HEWISH: Exeter. Gold and silversmith, watch and clockmaker; son of John and Ann Adams whose business he continued at 38 High St and 1, 2 and 3 St Martin's Lane (now Midland Bank), firstly in partnership with his mother until 31 Oct 1814, then with brother Charles as Adams & Sons until 25 Mar 1817, finally in own name until 1852. Baptised at St Martin's, 13 Aug 1786; at Dartington in Feb 1814 married Charlotte, second daughter of Major S. South, 20th Regiment of Foot. Supplied a new clock (since destroyed) for St Sidwell's parish church, Exeter, 1837–8.

ADAMS, HAROLD: Dartmouth, Duke St. Listed in directories, 1893, 1902 and later.

ADAMS, JOHN: Exeter. Born *c*1756, died Mar 1806. Exeter Museum has pair-cased verge watch, 'John Adams Exeter No 114'. Succeeded Matthew Skinner who announced in *Exeter Flying Post*, 25 June 1779, that 'having on account of his ill state of health declined business . . . he has disposed of his stock in trade, together with his curious tools, in the goldsmith, jewelling, and engraving branches to Mr John Adams who succeeds him in the same shop, at the corner of St Martin's Lane'. A 1789 billhead states that watches were sold and repaired at his shop. Was an active member of the Exeter Goldsmiths Company.

ADAMS, JOHN: Teignmouth, 9 Hollands Rd East, 1889–90 (D).

ADAMS, THOMAS: Plymouth. Lantern clock *c*1685 with verge escapement and pendulum swinging between going and strike trains (Bellchambers).

ADAMS, WILLIAM: Exeter watchmaker. Son of John Adams, a barber, he was apprenticed *c*1761 to William Hornsey, premium £14. Moved premises several times, but was living in St Stephen's when admitted as an Exeter freeman in Sept 1780. The 1803 Militia List records him in the parish of Allhallows, Goldsmith St, aged 56, but willing to serve as a volunteer. 'On Sunday last Mr Adams, a working watchmaker of this city, who had been to visit a farmer at Heavitree, and was returning about eight o'clock in the evening on the high causeway, a little distance from the one-mile stone, fell over into the road and was killed on the spot' (*Exeter Gazette*, 21 Feb 1811). His first wife, too, died in unfortunate circumstances, being found drowned near Exeter Quay in 1790.

ADAMS, WILLIAM: Ottery St Mary. Watchmaker, 1838 (D).

ADAMSON, WILLIAM: Exeter. Clockmaker, Holy Trinity parish, 1803 Militia List.

ALEXANDER, ——: Chittlehampton. Clock reported (Tapley-Soper).

ALEXANDER, ALBERT: Barnstaple. Eight-day longcase clock, square silvered dial, *c*1800.

ALEXANDER, CHARLES: Barnstaple. Listed at High St in 1838, at No 81 in 1844 and at No 33, 1866–73 (D).

ALFORD (ALLFORD), GEORGE: Tavistock, 37 Brook St, 1856–89 (D).

ALFORD (ALLFORD), JAMES: Tavistock, 37 Brook St, 1893–1902 and later (D).

ALLEN, PETER: South Molton. Apprenticed *c*1758 to John Potter of South Molton, watchmaker, premium £45. Listed as freeman and watchmaker, *Universal British Directory*, 1798. A man of this name buried at South Molton 19 Mar 1799.

ALLFORD, GEORGE: Ottery St Mary, Silver St, 1878 (D).

ALLIN, JOHN: Exeter, 177 Cowick St, St Thomas, 1889–90 (D).

ALMOND, RICHARD P: Devonport, 14 Pond Lane, 1850 (D).

ALWOOD (ALWORDE), WILLIAM: Tiverton. Probably a blacksmith; keeper of St Peter's Church, Tiverton, clock and chimes 1611–12. Paid in 1607–8 for 'amendinge of the clock and cheyme' and for supplying bell ropes.

ANGEL, FRANCIS (FRANK): Exeter. Recorded at North St in 1851 Census as watchmaker, aged 26, born Totnes, with wife Emma and infant daughter. Address 7 North St in 1852 (EPJ); at Paris St *c*1856–72 (D); listed also as 'naturalist'.

ANTILL, SAMUEL: Plymouth. 69 Union St, Stonehouse, 1889; the business listed under Mrs Sarah Antill in 1890 (D).

ANTONEY, WALTER: Ashburton. Parish-church sexton, clock winder and handyman; in 1499–1500 paid 4d for 'helping the man who mended le clocke' (churchwarden's accounts). Some books state erroneously that Antoney made a clock for the church.

ARCH, SAMUEL: Tiverton. On inside of wood and glass case of St Peter's Church, Tiverton, clock, among names of clock winders, is pencilled 'S. Arch Nov 17 1883'. Listed as watch and clockmaker, 41 Bampton St, 1897–1914 (D).

ARCHIBALD, ALEXANDER: Tiverton. Recorded in the 1861 Census as aged 46, clockmaker, born Scotland, together with wife Ann, 42, born Budleigh and children Louis, 14, Torquay; John, 6, Exeter; and Alfred, 3, Tiverton.

ASH, HENRY: Dartmouth. Watch and clockmaker, Foss St, 1838–50 (D).

ASHFORD, JAMES: Exeter. Watchmaker, St Olave parish. Freed on bail, 18 Dec 1786, to appear at Quarter Sessions to answer a charge of assault. William Upjohn, the Close, watchmaker, and Christopher Furze of St George, carver and gilder, provided sureties (DRO).

ATWILL, WILLIAM: Plymouth. 13 Union St, East Stonehouse, 1883 (D).

AVENT, THOMAS: Exeter. Watch and clockmaker, New Bridge St, 1836–8 (EPJ). Listed later as goldsmith and jeweller.

AVIOLET, SAMUEL ANTHONY: Exeter. Watchmaker, jeweller and goldsmith; successor to Upjohn & Co, 39 High St, in 1877; later at Queen St. The Aviolet family looked after the Exeter Cathedral clock for some sixty-five years until the cathedral was damaged in the 1942 blitz.

BABIDGE, WILLIAM HENRY: Torquay, 55 Market St, 1897; & Sons, 1902 and later (D).

BADCOCK, WILLIAM: Holsworthy, Station Rd, 1902 and later (D).

BAILEY, FREDERICK: Torquay, 53 Market St, 1889–90 (D).

BAKE, ROBERT: Plymouth and Exeter. Watch and clockmaker and silversmith; possibly the son of Robert Bake, of Camelford, Cornwall (see

*Sherborne Mercury*, 13 Apr 1767 – advertisement for 'a partner in trade in a shop where is two forges, and constant employ for two men and a boy'). Working in Plymouth June 1783 when he took out a licence to marry Susanna Wolland of Heavitree, Exeter. In 1785 took over Thomas Upjohn's shop on Bell Hill, South St, Exeter; but soon found himself in financial difficulties. May 1788 it was announced that 'a dividend of the effects of Mr Robert Bake, Jun, late of Exeter, watchmaker', would be paid. His name has been recorded incorrectly as Robert Bates in some published lists.

BAKER, JOHN: Exeter. St Martin's parish, jury list Jan 1777.

BAKER, JOHN: Plymouth Dock. Clock and watchmaker; married Elizabeth Tripe at Stoke Damerel parish church, 14 May 1798 (register). Listed in the *Universal British Directory*, 1798.

BAKER, JOHN HENRY: Bideford, 28 Mill St, 1883 (D).

BAKER, JOHN HENRY: Cullompton, 1889–1902 and later (D). Took over Mayne's watchmaking shop in Fore St, next to White Hart Hotel. Turret clock at old police station, Cullompton, has large setting dial inscribed 'J. H. Baker Cullompton' (Scott). Photograph of shop in Tiverton Museum collection.

BAKER, W: Plymouth, 1 Saltash St, 1856 (D).

BALHATCHET, FRANCIS JAMES: Newton Abbot, 23 Courtenay St, 1902 and later (D).

BALL, ALFRED ERNEST: West Hooe, Plymstock, 1890 and later (D).

BALL, CHRISTOPHER: South Brent, 1889–90 (D).

BALL, THOMAS BECKENHAM: Watch and clockmaker; his son, William Thomas Ball, engine driver, married Elizabeth Broom, tailoress, at Topsham 1 June 1852 (parish register).

BALLE, JOHN: Exeter. Longcase clocks signed 'Jno. Balle Exeter'. Apprenticed *c*1764 to Richard Jenkins, jeweller, premium £31 10s;

Plate 24 Square 30cm (12in) dial from an early Stumbels oak longcase clock, made pre-1730 while he was living at Aveton Gifford and signed with the name 'Stumbles'. The eight-day movement has pins around the back of the main-wheel to control the strike, as at Forde House (Plate 23). The dial has diamond-shaped half-quarter markings and bird and fruit basket decoration around the calendar aperture

Plate 25 Square 28cm (11in) dial of thirty-hour longcase clock signed 'Willm Stumbels Totnes'. It dates from after 1730. Unusually-shaped nameplates are a feature of Stumbels's work

William Stumbles
Aveton Gifford

Willm Stumbels
TOTNES

Plate 26 Stumbels, of Totnes. The dial of the equation clock in the marble hall at Powderham Castle. The layout is masterly and the workmanship of the very highest class. Notice the engraved border around the edge. The movement has an eleven-bell musical train fitted with double hammers

admitted as member of Exeter Goldsmiths Company in 1780, serving later as warden and accountant; Exeter's mayor in 1795. A member of the Chamber of Exeter, the city's ruling body, but removed 21 Nov 1808 because of his poverty. However, after legal argument, resignation on grounds of ill-health accepted 18 July 1809. On 10 Apr 1813 was granted £30 per annum 'for his better circumstances and support' (Robert Newton: 'The Membership of the Chamber of Exeter', DNQ Autumn 1977). Died in 1814. Shop was at 199 High St.

BANKS, GEORGE: Devonport, 45 St Aubyn St, 1830 (D).

BANKS, RALPH: Plymouth Dock. Watchmaker and silversmith, Catherine St; married Rebecca Palmer Collings at Stoke Damerel parish church on 10 Dec 1782 (register). His clocks and watches are of good quality, some having been bought ready-made from London makers; among his suppliers 1796–9 was John Thwaites, of Clerkenwell. Silver watches lost (EFP, 15 Jan 1795). An 8-day mahogany longcase clock by Banks fetched £790 in a 1977 Plymouth sale of antiques. The business listed under Rebecca Banks, silversmith, Fore St, Dock, in *Picture of Plymouth*, 1812.

BANNISTER (BANISTER), RICHARD: Axminster. A daughter Sarah baptised at the parish church in 1728 and a son Richard in 1730 (parish register). Typescript at Musbury Church states: 'The clock is dated 1729 and was made by Richard Banister, of Axminster'.

BARBER, BENJAMIN: Cullompton, Fore St, 1844 (D).

BARNES, REUBEN THOMAS: Dawlish. 5 Park Rd, 1883 (D).

BARNETT, JOHN: Tavistock in late eighteenth century/early nineteenth; at Higher Market St, 1823 (D). Inlaid oak longcase clock, brass dial with rocking ship, signed with maker's name around top of arch; Tavistock churchwarden's accounts have entry: 'April 16th 1781. Agreed to repair the Clock & Chymes from the present date to Easter next finding all Materials and leave it in good repair for the sum of Two Pounds and Two Shillings per year [Signed] Barnett'. 'There was a famous clockmaker, named John Barnett, at Tavistock, rather less than two hundred years ago' (*Transactions of the Devonshire Association*, vol 21, p332).

BARNETT, JOHN: Plymouth Dock. Clock and watchmaker, hatter, etc, Fore St, 1822 (D). Watchmaker of same name worked at Kingsand; his wife was called Susanna and two sons and a daughter were baptised at Maker, 1814–19. 'From Tuesday night's Gazette. Bankrupts . . . John Barnett, of Plymouth, watchmaker' (EFP, 9 Sept 1819).

BARNS, WILLIAM FRY: Ilfracombe, 1 St James Place, 1883 (D).

BARROW, WALTER PHILIP: Torrington, 9 South St, 1890 (D).

BARTLETT (BURTLETTE), ALBERT: Winkleigh, 1878–1902 (D).

BARTLETT, GEORGE: Devonport. Listed at 36 Fore St and at Tavistock St, 1873–89; at 104 Fore St, 1893 (D).

BARTLETT, HUBERT: Torrington. Watchmaker and photographer, High St, 1866 (D). Died 16 Oct 1869, aged 30 (EFP).

BARTLETT, MOSES: Exeter. Probably an instrument maker or engraver; Exeter Museum has a brass nocturnal dial, 'Moses Bartlett Exon fecit'. A man of this name was buried in St Mary Major parish 14 Oct 1709 (register).

BARTLETT, WILLIAM: Plymouth, 2 Russell St, 1850 (D).

BASCH, EMANUEL: Plymouth, 18 Whimple St, 1890 (D).

BASSETT, WILLIAM: Okehampton. Notice dated 3 Nov 1800 states: 'The creditors of Mr William Bassett, of Oakhampton, in Devon, clock and watchmaker, are requested to send their accounts forthwith to John Colling, of Oakhampton aforesaid, attorney at law; and all persons indebted to the said William Bassett, are requested to pay their debts to the said John Colling, or they will be prosecuted. N.B. The stock in trade of the said William Bassett, consisting of clocks, watches, guns, and all sorts of hardware, will be disposed of by private contract; and the person who purchaseth the same, may be accommodated with the shop and dwelling house' (EFP).

BASSETT, WILLIAM: Okehampton. Watchmaker, 1830–56 (D).

BASTOW, JOHN: Devonport. Watch and clockmaker, 30 Cross St, 1830; 26 Marlborough St, 1838 (D).

BASTOW, THOMAS: Plymouth. Watch and clockmaker, Finewell St, 1822–30; 18 Buckwell St, 1844 (D).

BATE, ALBERT: Plymouth. Manor St, Stonehouse, 1878; Chapel St, Stonehouse, 1883–9; 23 Ebrington St, 1893 (D).

BATE, S: Holsworthy, 1856 (D).

BATES *see* BAKE

BATHO, THOMAS: Plympton St Maurice. Clockmaker and repairer, 1889 (D).

BAYLY (BAYLEY), RICHARD: Exeter. Watchmaker and goldsmith from London; worked for Richard Upjohn, Exeter, whose business he carried on for a brief spell after latter's death in May 1778. A Richard Bayley was later in partnership with Peter Upjohn in London; one of their clocks, a high grade regulator-style mahogany longcase with gridiron pendulum, is illustrated in Cescinsky & Webster's *English Domestic Clocks*, Fig 254.

BEALEY, BERTRAM C: Exeter. Nicknamed 'Old Clockie', Bealey was for many years a full-time clock winder. Born in 1873, he served his apprenticeship with Milford K. Moxey, Exeter, and later worked for W. A. Gregory as a clock jobber before commencing business on his own at Paris St. During his busiest years he wound 400 clocks every week. He retired in 1942 after his premises and all their contents were destroyed by enemy action.

BEARE, JOHN: Pilton, Barnstaple. Longcase clock, 'Jno. Beare Pilton', its bell stamped with date 1780, reported in letter to *North Devon Journal*, 14 Nov 1915.

BEDFORD, STEWART H: Exeter. Watchmaker and jeweller, 48 Southernhay East, 1866 (D). Died, aged 24, 7 Nov 1866 at his father's house in Coombe St, Exeter (EFP).

BEER, BENJAMIN: Plymouth. Gold and silversmith, clock and watchmaker; agent to the Birmingham Insurance Office and to Richardson, Goodluck & Co's Lottery Office; in Old Town, 1814 (D). George III bracket clock with silvered dial and arched mahogany case with brass ball-finials sold at auction in 1976; longcase clock *c*1790.

BEER, JOHN R: Plymouth, 29 Buckwell St, 1844 (D).

BEER, WILLIAM: Tiverton, 42 Gold St, 1914 (D).

BEIRLEY BROS: Exeter, 105 South St, 1893 (D).

BELLRINGER, FRANCIS: Crediton, High St, 1850 (D).

BENHAM, Mrs JANE: Cullompton. Clock and watchmaker, Fore St, 1866–73 (D).

BENHAM, JOHN: Cullompton and Bradninch. Watch and clockmaker and jeweller; died at Cullompton, 21 May 1866, aged 56 (EFP). Listed in directories at Cullompton from 1838 and at Bradninch in 1844 and 1850. In Sept 1860 offered £50 reward after his shop in Fore St, Cullompton, was burgled; some 10 gold watches and about 100 silver watches, rings and other items were stolen, total value estimated at £500 (EFP, 12 Sept 1860). According to note at Exeter Library, the dials of Benham's clocks were made in Bristol, but the movements were made locally, as were the cases – by 'Cherry Top' Baker (Tapley-Soper). White dial clocks noted.

BENNET, JOHN: Plymouth. Quarter-striking lantern clock with date hand, moon dial, tide dial etc, dated 1668 (*Antiquarian Horology*, Sept 1982, p486).

BENNETT (BENNET), ——: Plymouth. 'Died . . . At Plymouth, Mr Bennet, aged 65, father of Mr Bennet, watchmaker, Buckwell St' (*Western Times*, 9 Apr 1836). Month regulator signed 'Bennett Plymouth' but with 'George Jamison' stamped on the bottom of the plate and pulley (*Antiquarian Horology*, Mar 1965, p313). Baillie lists George Jamison, London, Portsea and Portsmouth, 1786–1810.

BENNETT, CHARLES: Totnes, High St, 1883–1902 and later (D).

BENNETT, EDWIN: Tavistock, 21 West St, 1878–83 (D).

BENNETT, JAMES: Kingsbridge. Clock and watchmaker, Fore St, 1823–56 (D). Daughter married (EFP, 7 Dec 1854).

BENNETT, JOHN: Plymouth. Longcase clock *c*1740 (Bellchambers); another, 'John Bennett Plimouth', slightly earlier in period.

BENNETT, JOSEPH: Plymouth. Watchmaker, 25 Bilbury St, 1830–56 (D).

BENNETT, JOSEPH: Totnes. Watchmaker, Fore St, 1850–73 (D).

BENNETT, JOSEPH WEEKES: Kingsbridge, 49 Fore St, 1866–93 (D).

BENTLEY, SAMUEL: Kingsbridge. 'Many longcase clocks, late 18th century' (Bellchambers). Examples include 8-day and 30-hour movements, some with engraved scenes in the dial centre. One such, in an earlier marquetry case, fetched £1,550 at auction in Torquay in 1976. Watches also noted. Listed at Fore St, 1823–30 (D). 'A gold watch lost at the Eugene Inn, in Totnes, on Tuesday the 4th inst. together with a metal chain, gold seal with Hercules cut in a cornelian stone, also a metal key, and an enamelled hook, much worn, maker's name "Wightwick, London". If found and delivered for the owner to Mr Philip Brockedon, Totnes, or Mr Samuel Bentley, Kingsbridge, a reward of two guineas will be paid: or if information is given to either of them where the said watch is, so as the same may be recovered, a like reward will be paid on such recovery' (EFP, 13 July 1797).

BERNSTEIN, PHILIP: Plymouth, 2 Old Town St, 1890 (D).

BESLEY, THOMAS: Tiverton. Watch and clockmaker, St Peter St, 1823 (D).

BEVAN, THOMAS: Okehampton. Silversmith, clock and watchmaker; reported bankrupt (EFP, 17 Nov 1814). A clockmaker of the same name was in business later at Stratton, Cornwall, to about 1845 (Miles Brown).

BEYER, GEORGE CHRISTIAN: Listed at East St, Newton Abbot, 1856; 11 Fleet St, Torquay, 1866 (D).

BICKELL, JOHN: South Molton. Watchmaker, silversmith and gunsmith, 19 Broad St, 1883 (D).

BICKELL, RICHARD JOHN: South Molton. Clock and watchmaker, jeweller, silversmith and gunsmith, South St, 1838; Broad St, 1844–78 (D). Was mayor of South Molton in 1852. Entries in the borough treasurer's accounts record that he was paid £101 in 1851–2 for the town-hall clock. His wife Harriet died 8 May 1863, aged 49 (EFP). The business at 19 Broad St listed under John Bickell in 1883, R. J. Bickell's address in that year being given as Sunnyside.

BICKLE, R. H: Bishop's Nympton; name recorded (DNQ, vol 9).

BICKLE, THOMAS: Plymouth, 42 Union St, 1893–1902 and later (D).

BIDGOOD, GEORGE: Cullompton. Clock and watchmaker, listed in 1838; at Exeter St, 1844; at Fore St, 1856–66 (D). White-dial longcase clocks.

BIDGOOD, WILLIAM: Plymouth, 31 Claremont St, 1889; & Son, 4 Regent St, 1893; 79 Regent St, 1897–1902 and later (D).

BIDGOOD, WILLIAM: Ivybridge, Fore St, 1902 and later (D).

BIDLAKE, THOMAS: Ashburton. Longcase clocks, including early George III example with 305mm (12in) arched dial in burr-elm case. Another brass-dial example seen in Munich. A man of this name died, aged 83, and was buried at Ashburton 5 Jan 1821 (parish register).

BIERLY (BIERLEY) *see* BRIERLY

BILBIE, THOMAS: Cullompton. Bellfounder, clock and watchmaker; listed in the *Universal British Directory*, 1793. Will proved in Exeter in 1798. Several longcase clocks reported. He or a predecessor of same name kept Cullompton Church clock and chimes in repair at various times between 1750 and 1796–7; also Rewe Church clock from 1777 to 1798 (churchwardens' accounts). A chime barrel at Beaminster Church, Dorset, has brass plate:

Mr Tho Harris WARDENS
Mr John Hearn 1767
Tho[s] Bilbie CULLUMPTON fecit

BILBIE, THOMAS CASTLEMAN: Cullompton. Bellfounder and clockmaker. Son of Thomas Bilbie, he was born on 5 July 1759 and privately baptised the following month at Cullompton parish church; died in 1813. Will, in DRO, proved 4 Jan 1814, mentions widow Elizabeth, son John and daughter Ann Watson, wife of William Watson. His effects were sworn at under £450. The foundry was sold to William Pannell.

BILLING, JOHN: Landscore, Crediton. Repairer, 1902 and later (D).

BISHOP, JOHN: Topsham. Listed in *Time in Exeter* as clockmaker; but closer examination of 1796 document reveals he was, in fact, a blockmaker.

BISHOP, THOMAS: Axminster. 30-hour longcase clock, brass dial, *c*1770; another, 8-day, in oak and mahogany case.

BLACKBEARD, H. J: Exeter. Watchmaker, 82 Paris St, 1890 (D).

BLACKBEARD, JOHN: Exeter. Watchmaker, Paris St. Found drowned in Exeter Canal (EFP, 20 Feb 1884). Inquest told he suffered from dizziness. Evidence given by his son, Henry Blackbeard.

BLACKBEARD, WILLIAM J: Exeter. Recorded at Heavitree with family in 1861 Census, aged 37, watch and clock finisher, born Islington, Middlesex.

BLACKBEARD, WILLIAM JOHN: Newton Abbot, 24 Fairfield Terrace, 1893 (D).

BLACKFORD, JOHN: Plymouth Dock (Devonport). Watch and clockmaker, Catherine St, 1812; 53 Fore St, 1830 (D). Reported bankrupt May 1832 (EFP). 'Exeter Assizes. Benjamin Baker was charged with having on the night of the 1st November last, burglariously broken the shop window of John Blackford, of Devonport, watchmaker, and stolen therefrom one gold and two silver watches, value £32 and upwards. The prisoner was a private in the 94th regiment, stationed at Plymouth. Guilty. Sentence of death recorded' (EFP, 26 Mar 1829).

BLACKFORD, SYDNEY (SIDNEY): Lynmouth, 1893; Lynton, 1897–1902 and later (D).

BLACKMORE, JAMES: Sidmouth. Received £1 1s on 13 May 1783 for repairing East Budleigh Church clock and entered into agreement with the churchwardens 'to keep the said parish clock in good and sufficient repair for one whole year . . . without any manner of charge . . . And after the said year has expired . . . further agrees to keep the said parish or church clock in good and sufficient repair for the sum of five shillings a year for seven years successive'. Died 2 Apr 1828, being described in his will as a plumber and leaving estate valued at under £600. His son, James Blackmore jun, clock and watchmaker, was appointed executor (DRO).

BLACKMORE, JAMES (ii): Sidmouth. Watch and clockmaker and stamp distributor, Fore St; listed in directories, 1798–1844. Shop broken open and seven watches stolen (EFP, 23 Apr 1795). Married Agnes Hutchings at the parish church on 13 Dec 1799 (register). 'To be sold for ten guineas, an excellent secondhand church clock, warrented to keep good time. Apply to Mr James Blackmore, Jun, Sidmouth' (*Exeter Gazette*, 21 Mar 1818). Mahogany bracket clock with silvered dial, two subsidiary dials, moon's phases, and calendar, repeating hours; also silvered-dial longcase clock.

BLACKMORE, JOHN: Exeter. Watchmaker; son of John Blackmore, apothecary, admitted as Exeter freeman 6 Feb 1698–9. Joseph Purchase was apprenticed to him the following January, but in March 1701–2 it was stated at Exeter Quarter Sessions that Blackmore 'hath deserted his dwelling house and left his said apprentice wholly destitute of maintenance and made noe provision for the instructing of his said apprentice'.

BLACKMORE, WILLIAM HERBERT: Plymouth. 8 Townsend Hill, Mannamead, 1893; 3 Cornwall St, 1897 (D).

BLACKWELL, ROBERT: Made a church clock for Colebrooke, near Crediton, in 1659 for £6 and was paid for repairs the following year (churchwardens' accounts). The clock was replaced in 1698.

BLIGHT, EDWARD (EDWIN) GEORGE: Plymouth, 118 Exeter St, 1883–1902 and later (D).

BLIGHT, JOHN: Torrington, High St, 1878 (D).

BLOODWORTH, JOHN: Exeter and Torrington. Clock and watchmaker; a debtor seeking release from imprisonment in the Sheriff's Ward in St Thomas (EFP, 9 July 1801).

BLOODWORTH, RICHARD: Exeter. Clock and watchmaker, Bartholomew St, 1801 (EPJ).

BOLITHO, SAMUEL: South Brent, Station Rd, 1902 (D).

BOLT, Mrs EMILY: Dawlish. Continued William Bolt's business at 13 Strand. Listed 1889–1902 and later (D).

BOLT, RICHARD: Teignmouth. Watch and clockmaker, Northumberland Place, 1830–8; Regent Place, 1844–56 (D). Mahogany longcase clock, silvered dial with bird and *tempus fugit* in arch (Tapley-Soper).

BOLT, WILLIAM: Dawlish, Brunswick Place, 1844–50; 13 Strand, 1856–83 (D). Longcase clock 'W. Bolt Dawlish'.

BOLT, WILLIAM: Teignmouth. Watchmaker, Regent Place, 1866 (D).

BOND, C: North Tawton and Okehampton. Marriage reported in *Exeter*

*Flying Post*, 15 Sept 1853 – 'Sept. 4, at Okehampton, Mr C. Bond, watchmaker, of Northtawton, to Mary, only child of Mr William Seymour, ironmonger, of Okehampton'. Listed at Fore St, Okehampton, 1856 (D). *See also* Seymour & Bond.

BONE, GEORGE: Plymouth. Watch and clockmaker, 76 James St, Plymouth Dock, 1822; 10 Treville St, 1830 (D).

BONEY, CALEB: Plymouth. Report of theft of watches, clocks, jewellery and plate from shop of Mr Caleb Boney, silversmith etc, Bedford St, Plymouth; value of property stolen £900-£1,000; £50 reward offered (EFP, 7 Oct 1874). Listed at 20 Bedford St, 1878 (D). See Padstow and Saltash entries in *Cornish Clocks and Clockmakers* (H. Miles Brown).

BONEY, F. H: Plymouth. 49 Union St, Stonehouse, and Bedford St, 1890 (D).

BONEY, JOHN: Plymouth, 108 Exeter St, 1883–1902 and later (D).

BOON, ——: North Molton. Inscription 'Boon Northmolton made this clock 1841' noted on 30-hour movement of longcase clock, the dial signed R. J. Bickell, South Molton.

BORELLI, DOMINICK: Plymouth. Watch and clockmaker and silversmith; Edgcumbe St, Stonehouse, 1838–56 (D). Watch movement, 'D. Borelli Stonehouse', at Torquay Natural History Museum.

BOTELER, HENRY: Repaired the Exeter Cathedral clock at various times 1405–29 (fabric accounts).

BOUTCHER, GEORGE: Broadclyst. Exeter Museum has walnut longcase, 6-tune musical clock *c*1740, the chapter-ring signed 'Geo: Boutcher Broadclist' (Plate 44). A man of this name buried at Broadclyst on 14 Mar 1744–5 (register) and his will proved shortly afterwards in Exeter.

BOWDEN, AMBROSE: Plymouth. Longcase clock reported (DNQ, vol 8).

BOWDEN, FREDERICK E: Plymouth, 9 George St, 1873–1902; listed later as F. E. Bowden & Son, 72 George St (D).

BOWDEN, JOHN: Exeter. Gunsmith, St Edmund's parish in jury list Apr 1747. Repaired Exeter church clocks.

BOWDEN, MATTHEW: Exeter. St Mary Steps churchwarden's accounts have: 1797, 'May 22nd paid Mr Matw. Bowden for the clock & chimes to this day by order of the parishioners from Lady day last 6s 6d . . . for a collar to ye locking wheel to ye clock . . . 1s'

BOWDEN, SAMUEL: Meeth. 30-hour longcase clock, 'Samuel Bowden

Meeth 1786', the dial centre engraved with house, castle and ship (Tapley-Soper).

BOWDEN, WILLIAM: Paignton. Watch and clockmaker, 1830 (D).

BOWRING, BENJAMIN: Exeter. Born 1778; died Liverpool 1 June 1846. Apprenticed to the elder John Tucker in Tiverton. In Oct 1803 opened a shop as watchmaker, silversmith, jeweller and engraver at 225 High St, Exeter, moving to 199 High St in 1806. Later established a business in St John's, Newfoundland, being succeeded in Exeter by Henry Ellis in 1816. Good-quality longcase and bracket clocks.*

BOYCE, ARTHUR & WILLIAM: Teignmouth, 2 Wellington St, 1873–93 (D).

BOYCE, EDMUND: Exmouth. Watchmaker, silversmith and jeweller, 20 Strand, 1856–1902 and later (D). Wall clocks noted.

BOYCE, JOHN: Teignmouth. Watch and clockmaker; listed in 1830; at 2 Wellington Row, 1844–50; 2 Wellington St, 1856 (D).

BOYCE, JOHN HOCKINGS: Moretonhampstead, Cross St, 1893 (D).

BOYCE, WILLIAM: Teignmouth, 44 Northumberland Place, 1902 and later (D).

BOYNE, CHARLES HENRY: Tiverton. In 1871 Census aged 32, watchmaker, born Barnstaple, with wife and three children, Bampton St. Listed in directories at St Paul's St, 1866; Bampton St, 1873–8.

BOX, FREDERICK: Cullompton, Fore St, 1883 (D).

BRADDEN, THOMAS: Plymouth Dock. Clock and watchmaker; listed in *Universal British Directory*, 1798.

BRADEY, JOHN: Watchmaker; married Hannah Jago at Stoke Damerel, Plymouth, 29 Jan 1799 (register).

BRADFORD, BROTHERS: Torquay, 4 Victoria Parade, 1856–66 (D).

BRADFORD, DENNIS: Torquay, 4 Victoria Parade, 1878 (D).

BRADFORD, EDWIN G: Teignmouth and Torquay. 'Edwin G. Bradford, goldsmith, watchmaker, and malachite and Madrepore ornament manufacturer, has removed from Bank St, Teignmouth, to 8, Strand, Torquay, on the premises lately occupied by Mr Braham, whose business he has purchased, and where greater facilities are afforded for supplying a constant change of novelties' (EFP, 14 Nov 1850).

BRADFORD, J: Tiverton. Clock reported; no further details.

BRADFORD, JAMES: Crediton. Clock and watchmaker; took out a licence on 27 Mar 1806 to marry Sarah Tar of Crediton.

BRADFORD, JAMES: Newton Abbot. Watch and clockmaker; took out licence on 23 Mar 1813 to marry. Listed at Wolborough St, 1823; East St, 1830; Bridge St, 1844 (D).

BRADFORD, JAMES: Totnes. Watchmaker, Fore St, 1844–50 (D).

BRADFORD, JAMES: Torquay. Watch and clockmaker and Madrepore manufacturer, 7 Victoria Parade, 1850–66 (D).

BRADFORD, JOHN: East Anstey. 'Devon Assizes: George Harris, 20 (second offence), was charged with burglary in the house of John Bradford, clock and watchmaker, at East Anstey, on the night of the 21st June last, and stealing therefrom 5 silver watches, a silver spoon, bread, cheese &c. The prisoner acknowledged to the constable that his brother broke the house, and assisted him in getting in at the window; he first took the bread and cheese, when his brother asked if there were no watches in the windows – prisoner replied, yes. Upon which he said, "Go back and take the heaviest:" – he did so, and selected five. Guilty – Death recorded' (EFP, 13 Aug 1829).

BRADFORD, JOHN and JAMES: Crediton. Clock and watchmakers, linen drapers and grocers; partnership dissolved 25 Dec 1805, James Bradford to carry on (EFP, 22 May 1806).

BRADFORD, PHILIP: South Molton. Watch and clockmaker, South St, 1830–44 (D).

BRADFORD, WILLIAM: Cheriton Fitzpaine churchwarden's accounts include: 1744–5, 'pd Wm Bradford for Repairing the Clock £1 11s 6d'; 1749–50, 'To Mr Bradford bellfounder two years keeping the clock 10s'.

BRADFORD, WILLIAM: Drayford, Witheridge. 30-hour brass-dial longcase clocks, one illustrated in C. N. Ponsford *et al*, *Clocks and Clockmakers of Tiverton*; another has hour hand only. Baptisms at Witheridge in 1776 included William, son of William Bradford, clockmaker, and Elinor. This clockmaker is also mentioned in the will, dated 4 June 1779, of James Bradford of Stockleigh English, yeoman (Moger).

BRADFORD, W: East Anstey. 30-hour longcase clock, post-framed movement inscribed 'April 2nd 1794 No. 705'; countwheel with raised teeth, dial with separate chapter ring and engraved pictorial centre; attractive case.

BRADFORD, WILLIAM: Tiverton and Exeter. East side of Bampton St (1808 church rate). Made a public apology to Mr Samuel Wotton for slander

(EFP, 24 Aug 1809). His wife Jane died 29 Nov 1826 (EFP). He moved to Exeter some time after 1830 and can be traced in Mary Arches St in the 1840s. In 1846 entered into an agreement to repair St Thomas Church clock at an expense not exceeding £10 (churchwarden's accounts). He died 3 July 1849; will in DRO. Longcase clocks and watches. A silver verge watch, 'Wm Bradford Tiverton, no. 184', is hallmarked 1826.

BRAHAM, JAMES: Torquay. 'To Her R. H. the Duchess of Clarence', 1830 (D). White-dial longcase clock 'James Braham Torquay No. 19' (Tapley-Soper); bracket clock noted in South Devon country house. Braham signed a petition in 1836 'for the making of the Torbay Road' along the seafront. 'Mr Braham, goldsmith, jeweller and watchmaker, 8 The Strand, Torquay, returns his best thanks to the nobility, gentry, and the public, for the very high and distinguished patronage received during the twenty-two years he has been in trade, and takes the present opportunity of informing them, that having decided on retiring from business, he begs to offer his valuable stock of gold and silver ornaments, silver and plated goods, watches, clocks and inlaid Madrepore work, at a considerable reduction in price . . . NB Until disposed of, the working and manufacturing department will be carried on as usual, and all orders executed with Mr Braham's accustomed punctuality' (*Exeter Gazette*, 6 Jan 1849). The business was taken over in 1850 by Edwin G. Bradford.

BRAHAM, JOHN: Torquay. Watchmaker, Abbey Place, 1823 (D).

BRAITHWAITE, WILLIAM: Dartmouth, The Quay, 1897 and later (D).

BRAUND, JOHN: Hatherleigh. Clock and watchmaker, High St, 1823–83 (D). 'On the night of the 4th inst. as Mr John Braund, watchmaker, Hatherleigh, was, about 9 o'clock, on his return home from Monk-Okehampton, in crossing a clapper-bridge, from its slippery state, from the frost, he slipped and sustained a severe fracture of the leg . . . The place was retired and no one passed within hail until five o'clock the following morning, when two carters passed . . . He implored them to stop and assist him into one of their carts; the selfish and inhuman fellows however refused, on the plea that they had come that way to evade the turnpike toll, and if they assisted him should be discovered . . . and otherwise than by saying they would apprise some cottagers not far off, positively refused to meddle with him and drove off. He now gave up all for lost . . . when from the carters having fulfilled their promise, some labouring men drew near and with difficulty relieved and took him in a cart to his home, where he still lies in a very precarious state' (EFP, 17 Feb 1831).

BRIDGES, JAMES COLMAN: Okehampton. Clock and watchmaker; name is on a deed dated 1 Mar 1790 in borough records (DRO). Listed in *Universal British Directory*, 1798.

BRIERLY (BIERL(E)Y), JOSEPH: Exeter, 2 Milk St, 1889; 100 Paris St, 1897 (D).

BRIGHT, JAMES: Hartland. Listed 1866–90; & Sons, The Square, 1897–1902 and later; also at Bradworthy, 1902 and later (D).

BRIMICOMBE, ROBERT: Bradninch, 1878–1902 (D).

BRISSINGTON, WILLIAM: Plymouth. 31 St Mary St, Stonehouse, 1850–66 (D).

BRISTOW, WILLIAM GEORGE: Skilled watchmaker from London employed in Exeter by Henry Ellis in 1832. Lived in St Thomas and died, aged 42, on 17 Dec 1836.

BROAD, WILLIAM: Crediton, 2 Bowden Hill. Watchpapers dated 1876–85: 'Broad watch & clock maker Crediton'.

BROCK, CHARLES: Plymouth, 165 King St, 1883–1902 (D).

BROCK, JOHN: Plymouth, Frankfort St, 1889–93 (D).

BROCKEDON, PHILIP: Kingsbridge and Totnes. Married 1784; buried at Totnes, 17 Sept 1802.*

BROCKEDON, WILLIAM (1787–1854): 'Subject and history painter' and Alpine traveller. Son of Philip Brockedon, Totnes; worked in the family clock and watchmaking business when young.*

BROOKING & SON: Exeter, 270 High St and Gandy St, 1878 and later (D). Also pawnbrokers.

BROOKING, JOHN: Apprenticed *c*1759 to James Pike of Newton Abbot, premium £31 10s.

BROOM, CHARLES: Tiverton, 104 Westexe South, 1914 (D).

BROOM, JOHN: Exeter. Recorded in St Thomas in 1815; by 1825 had moved from Okehampton St to 160 Fore St, moving to Market St about 1849. The *Exeter Flying Post* recorded his death in 1852: 'Feb. 25, after a lingering illness, Mr John Broom, watchmaker, aged 57, respected by all who knew him.' Eight-day and 30-hour longcase clocks, also watches; dial of 30-hour clock at Totnes Museum, decorated with painted bird of paradise. Exeter Museum has pair-case verge watch by him, bearing London hallmarks for 1829–30. He was the 'Mr John Broom, of this city', referred to by Dr Oliver, the nineteenth-century Exeter historian, as having an 'excellent table clock' by Jacob Lovelace.

BROWN, H. A: Exeter, Market St, 1844–6; 168 Fore St, 1847–50 (EPJ).

BROWN, JOHN: Plymouth. Watchmaker, goldsmith and jeweller of St Andrew's parish. Stepson of Nathaniel Upjohn, he married Patience Kimber at Bideford parish church 8 Feb 1764 (register). Recorded as sending wares to the Exeter Assay Office between 1765 and 1783 (*Catalogue of Exeter Silver* in Exeter Museum).

BROWN, JOHN: Honiton. A 30-hour longcase clock, 'Ino. Brown Honiton', sold at auction in 1976 for £310.

BROWN, JOHN: Modbury. Watch and clockmaker, 1838 (D).

BROWN, JOHN: Plymouth. 114 Union St, Stonehouse, 1878 (D).

BROWN, JOHN: Exeter. Clockmaker, John St, 1852 polling list.

BROWN, JOHN and ANTHONY: Exeter. Watch and clockmakers, 168 Fore St, 1850 (D).

BROWN, MATHIAS & CO: Exeter. Clockmakers, 1 Market St, 1838–43 (EPJ).

BROWN, NICHOLAS: Kingsbridge. Watchmaker and jeweller, Fore St, 1850–83 (D).

BROWN, NICHOLAS EDWARD: Plymouth, 8 Tavistock Rd, 1883 (D).

BROWN, WILLIAM: Exeter. Watchmaker; married Jane Upjohn, daughter of the elder William Upjohn (qv), at St Mary Arches Church, Exeter, 25 July 1781. Described as 'of the Close, watchmaker', jury list Apr 1783. A witness to the will dated 20 Aug 1783 of Thomas Upjohn, watchmaker.

BROWN, WILLIAM: Brixham, 70 Bolton St, 1878; 19 Fore St, 1883 (D).

BROWNE, THOMAS: Apprenticed *c*1721 to Daniel Gossier, 'Hamoze Dock', Plymouth, premium £20.

BRUFORD, WILLIAM & SON: Exeter. 'Late S. and E. Piper' (advertisement, Besley's *Exeter Directory*, 1906).

BRYAN, THOMAS: Honiton, New St, 1889–93 (D).

BUCKINGHAM, JOHN: Plymouth Dock (Devonport). Clock and watchmaker and silversmith, Catherine St, 1822–38 (D).

BUCKLOW, ALEXANDER: Moretonhampstead, Station Rd, 1897–1902 and later (D).

BUCKNELL, JAMES: Crediton. Clock and watchmaker; apprenticed *c*1756 to George Newman of Crediton, premium £8 8s. His own apprentices included Simon Davy and John Trobridge. Buried at Crediton 7 Jan 1782 (parish register).

BUCKNELL, JAMES (ii): Crediton. Son of the above; clock and watchmaker and silversmith. In notice dated 9 Jan 1782 he states that 'he carries on the business in partnership with his mother, in the same house in which his late deceased father lived' (EFP). Took out licence on 27 Feb 1792 to marry Sarah Webb of Sandford; died, aged 63, on 18 Oct 1827, leaving estate sworn at under £6,000. Will in DRO, dated 7 May 1818; witnesses John and Elizabeth Traies. Verge watch, 'Jas. Bucknell Crediton', at Exeter Museum; longcase clocks. Was paid £8 8s in 1819 for supplying a new dial, motion work and hand to the church clock of Colebrooke (churchwarden's receipts). Premises were at Market Place.

BUCKNELL, JAMES (iii): Crediton. Son of above; resigned business in favour of George Webber (EFP, notice dated 9 Nov 1833).

BUCKNELL, WILLIAM: Crediton. Watch and clockmaker; listed in *Universal British Directory*, 1793. Possibly William, son of James and Sarah, baptised at Crediton 26 Jan 1774 (parish register).

BULFORD, JOHN ROGERS: Devonport, 14 Marlborough St, 1856–89; business listed under Mrs Mary Bulford in 1890 (D).

BULKELEY, JAMES: Starcross. Kenton baptisms include: 'Bulkeley, Isabella Hunter, daughter of James, watchmaker, and Isabella, Starcross, 4 Apr 1816' (register).

BULKELEY, JOHN: Starcross, Dawlish and Newton Abbot. Clock and watchmaker; born *c*1785 and apprenticed *c*1802 to William Welsford at Teignmouth. Recorded at East St, Newton Abbot, in 1823 (D). A pauper in Kingsbridge Union Workhouse in 1838 (DRO).*

BULLGLASS, WILLIAM: Chudleigh. Eight-day longcase clock with deadbeat escapement and maintaining power, *c*1795 (Bellchambers).

BURGE, ARTHUR HENRY: Newton Abbot, 97 East St, 1883; 11 Highweek St, 1889-1902 (D).

BURGOINE, RICHARD: Dartmouth, Newcomen Rd, 1883–1902 and later (D).

BURNECLE, WILLIAM: Appledore. White-dial longcase clock, 'W. Burnecle Appledore'. A brazier of this name listed in 1850 (D).

BURNELL, WILLIAM: Lynton, 7 Alford Terrace, 1856–73 (D).

BURNETT (BURNET), ABRAHAM FILMORE: Exeter and Topsham.

# £50 REWARD.

**The Shop of Mr. BURRINGTON, Watch Maker, High Street, Exeter, was burglariously entered on the night of the 14th inst., and from 60 to 70 GOLD and SILVER WATCHES, 30 to 40 GOLD GUARD CHAINS, 30 to 40 ALBERT CHAINS, a number of Second-hand SILVER WATCHES, &c., &c., Stolen, of which at present only the following can be particularized:**

Lever Three-quarter Gold Dial, raised figures, seconds, 1,165.
Ditto ditto ditto not seconds.
Ditto ditto ditto.
Ditto Enamelled Dial, sunk seconds, *Burrington.*
Ditto ditto ditto *No Name.* No. 140.
Ditto Gold Dial, No. 260.
Gold Dial Vertical, second-hand. *Burrington,*
Ditto ditto ditto. *No Name.*
Four Gold Watches, ditto.
Gold Watch Case.
Six Gold Genevas, horizontal.
Gold Geneva Lever. *Stauffer.* 54,299.
Gold Horizontal, gold dome, 66,828.
Gold ditto ditto 58,372.
Small Gold Watch, Geneva, size of shilling.
Gold Case and Dial only.
Thirty to Forty Gold Guard Chains.
Ditto ditto Alberts, various patterns.
A number of Silver Second-hand Watches.

**The above Reward will be paid by Mr. BURRINGTON, upon the apprehension and conviction of the Offender and Offenders.**

**Information to D. STEEL, Superintendent of Police, Exeter.**

Dated 15th January, 1856

Latimer, Printer, Exeter

Fig 9 Reward poster concerning theft of watches

Longcase clocks, one with Adam and Eve in the arch of the dial, another, 30-hour, with a 280mm (11in) square brass dial. Described as watchmaker, Holy Trinity parish, when he married Elizabeth Harding at St John's Church, Exeter, 1 June 1783. Moved to Topsham where he kept the parish clock in repair from 1786 to 1795 (churchwarden's accounts); had a share in the lighter *Good Intent*, registered 26 Nov 1792. Died 15 Aug 1840; will in DRO mentions wife and daughter, Elizabeth (or Betsey) Harding Burnett; effects sworn at under £20.

BURRINGTON, JOHN: Exeter. Recorded aged 30 in 1841 Census, with wife Elizabeth; died in 1878. 'Burrington, chronometer and watch maker, No 2 High St (adjoining the Grammar School), begs to inform . . . of having commenced as above, & trusts that a long previous experience in this city, added to a valuable acquaintance with the detail of watch manufacture, acquired in one of the first houses in London, enables him with confidence, respectfully to solicit a share of their patronage. A particular study of Geneva and foreign watches in general, renders him perfectly conversant with that the least understood, and more difficult branch of the art. Jewellery and plate' (EFP, 25 Apr 1839). Was a member of Exeter City Council and in 1851 seconded Dr Shapter's proposal to adopt Greenwich Mean Time in Exeter. In Jan 1856 offered £50 reward for information following a burglary from his shop in which some seventy watches and other items were stolen (Fig 9). Listed at 2 High St until 1868 (EPJ); at 3 Bampfylde St later.

BURRINGTON, THOMAS (the younger): Exeter, St Mary Major parish, watchmaker. Took out licence on 26 Oct 1764 to marry Martha Webb, widow, of Holy Trinity parish.

BURROW, HENRY MARSHALL: Bideford, Mill St, 1873–1902 (D).

BURROW, JAMES: Uffculme, 1850 (D).

BURROW, SAMUEL EDWIN: Totnes, 15 High St, 1883–93 (D). Tower clock formerly at Totnes parish church, now at Tuckenhay Mill, has inscription 'Reconstructed by Sainsbury Bros., Walthamstow, for S. E. Burrow, Totnes'.

BURSTON, JOHN: Teignmouth. Watchmaker, Fore St, 1878 (D).

BURSTON, JOHN: Ashburton, North St, 1878; East St, 1883–9. Same name listed at 63 Queen St, Newton Abbot, 1902 and later (D).

BURT, HENRY: Chudleigh, 1883 (D).

BURT, ROBERT HENRY: Devonport, Queen St, 1866–78 (D).

BURT, WILLIAM HENRY: Devonport, 4 Marlborough St, 1878–93 (D).

BURTON, HENRY: Torquay, Braddons Row, 1850 (D).

BUSSELL, THOMAS: Cullompton. Silvered-dial longcase clocks, one illustrated in *The White Dial Clock* (1974 edition) by Brian Loomes, p17. 'All persons indebted to Thomas Bussell, late of Cullompton, in the county of

Devon, watchmaker and ironmonger, are requested immediately to pay their respective debts to Mr Wm. Leigh, of Cullompton, aforesaid, solicitor; and all persons having any claims or demands on the said Thomas Bussell, are desired to transmit them the particulars thereof to the said Wm. Leigh, that the same may be arranged for settlement. Dated 28th August 1804' (EFP).

BUSSEY (BUSSY), GILES: Exeter. Locksmith and turret-clock repairer. St Mary Steps churchwarden's accounts state: 1723, 'Paid Giles Bussy half a year's salary for keeping ye clock £1 5s; for mending the hammer of the clock 1s; for a rope for the chimes 5s 9d; for a clock weight 2s.' He also looked after the St Edmund's Church clock from 1704 until the 1720s, and carried out work on the bells.

BUTLAND, BEN: Plymouth, 15 Old Town St, 1883–90 (D).

BUTTALL (BUTTAL), JOHN: Plymouth. Silvered-dial longcase clocks of the late eighteenth/early nineteenth centuries. One has Britannia in arch of dial and well-engraved full-length portraits in spandrels; another has rocking ship in arch with eight 'Armada' ships depicted in action behind it.

BUTTRESS, WILLIAM: Crediton. Clock and watchmaker; took out a licence on 1 Dec 1809 to marry Susanna Thomas of Crediton. Unusual oak longcase regulator, 'Buttress Crediton' 8-day movement with tapered plates, rack-striking, deadbeat escapement, Harrison maintaining power and massive pendulum bob, 330mm (13in) circular brass dial with silvered chapter ring and subsidiary seconds dial – 2m (6ft 4½in) tall.

CAINE, JOHN: Apprenticed *c*1755 to Matthew Sayer of Exeter, clock-maker, premium £20.

CALLE, SAMUEL: Exeter. Clockmaker; mentioned as surety in Exeter Quarter Sessions records, 21 Aug 1685.

CALLEY, WILLIAM: Brixham, Burton St, 1893–7; 53 Bolton St, 1902 (D).

CANN, STEPHEN: Morchard Bishop. Watchmaker, 1850–66 (D).

CANON, C. E: Exeter, 30 New Bridge St, 1902 (D).

CAPON, FRANCIS W: Ottery St Mary. Watchmaker, 1838 (D).

CAPPS, WILLIAM GEORGE: Plymouth. Union St, Stonehouse, 1889–93; also at 25 Morley St, 1893; 35 Frankfort St, 1902 and later (D).

CARLETON, WILLIAM: Torquay. Watchmaker, 2 Braddons Row, Fleet

St, 1856; 14 Torwood St, 1866; Park Crescent, 1873–90 (D). Carried out modifications to the tower clock at Torre Abbey in 1875.

CARNE, GEORGE: Plymouth. Watchmaker; signed letter dated 20 June 1833, stating willingness to have his stock of foreign watches stamped (Clockmakers' Company archives).

CARPENTER, G: Plymouth Dock. Clock and watchmaker, Boot Lane, 1822 (D).

CARPENTER, WILLIAM: Described as watchmaker of East Budleigh, when he took out a licence 4 Jan 1822 to marry Charlotte Cox of Woodbury. Listed at Budleigh Salterton, 1838–56 (D).

CARTER, EDWARD: Exmouth, Strand, 1844; Parade, 1850–6 (D).

CARTER, JOHN: Totnes. Watchmaker, listed in 1823; Lower Main St, 1830 (D). Arched dial from longcase clock at Totnes Museum – brass, one-piece with narrow curved calendar aperture just below the centre, signed 'Jno Carter Totnes', late eighteenth/early nineteenth centuries. A 30-hour movement, 280mm (11in) dial, 'I. N. Carter Totnes'.

CARY, RICHARD: Torquay, St Marychurch, 1878; 19 Abbey Rd, 1890 (D).

CAVILL, WILLIAM JOHN: Exeter, 12 Russell Terrace, 1890 (D).

CAWDLE, HENRY: Torquay, 58 Lower Union St, 1878–83; 12 Bexley Terrace, Bampfylde Rd, 1890 (D).

CAWDLE, WILLIAM: Torquay. Watch and clockmaker, 26 Lower Union St, 1850–73 (D). In 1863 supplied clock for St Mary Magdalene Church, Upton, Torquay.

CELLER, ——: German clock seller who visited Exeter. The *Flying Post* has advertisement dated 22 Jan 1795: 'Musical clocks to be seen or sold by Mr Celler, from Germany. Mr Celler begs leave to inform the ladies and gentlemen of Exeter, and the public in general, that his stay will be but a few days longer, he hopes they will not lose the present opportunity of viewing those curious pieces of mechanism, at the Half-Moon Inn. Admittance the forenoon, One Shilling each, Children threepence. After four o'clock sixpence each, servants and children threepence.'

CHAMBERLAIN, HENRY: Tiverton. Clock and watchmaker; possibly working in the late-eighteenth century. Listed at Bridge St in 1830 (D). Brass-dial oak longcase clock, 'H. Chamberlain Tiverton', advertised for sale in 1979 for £675. A clock with musical chimes reported, also white-dial longcase clocks signed 'Chamberlain Tiverton'. Dial at Tiverton Museum.

CHAMBERLAIN, T. F: Tiverton. Watch and clockmaker, Bridge St, 1823 (D).

CHAMBERLAIN, WILLIAM: Exeter. In 1861 Census at South St, aged 35, watchmaker, born London.

CHAMBERLAYNE *see* CHAMBERLAIN

CHAMBERS, WILLIAM HENRY: Exeter, Fore St, 1870; & Son, 1893 and later (D).

CHANT, LLEWELLYN JOHN (or LOUIS): Lynmouth, 1889–90 (D).

CHAPMAN *see* Morrison, James

CHARLTON, CORNELIUS: Exeter. In 1851 Census in St Thomas, aged 27, watchmaker, born Exeter.

CHASTY, CHARLES: Dawlish. Watchmaker, 1838 (D).

CHASTY, CHARLES HEARD: Watchmaker. Listed at Shaldon in 1850; at 25 Bitton St, West Teignmouth, 1883 (D).

CHASTY, ROBERT: Hatherleigh. Watch and clockmaker, Market Place, 1823–66 (D). A correspondent *c*1916 writes that Robert Chasty 'often permitted me to see him at work cutting cog-wheels, etc.', and that he 'was an aged man when I knew him in the forties' (DNQ, vol 9). Longcase clock with Adam and Eve automata.

CHASTY, Mrs SUSAN: Teignmouth. Watchmaker, Fore St, 1856–66 (D).

CHASTY, WILLIAM: Teignmouth. Watch and clockmaker and working jeweller; goods stolen (EFP, 28 Oct 1824). Listed at Regent Place, 1830; Fore St, 1844–50 (D). Died 'Oct. 2, at West Teignmouth, Mr W. Chasty, watchmaker' (EFP, 14 Oct 1852).

CHASTY (CHASTEY), WILLIAM: Totnes. Watchmaker, 1823 (D). Eight-day and 30-hour longcase clocks.

CHASTY, WILLIAM: Torquay. Watch and clockmaker, Torwood St, 1830 (D).

CHAUNTER, JOHN: Exeter. A witness to will dated 21 Jan 1768 of the elder William Upjohn, of Exeter. Described as watchmaker, St Kerrian's parish, in jury list, Oct 1775. 'Next of kin. If John Chaunter, watchmaker, formerly living in the city of Exeter, but since in Gosport, near Portsmouth, and at Calne, in Wiltshire; and also William Chaunter, seaman, formerly of Exeter, will apply to the printers of this paper, they will hear of something to their advantage' (EFP, 24 Apr 1806).

CHAUNTER, THOMAS: Exeter. Watchmaker; took out licence on 9 Dec 1749 to marry Susanna Ley of Exeter.

CHENHALL, ——: Plymouth. 'In 1858, a watchmaker named Chenhall, of Drake St, Plymouth, exhibited in his shop window a clock of the size of an ordinary 8-day clock, with a novel and very simple movement, which was said to be capable of going as long as the durability of the materials permitted, without the aid of weight or spring, and in short without any manual assistance whatever' (E. J. Wood: *Curiosities of Clocks & Watches*).

CHENHALL, Mrs GRACE & SONS: Tavistock, 24 West St, 1889–1902 (D). The business, formerly James Chenhall's, also listed in directories under style Mrs J. Chenhall & Sons.

CHENHALL, J: Plymouth, 33 Kinterbury St, 1856 (D).

CHENHALL, JAMES: Buckfastleigh, 1850 (D).

CHENHALL, JAMES: Tavistock. Listed at West St in 1866, at No 2 in 1873 and at No 24, 1878–83 (D).

CHENHALL, JOHN: Totnes, Fore St, 1873 (D).

CHERITON, GEORGE: Exeter. Clockmaker; witness to a marriage document dated 7 July 1705.

CHING, WILLIAM HENRY: Devonport, Duke St, 1850–83 (D).

CHING, WILLIAM THOMAS: Kingsbridge, 28 Fore St, 1889–90 (D).

CHOPE, WILLIAM: Hartland. White-dial longcase clock (DNQ vol 8).

CHORLEY, JOHN: Tiverton. Clock and watchmaker and taxidermist (bird preserver, naturalist etc), 1856–73 (D). Watchpaper at Tiverton Museum: 'J. Chorley, Clock & Watch Maker, Working Jeweller, Westexe, Tiverton. Clocks attended to in the Country. Horizontal, Duplex, Lever & Repeating Watches, Musical Boxes, Plate, Jewellery, etc Repaired'.

CHUBB, JOHN: Apprenticed *c*1759 to Francis Pile, Honiton, premium £20. Was a witness to Pile's will, dated 22 June 1763 (DRO).

CHUDLEIGH, WILLIAM JOSEPH: Newton Abbot, 4 Gloucester Rd, 1897 (D).

CLAMPITT, CHARLES H: Exeter, 46 Paul St, 1897 (D).

CLARK, ALFRED: Of Honiton. Bound apprentice to John Compton, of Honiton, watchmaker, in 1885 (DRO).

CLARK, EDWARD: Northam, 1873 (D).

CLARKE, ——: Newton Abbot. Supplied the 1863 church clock at Leusdon. Possibly John Clarke, listed as a gunsmith at Newton Abbot, 1844–70 (D).

CLARKE, CHARLES: South Molton, 1 Barnstaple St, 1878; Broad St, 1883–1902 and later (D).

CLARKE, GEORGE P: Plymouth. 30 William St, Morice Town, Devonport, 1890 (D).

CLARKE, JOSEPH: Exeter. Watchmaker, took out licence on 2 Dec 1724 to marry Elizabeth Potter of Bovey Tracey, widow. At Exeter Quarter Sessions, 16 Feb 1727, informed against Thomas Tilly for stealing a watch. Clarke's name occurs again in the Quarter Sessions records in 1740.

CLEAK, ADAM (1732–1809): Luppitt, Bridport (Dorset), and Upottery. Made the church clock at Upottery which has a brass plate inscribed: 'The Gift of The Right Hon. Henry Addington. A. Cleak, BRIDPORT Fecit May 1$^{th}$ 1794'. Son of William and Temperance Cleak; baptised at Churchstanton, 20 Dec 1732; married Joan Quick of Luppitt on 11 Apr 1757, by whom he had four children. Joan died 1768 and four years later he married Polly Culverwell (died 1814). Established himself as a clockmaker in West St, Bridport, *c*1782; in 1804 moved to Upottery where he owned a property called Stylinghayes with nine acres of ground. Was buried in Upottery churchyard on 12 Feb 1809. Will in DRO among Addington Papers. 8-day and 30-hour longcase clocks, including a 'one hander' made when he was at Luppitt. 'Mr Cleak for cleaning the clock £2 12s 6d' (entry dated 18 Nov 1790, Colyton churchwarden's accounts). The Bridport business was continued by his son John Cleak (1763–1833).

CLEAK (CLEAKE), EZEKIEL: Exeter. Locksmith; son of Jonathan, baptised in St Mary Major parish in 1639; recorded there in 1671 (Hearth Tax). 'To Ezekiell Cleake for Mending the Clock £1 15s. For a horse & a boy to fetch and carry him home and expenses on him 4s 8d' (East Budleigh churchwarden's accounts 1668).

CLEAK (CLEAKE), EZEKIEL (ii): Exeter. Locksmith and turret-clock repairer. Apprentice of his father, Ezekiel Cleak, locksmith; admitted as Exeter freeman 31 Dec 1688. Burial recorded in St Kerrian's register, 1 Aug 1709.

CLEAK (CLEAKE), EZEKIEL (iii): Exeter. Clockmaker, son of the above, admitted as Exeter freeman by succession, 18 Sept 1710. Was churchwarden of St Kerrian's 1714–15 and his burial was recorded there on 30 Jan 1728 (register). 'Watch wheel engine seen 1971' (Osborne Index, Guildhall Library, London).

CLEMENT, EDWARD: Exeter. Clock and watchmaker, working in the city in 1684, died 1720.*

CLEMENT, WILLIAM: Totnes. Clock and watchmaker, working in the town *c*1700, died 1736.*

CLEMMOW, CHARLES: Ilfracombe, Market Row, 1873 (D).

CLOCKMAKER OF EXETER: 'Item to the Clocke maker of Exetter for the mendyng of the Clocke 6s 8d' (Woodbury churchwarden's accounts 1541–2).

CLOCKMAKER, PETER: Exeter, South Quarter, 1537 (DRO).

CLODE, HAROLD: Exeter, 134 Fore St Hill, 1878 (D).

CLUTTERBUCK, H: Devonport, 71 St Aubyn St, 1866 (D).

COCKEY, EDWARD: Warminster, Wilts. A famous maker of astronomical and other longcase clocks. Had a connection with Totnes where his son, Christopher, was baptised 26 Aug 1696 (parish register). A number of craftsmen named Cockey worked at Totnes in the metal-working trades, but not apparently as clockmakers.

COCKREN, EDWARD: Exeter, 30 North St, 1862; Broad St, 1865; 113 Fore St, 1870 (D).

CODNER, PHILIP: Totnes. Entered into agreement in 1672 to keep and repair the town clock and chimes for seven years (borough records, DRO). Buried at Totnes on 5 Nov 1686 (register).

COE, ALFRED I: West Teignmouth, 11 Bank St, 1889–97; the business listed under Mrs A. I. Coe in 1902 (D).

COFFIN, JAMES: Apprenticed *c*1752 to William Upjohn, Exeter, clock and watchmaker, premium £12 12s. 'James Coffin to take care of clock, chimes and quarter-clock for 5 years in place of John Cole deceased' (entry dated 7 June 1764, Barnstaple parish church vestry minutes).

COHEN, D: Plymouth. Watchmaker, Frankfort Place, 1814 (D).

COHEN, HYAM: Exeter. Clothier and watchmaker, 98 Fore St. Bankrupt in 1847 (DRO).

COHEN, L: Plymouth Dock. Watchmaker, Market St, 1814 (D).

COHEN, Mrs: Plymouth Dock. Watchmaker, Market St, 1814 (D).

COHIN, SAMUEL: Plymouth. Watchmaker, listed in the *Universal British Directory*, 1798.

COLE (COLES), JOHN: Barnstaple. Described in parish register as clockmaker when buried on 11 Jan 1757. A church clock he made for Combe Martin in 1734 can be seen at Ilfracombe Museum; also made church clocks (since replaced) for Northam and Ilfracombe and kept the Barnstaple parish church clock and chimes in repair, his widow having a salary for the same in 1760. In 1742 he and his brother were paid for mending High Bickington Church clock (churchwarden's accounts). Was maker of 8-day and 30-hour longcase clocks, including a one-hander with rear rack strike (see *Antiquarian Horology*, Winter 1977, p623); and also of a good musical clock with astronomical work and tidal dial (Bellchambers).

COLES, WILLIAM D: Exeter, 6 Bedford St, 1870; 44 Longbrook St, 1883 (D).

COLLEBER, JOHN: Exeter. Watchmaker; provided a £20 surety at the city Quarter Sessions, 29 Mar 1711.

COLLIER, JOHN: Exeter. Apprenticed *c*1742 to Edward Upjohn, watch and clockmaker, premium £30.

COLLINS, AUGUSTUS: Torrington, High St, 1883; 13 Fore St, 1890; & Son, 1902 and later (D).

COLLINS, HENRY: Devonport, 23 Tavistock St, 1897 (D).

COLLINS, JOHN: Probably a smith. Keeper of St Peter's Church, Tiverton, clock and chimes 1668; also paid for 'wyer for the chymes' and for work about the tower and bells.

COMMINGS family: Thorverton. Surname originally COMMINGS (COMMINS), but was changed by deed poll in the mid-nineteenth century to CUMMINGS.

COMMINGS (COMMINS), JAMES: Thorverton. Clockmaker; son of William, blacksmith, and Mary, baptised 1791. Married Jane Lipscombe at Silverton in 1817 and died in 1853. Tomb at Thorverton: 'James Commings, 62, clock manufacturer'. 'Paid James Commins for cleaning the church clock £1 6s' (Brampford Speke churchwarden's accounts, 1844). Business listed under Mrs JANE COMMINGS in 1856 (D); she died in 1864, aged 73.

COMMINGS (CUMMINGS), JAMES LEVIL: Thorverton. Clockmaker and bellhanger; recorded in 1851 Census. Son of James and Jane and an older brother of William L. Spent most of his working life in London.

CUMMINGS, JAMES RICHARD: Thorverton. Son of William L. and Charlotte, baptised 1868; died 1955. Brass plate in Thorverton parish church states: 'Church clock improved. Outer dial and dial work were added by W. L. Cummings & Son, Clock Makers of this Parish 1903'. Recorded at 8 Ironbridge, Exeter, in 1906 (D); at Thorverton later.*

CUMMINGS, WILLIAM LIPSCOMBE: Thorverton. Clock and watchmaker and bellhanger; son of James, watchmaker, and Jane, baptised in 1833. Married Charlotte Risdon at Virginstowe in 1865, died in 1913. Continued the family's business at top of Bullen St. In 1870 was successful in application to become the new postmaster; the 1871 Census describes him as watchmaker finisher and his wife as postmistress. Some time afterwards they moved to the present Post Office at the bottom of the street. Square-dial longcase clock movement, 'W. Commings Thorverton'. Several family timepieces preserved by his descendants; also account books with references to visits to the houses of the aristocracy and gentry – Aclands and Bullers amongst them – on clock business.

COMPTON, JOHN: Honiton, High St, 1889–1902 and later (D). Alfred Clark was apprenticed to him in 1885.

CONIBEAR, GEORGE HENRY: Exeter. Served apprenticeship with John Parrish; listed at 14 New Bridge St, 1889; then at 13 New Bridge St. The business still at the same address in 1956 (D).

CONNEYBEAR, JAMES: Ashburton. Clock and watchmaker and plumber. Address given as Buckland when he married Betsey Hannaford at the parish church on 14 Dec 1819. 'Committed to the Devon County Gaol . . . Charles Newell, for stealing three watches, chains and seals from the shop of James Coneybear, watchmaker, of Ashburton' (EFP, 7 Aug 1823). Before Insolvent Debtors Court but discharged (EFP, 13 July 1826). Listed at North St, 1823–30; at West St, 1838–66 (D).

CONNEYBEAR, SAMUEL: Ashburton. Clock and watchmaker; surname variously spelt in different references, eg Connaybear, Connibear and Conibere. 'Paid Coneybeer for cleaning and repairing ye clock £6 6s' (Paignton churchwarden's accounts, 1781–2). A brass-dialled movement, 'Saml Connibear Ashburton', seen in Exeter shop in 1921 (Tapley-Soper). Listed in the *Universal British Directory*, 1793.

COOK, JOHN: Apprenticed *c*1736 to John Osmond of Pilton, Barnstaple, clockmaker, premium £4 4s.

COOKSON, WILLIAM: Tiverton. Goldsmith, watchmaker and optician, 13 Fore St, 1894 (D). Successor to James Grason.

COOMBE, EDMUND JOHN: Newton Abbot, Market St, 1883 (D).

COOMBE, EDMUND JOHN: Honiton, New St, 1889–97 (D).

COOMBE, WILLIAM: Exeter, St Mary Major parish; recorded as watchmaker, aged 38, son of William Coombe, tinplate worker, when admitted as Exeter freeman by succession, 2 Jan 1790.

COOMBES, GEORGE: Exmouth, Chapel St, 1883–1902 and later (D).

COOMBES, GEORGE: Budleigh Salterton, 56 High St, 1902 and later (D).

COPPLESTONE, WILLIAM: Plymouth. Clockmaker, married Susanna Hooper at Stoke Damerel 4 Dec 1782; on 15 July 1791 married Ann Penney, also at Stoke Damerel (register). Silvered-dial longcase clock reported; another, painted dial, 8-day, in oak and mahogany case.

CORNELIUS, WILLIAM: Dawlish. Watchmaker and jeweller, Strand, 1866; High St, 1878 (D).

CORNELIUS, WILLIAM JOHN jun: Plymouth, 21 Cecil St, 1893–7 (D).

CORNELIUS, W. J: Plymouth, 11 Mutley Plain, 1897 (D).

CORNISH, CHARLES HAROLD: Plymouth, 1 St Andrew's Place, 1893–1902 and later (D). Restored several Devon church clocks in a drastic manner; an example is North Tawton, to which he fitted a number of new parts in 1899.

CORNISH, WILLIAM HENRY: Okehampton, Fore St, 1893–1902 and later (D).

CORYNDON, WILLIAM: Watchmaker. Listed at St Keyne, Cornwall, in 1765, 'late of Plymouth' (Miles Brown).

CORYTON, GEORGE: Watchmaker. Married Mary Patrick, 'spinster and minor with consent of her mother', at Stoke Damerel, Plymouth, 16 Dec 1773 (register).

COTTEY, ABELL (ABEL): Crediton, Exeter, and Philadelphia, USA. Died 1711.*

COTTEY, JOHN: Exeter. Clockmaker; accused at Exeter Quarter Sessions, 3 Oct 1698, of stealing deeds and writings from the house of his father Abell Cottey.*

COTTIE, ——: Tiverton area. 'Itm payd for mendinge of the clocke 13s 4d. Itm payd to Cottie for mendinge of the pullie of the clocke 1d' (Halberton churchwarden's accounts, 1581).

COTTLE, JOSHUA: Bideford, Market Place, 1902 and later (D).

COTTRELL, Mrs MARY: Tiverton. Watch and clockmaker, Bampton St, 1890–3 (D).

COTTRELL, THOMAS: Tiverton. Watchmaker and jeweller, Bampton St, 1878–89 (D).

COUCH, CONROY: Torquay, 54 Lower Union St, 1883–93; 128 Union St, 1897–1902 and later; also at 1 Strand (D).

COUCH, RICHARD: Brixham. Watch and clockmaker and pawnbroker; The Quay, 1823; Market Place, 1830; Fore St, 1838–73 (D).

COULDRIDGE, WILLIAM HENRY: Totnes. Watch and clockmaker, New Walk, 1866 (D).

COULDRIDGE, WILLIAM HENRY: Exeter, 177 Cowick St, 1893 (D).

COURTI, PAUL jun: Exeter. Working watchmaker and jeweller; moved premises eight or more times but was recorded at 7 High St (1843 voters' list) and at 1 Blackboy Rd (1881). Paul Courti (Corti) sen, barometer maker and tobacconist, died 1876, aged about 75.

COVENTON, JOHN G: Lympstone, 1878 (D).

COX, WILLIAM C: Plymouth. 87 Fore St, Stoke, 1830 (D).

COXWORTHY, THOMAS: Plymouth Dock. Clock and watchmaker, listed in the *Universal British Directory*, 1798.

CREWS, CHARLES FREDERICK: Plymouth, 45 Treville St, 1838; 23 Bedford St, 1844–50 (D).

CROFT, JOHN: Plymouth Dock. Clock and watchmaker and silversmith, Catherine St, 1812–14 (D).

CROFT, ROBERT: Plymouth Dock, *c*1790 (Baillie).

CROFTS, JOHN: Torquay, 1 Union Road, 1838 (D).

CROOK, HENRY: Plymouth, 72 Regent St, 1897 (D).

CROSS, CHARLES: Exeter. Served his apprenticeship with Charles Price at Wiveliscombe, Somerset; then worked in Exeter, making clocks for Benjamin Bowring (pre-1816). Was similarly employed by James Bradford at Newton Abbot and, later, by Henry Ellis, of Exeter. 'Died – Nov. 25, in Paul St, in this city, Mr Charles Cross, watchmaker, in his 73rd year, much respected' (EFP, 5 Dec 1860).

CROSS, CHARLES (ii): Exeter. Recorded in 1861 Census in Paul St, aged 39, watchmaker, born Exeter; together with his son, William, 19, watchmaker. Listed there until 1867 (EPJ).

CROSS, CHARLES: Newton Abbot, 110 Queen St, 1873–89 (D).

CROSS, HENRY: Tiverton. 24 Bampton St, 1893–1902 and later (D). Took over the Pinkstone family business in which he served his apprenticeship. Tiverton Museum has ornate glass panel from the shop door, 'H. CROSS Jeweller', and billheads dated 1906 and 1913. Described himself as practical watch and clockmaker, jeweller, silversmith, optician, goldsmith and diamond merchant.

CROSS, HENRY JOHN: Exeter, 51 Longbrook St, 1889 (D).

CROSSE, EDMUND: Locksmith, of Thorverton. Admitted as an Exeter freeman by fine, 28 Aug 1654; repaired Exeter church clocks.

CROSSMAN, FREDERICK: Newton Abbot. Watch and clockmaker and engraver, East St, 1844 (D).

CROSSMAN, FREDERICK: Newton Abbot, 39 Queen St, 1866–78 (D).

CROSSMAN, FREDERICK HUGH: Chudleigh, 1850–6 (D).

CROW, T. LEONARD: Tiverton. 'From Mappin & Webb's'; he took over Munford & Son's business at 21 Fore St in the early 1900s; also at Crediton. An advertisement has two black crows and states: 'Come Round Our Way for Clocks, Rings, Optics, Watches'.

CROWDER, JOHN: East Budleigh. Repaired church clocks over a wide area of the county in reign of Elizabeth I; 'pd John Crowder of Budligh for keping of the clocke' (South Tawton churchwarden's accounts 1585–6). At Woodbury in 1561–2 entered into agreement with the parish to keep the church clock in repair 'for the yerelie fee of ij s w[ith]out meat or drinke as often as nede requyrithe' (Fig 10). Looked after church clock at Chudleigh from about 1582 to 1593; and at Wolborough, Newton Abbot, in the year of the Armada, 1588, was paid 5s 'for mendinge of the Cloke & the Orgones'.

CROWTHER, HENRY WILLIAM: Plymouth, 6 Mutley Plain, Tavistock Rd, 1890–3 (D).

CROYDON, CHARLES: Devonport. Watch and clockmaker and jeweller, Fore St, 1844–78; & Son, 20 Fore St, 1890 (D).

CRUNKHORN, WILLIAM JAMES. Devonport, 58 Pembroke St, 1883 (D).

CRUWYS, ——: South Molton. Clock reported (Tapley-Soper).

CUMMING, JOHN: Torquay. Plainmoor, St Marychurch, 1878 (D).

CUMMINGS, Thorverton. These entries are listed after COMMINGS.

Charge done upon the bell — Item paid to John Crowder for the amendement of the clock [illegible]
Item paid to Robert Smyth for two ropes for the clock [illegible]
Item paid for two bell ropes [illegible]
Item paid to Richard Yate for two new gogyns and mending of the bell whele [illegible]
Item paid to Richard Oke for keping of the clock this yere [illegible]

Md that hit is covenantid concludid and fully agreed betwene the parishe and the aforesaid John Crowder this yere that the same John shall and will yerelie during the lyff of the said John att his only coste and charges repare and amend the same clocke for the yerelie fee of ij s. wont mete or drinke as often as nede requyrithe and the yere began att Ester last past

The some of all paymentes and allowannces [illegible]

So that the said accomptantes doo owe unto the church the some of [illegible]

Fig 10 Extract from the Woodbury churchwarden's accounts for 1561–2 giving details of John Crowder's agreement with the parish to keep the clock in repair for two shillings a year 'as often as nede requyrithe'

CURTIS, WILLIAM: Exeter. Clock and watchmaker and silversmith; in business at New Bridge St from about 1790. Was paid £2 2s in 1793 for mending the St Edmund's parish chimes (churchwarden's accounts) and was a parish constable in 1803 (Militia List). Died, aged 88, on 21 Mar 1847 and was described in an obituary as 'the oldest tradesman in Exeter'. Will in DRO: mentions daughters Ann and Mary; estate sworn 'at under £3,000'.

His wife died in 1823. At one time an academy for drawing was conducted at his house by a Mr Ellis who 'instructed ladies in the beautiful art of painting on velvet' (EFP, 7 May 1807). Longcase clocks with well-engraved silvered brass dials in rich mahogany cases.

CURWOOD, Miss RHODA JEMIMA: Tiverton, Castle St, 1890 (D).

DANIEL, ABRAHAM: Plymouth. Miniature painter, watchmaker etc. Late eighteenth-century gilt pair-case stop seconds watch with subsidiary hour and minute dial, signed on dial 'Abrm Daniel Plymouth', movement signed also, plus number 9570 (illustrated DNQ, Spring 1976). Son of Nochaniah Daniel, of Bridgwater, he worked at Bath from time to time as a miniature painter: 'Died, at Plymouth, Mr Daniel, miniature painter, after a lingering illness, aged 43' (EFP, 20 Mar 1806). Samuel Hart was apprenticed to him in 1779. A brother, Joseph Daniel, was a miniature painter at Bath and a second brother, Phineas, was a watchmaker, silversmith and engraver in Bristol. For details of the family see Daphne Foskett, *British Portrait Miniatures*, pp 141–3.

DANIEL (DANIELS), SAMUEL: Colyton. Watch and clockmaker, listed in directories 1830–56. Received various payments, mostly of £4 per annum, for keeping Colyton Church clock in repair between 1842 and Lady Day 1868 (churchwardens' accounts). Several white-dial longcase clocks noted, including one in plain oak case with name 'Wright' stamped on back of dial.

DANIELL, ——: Plymouth. Clock and watchmaker, Drake St, 1814 (D).

DANIELL, ROBERT: Plymouth Dock. Watchmaker; married Jane Delacombe, a widow, at Stoke Damerel parish church, 4 Apr 1777 (register). Listed in the *Universal British Directory*, 1798.

DARKE, JOHN (*c*1743–1813): After working in London and then at Holsworthy (see Chapter 5) he can be traced in Barnstaple where in 1769 he entered into a 7-year agreement to keep the parish church clock, quarter-clock and chimes in repair. 'Wanted immediately by John Darke, Clock and Watch-Maker, in Barnstaple, and constantly Employed if approved of, a journeyman clock-maker. He must be a man of good character, and capable of clockmaking in all its branches' (EFP, 13 Mar 1772). Was a widower when he married Mary Harding at the parish church 14 Aug 1786; died, aged about 70, in August 1813 and was described in an obituary as 'a man who has passed through life with the fairest character' (EFP). Longcase and bracket clocks.

DARKE, JOHN (ii): Barnstaple. Son of the above, he continued the family clock and watchmaking and silversmith's business in the High St until his death in May 1825. Longcase clocks.

DASHPER, ALFRED: Torquay, South St, 1873–8 (D).

DASHWOOD, JOHN: Apprenticed *c*1762 to George Flashman, Exeter, watchmaker, premium £31 10s. Thirty-hour longcase clock, brass dial, by J. Dashwood, Bridport, noted by Tribe & Whatmoor (*Dorset Clocks and Clockmakers*, p82).

DAVEY, ——: Plymouth. Clock reported (Tapley-Soper).

DAVEY, JOHN: Torquay, 2 Somerton Terrace, St Elfrides Rd, 1893; 11 East St, 1897 (D).

DAVIES, FRANCIS: Plymouth. Watchmaker, of Stonehouse; took out licence on 21 Apr 1758 to marry Elizabeth Honey of Stonehouse, widow.

DAVIS, ADOLPHUS E: Axminster, West St, 1883 (D).

DAVIS, ARTHUR: Tiverton, Westleigh, Cullompton and Kentisbeare. A Quaker clock and watchmaker, recorded *c*1700–23.*

DAVIS, HENRY CHARLES: Newton Abbot and Moretonhampstead. 137 Queen St, Newton Abbot, 1893; 19 Wolborough St, Newton Abbot, 1897; Forder St, Moretonhampstead, 1902 and later (D).

DAVIS, J: Watchmaker, from Barnstaple. Married Elizabeth Brown at Great Torrington on 15 Apr 1854 (EFP).

DAVIS, RICHARD FREDERICK: Plymouth, 29 Whimple St, 1889; 96 Treville St, 1897–1902 and later (D).

DAVY, ROBERT: Exeter. Clockmaker, St Edmund's parish; provided a surety of £10 at the city Quarter Sessions, 5 Feb 1738.

DAVY, SIMON: Apprenticed *c*1764 to Jas. Bucknell, clockmaker, Crediton, premium £2.

DAVY, WILLIAM: Tavistock, West St, 1856 (D). Eight-day longcase clock, white dial, with painting of castle in arch, 'Wm Davy Tavistock', plain deal case (Tapley-Soper).

DAVYES, JOHN: Exeter. Clockmaker; documentary mention, Dec 1683 (Exeter Quarter Sessions).

DAWS, RICHARD: Devonport, 88 Fore St, 1844 (D).

DAWSON, MATTHEW: Plymouth Dock. A witness to the marriage of Thomas Beer, goldsmith, at Stock Damerel, 4 Dec 1780 (register). Made a longcase movement that now forms part of a massive organ clock with skeletons at the Red Lion Hotel, Salisbury.

DAY, CHRISTOPHER: South Molton. 'Paid Christopher Day his bill £3 9s 6½d' (South Molton churchwarden's accounts, 1733). Geo. Gold (or Gould) was apprenticed to him *c*1755, premium £10 10s.

DAY, HERBERT: North Tawton, 1873 (D).

DAY, THOMAS: Apprenticed *c*1762 to John Upjohn, Exeter, watchmaker, premium £26 5s.

DEBNAM, JOHN: From Tiverton. Ran away from Francis Redstall, clockmaker, of Overton (*Daily Post* 17 Jan 1726 – Buckley).

DEEME, HENRY: Luppitt. Thirty-hour square brass-dial longcase clock.

DEEME, HENRY: Honiton. Married Sarah Blew (or Blue) at Honiton 13 Sept 1756; died *c*1786, in which year letters of administration were obtained for his estate. Kept Kentisbeare Church clock in repair 1762–80 according to churchwarden's accounts. A sample entry reads: 1774 – 'pd Mr Henry Deeme for repairing the church clock £3 3s . . . a man & horse two days for carrying the clock to Honiton & bringing home again 4s'. Numerous longcase clocks reported. Exeter Museum has oak longcase, 30-hour with birds engraved on dial, on loan from the Liverpool Museum; a brass dial is at Totnes Museum. Bracket clock also reported (Bellchambers).

DELVE, JOHN: Barnstaple. Listed at High St in 1838 and at No 104 in 1844; at Boutport St, 1850 (D).

DELVE, JOHN: South Molton. Clockmaker; entered into agreement about repairs with North Molton churchwardens, 1852 (DRO). Listed at East St, 1856–73 (D).

DENBY, WILLIAM: Sidmouth. Watch and clockmaker, 1830 (D).

DENHAM, JOHN: Luton, Chudleigh, 1897–1902 (D).

DENNIS, ——: Torrington. Thirty-hour brass-dial longcase clock *c*1730.

DENNIS, JOHN: Torrington. Watch and clockmaker, silversmith, and ironmonger, High St, 1830–56 (D). Died 13 July 1856; will in DRO; estate under £200; widow Fanny Lewis Dennis appointed sole executrix. White-dial longcase clock at a north Devon hotel (Tapley-Soper).

DENT, GEORGE: Honiton, New St, 1873; High St, 1878–83 (D).

DEPREE, FREDERICK TEMPLER (1849–1926): Exeter. Son of a London solicitor, he became a partner *c*1878 in firm of Ellis & Tucker, watchmakers and jewellers, 200 High St. Business moved in early 1880s to new-built premises at 17 & 18 High St, on eastern corner of Bedford St, and traded as F. Templer Depree (late Messrs Ellis, Depree & Tucker). An advertisement states: 'Skilled workmen for the repair of watches, clocks & jewellery on the premises'. Depree became chairman in 1904 of gas engineers Willey & Co and took a prominent part in the commercial and social life of the city. The jewellers' business listed as Depree, Raeburn & Young in 1902, as Depree & Young in 1909, and as John Young & Co in 1920, being continued to *c*1930. Premises later became Barclays Bank, but were destroyed in the blitz in 1942. The firm supplied several church and turret clocks in the area. Regulator inscribed, 'F. T. Depree Exeter'.

DIETRICH, J. G: Exeter. *Exeter Gazette*, 5 Sept 1811, has this notice: 'Whereas, Joseph Kennett, late a private in the Marines, was intrusted by me, on Wednesday, the 21st August last, to sell goods on my account, at Crediton Fair, and had of me Clocks to the amount of several pounds, but has not since returned – this is to give notice, if the said Joseph Kennett does not return the said Clocks, or the amount of the same within one week from the date hereof, he will be prosecuted to the utmost rigour of the law; any person likewise that will give information where the said Joseph Kennett may be discovered, will greatly oblige. J. G. DIETRICH, Clock-maker, Mason's Arms, Back-lane, Exeter. N.B. J. Kennett is a Swiss; speaks bad English; had on a blue coat and grey pantaloons; 5ft 2ins high; 23 years old. Dated 3rd September 1811'.

DIETSCHEY, CHARLES: Devonport. 3 William St, Morice Town, 1897–1902 and later (D).

DILKE, EDWARD: Plymouth. Apprenticed *c*1751 to Walter Elliott of Plymouth, watchmaker, £30 premium. Described as watchmaker, of St Andrew's parish, when he took out licence 9 Dec 1760 to marry Mary Lake. Longcase clock in lacquer case with silvered-brass dial fetched £26 in Brooking Rowe sale at Plympton in 1917 (Tapley-Soper).

DILLIN, ROBERT: Plymouth Dock. Watchmaker, jeweller, goldsmith and engraver; listed in the *Universal British Directory*, 1798. Married Sarah Grills at Stoke Damerel parish church 31 Jan 1776 (register).

DINGLE, WILLIAM: Tavistock. Watchmaker; married Mary Ann Hawkins at Tavistock parish church 31 Aug 1822 (register).

DITCHET, SAMUEL: Hatherleigh. Watchmaker; married Joan Essery at Torrington parish church 30 Oct 1763 (register).

Plate 27 The hood of the 3.9m (13ft) Stumbels equation clock at Powderham Castle. The casework suggests the Channon family of cabinetmakers who were employed extensively at the castle

Plate 28 An S-shaped winding key from a Stumbels longcase clock

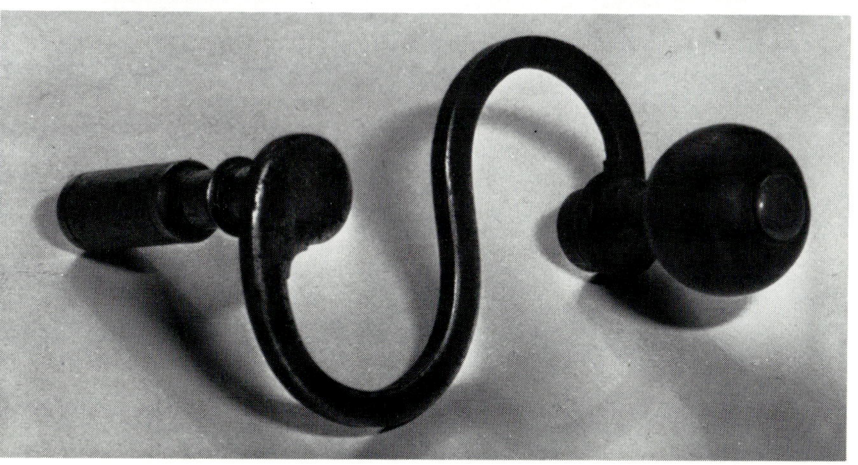

DOBBS, ALFRED REGINALD: Instow, The Quay, 1902 and later (D).

DODD, EDWARD: Devonport, 33 Marlborough St, 1850; 24 Canterbury St, 1873–83 (D).

DOMINY, R: Devonport, 37 Ker St, 1856 (D). G. H. Dominy listed as working jeweller, Mount St, in 1822 (D).

DONOVAN, JOHN: Exeter, Guinea St, 1865–6 (EPJ).

DORNING, ROBERT: Devonport, 37 Ker St, 1844–50 (D).

DOWN, ALBERT: Plymouth. 17 Manor St, East Stonehouse, 1883 (D).

DOWN, JAMES: Plymouth, Flora St, 1866; 22 King St, 1873–8; 22 Tavistock Rd, 1883–93 (D).

DOWN, ROBERT: Bideford. Watch and clockmaker and gunsmith, High St, 1830–56 (D). Died 22 June 1856, leaving son, Henry, and daughters, Ella and Ann Elizabeth. Will in DRO; estate valued at under £100.

DOWNING, JOHN: Hatherleigh, Bridge St, 1878 (D).

DOWNING, R. E: Tiverton. Watch and clockmaker, Newport St, 1838 (D). See next entry.

DOWNING, ROBERT EDWARD: Witheridge. Clock and watchmaker and brazier; listed 1856–78 (D). See next entry.

DOWNING, WILLIAM: Witheridge. Watch and clockmaker; listed with Robert Edward Downing in 1857 (D). A correspondent in 1978 states: 'The site of their house by Drayford Bridge is still recalled, and the remains of their sign'.

DRAYTON, JAMES: Possibly the son of Thomas Drayton of Chardstock. Apprenticed to Joel Spiller of Wellington, Somerset, clockmaker, for a premium of £16 paid in 1760.

DRAYTON, THOMAS: Chardstock. Clockmaker; son of Reverend John Drayton, vicar of Chardstock. He married Ann, daughter of clockmaker John Michell at Dorchester 7 Sept 1727, buried at Chardstock in 1788.

DREW, S. R: Okehampton, Fore St, 1873 (D).

DUFNER, LEOPOLD: Exeter, Friars Hill, 1878 (D).

Plates 29 & 30 Two superb tidal dial longcase clocks by William Stumbels, of Totnes. The mahogany one (left) has a regulator-type movement and is 2.3m (7ft 10in) tall; the other has a walnut case with a five-pillar movement and is 2.5m (8ft 5in) tall

DUGDALE, JAMES R: Devonport, 2 St Aubyn St, 1889–93 (D).

DUGDALE, L. H: Devonport, 2 St Aubyn St, 1897–1902 and later (D).

DUNDASS, ROBERT: Dartmouth. Maker of lost watch (*Public Advertiser*, 2 Jan 1788 – Buckley).

DUNN, RICHARD: Bideford, Market Place, 1878 (D).

DUNSFORD, MARTIN: Ashburton. Maker of longcase clocks and also of a clock with a very large dial; was living in 1787 (DNQ, vol 9).

DUNSFORD, WILLIAM: Plymouth. Silversmith and watchmaker; a signatory to registration of Baptist meeting house at Catdown, 3 Jan 1803. Eight-day silvered-dial mahogany longcase clock. An 1822 directory lists: 'Dunsford, Jas. N., jeweller and goldsmith, 43 Fore St, Plymouth Dock; and Dunsford, W. N., goldsmith, &c, agent for Carrol and Co. Lottery Office, Market St, Plymouth'.

DUPUY, ADRIAN: Plymouth. Watchmaker; took out licence 10 July 1734 to marry Florentia Footus of Charles, Plymouth, widow. An Audran Dupuy was apprenticed in London 4 May 1713 to Richard Glover (Clockmakers' Company register of apprentices); Odran Dupuy, Philadelphia, 1735 (Britten).

DURANT, WALTER: Plymouth. 120 Union St, East Stonehouse, 1897 (D).

DURCKHEIM, ISAAC MOSES: Newton Abbot. Watchmaker, East St, 1823 (D).

DYER, EPHRAIM: Bideford. Kept several turret clocks in repair, including those of Braunton (from 1689), Northam, and Hartland (1716 and 1723); and was a maker of 8-day and 30-hour longcase clocks. The late J. K. Bellchambers had 'a very pleasant 12in square longcase by him *c*1715. The case is made of pine, which must once have had a walnut veneer, and the hands are very fine. The clock has brass-covered weights and a bull's eye in the case'.

DYER, JOHN: Plymouth. Watch and clockmaker. 25 Edgcumbe St, Stonehouse, 1830 (D).

DYER, WILLIAM (*c*1746–1831): Barnstaple. Clockmaker, gunsmith and ironmonger; married Mary Heanes at Torrington parish church 28 Mar 1774 (register); listed in Bailey's *Western Directory*, 1783, and in *Universal British Directory*, 1793. 'Died – On the 11th instant Mr Wm Dyer, many years respectable ironmonger of Barnstaple, aged 85' (EFP, 19 May 1831). Longcase clocks include 30-hour example with posted frame movement, the dial engraved with maritime scene and flowers, and with rococo spandrels.

DYMOND, CHARLES: Devonport, St Aubyn St, 1850–83 (D).

EAMES, GEORGE: Tiverton. Watch and clockmaker and jeweller, Angel Hill, 1866–73 (D). Recorded in 1861 Census, aged 27, watchmaker, born Tiverton; together with wife Elizabeth, 30, and daughter.

EAMES, JAMES: Bradninch. Watch and clockmaker, 1856–66 (D).

EAMES, WILLIAM: Tiverton. Watch and clockmaker, West Exe, 1838–66 (D). Recorded in 1861 Census, aged 61, born Bradninch. Thirty-hour painted arch dial longcase clock with false winding squares, at Tiverton Museum, illustrated in C. N. Ponsford *et al*, *Clocks and Clockmakers of Tiverton.* Another reported from Nottingham, destroyed case of which was originally surmounted by unusually large and heavy ornaments – two large brass acorns and a brass ball topped by an engraved brass swan.

EARLY, W: Exeter. Watchmaker, 49 North St, 1876; 30 North St, 1877 (EPJ).

EARLY, WILLIAM HENRY: Tiverton. Clock and watchmaker, Church St, 1883; Bridge St, 1889; Angel Hill, 1893–7 (D). Henry Early listed at 12 Angel Hill in 1902.

EASTCOTT, RICHARD: Exeter. Clock and watchmaker; apprenticed to Jacob Lovelace in 1736 while a chorister at Exeter Cathedral; died 1795.*

EASTERBROOK, FRANK: Torquay. Cockington, 1893; Sherwell Rd, Chelston, 1902 and later (D).

EASTERBROOK, GEORGE PEPPERELL: Torquay, 9 Fleet St, 1889–1902 and later (D).

EASTERBROOK, HENRY: Dawlish. Watchmaker, Albert Place, 1866–73; Albert St, 1878–89 (D).

EASTERBROOK, HENRY: Chagford, 1883 (D).

EASTERBROOK, WILLIAM: Kingsbridge, 1850 (D).

EASTERBROOK, WILLIAM: Torquay, Park St, 1873–8; 21 Torwood St and 19 Wellswood Place, Ilsham, 1883; 7 Parkfield Rd, 1893–7 (D).

EASTMAN, HENRY: Exeter. Clock and watchmaker; listed at St Thomas, 1789; New Bridge, 1791 (EPJ); and at Catherine St a year or so later. Announced in Jan 1790 that he had moved 'from his late residence near Mr Ford's, to a house opposite the Island Steps'. He further stated that he worked 'three years and a half under direction of the nephew of the celebrated Mr Graham, watchmaker, London' (EFP).

EASTMOND, JOHN: Starcross, 1883 (D).

EASTON, JAMES & CO: Barnstaple, 34 High St, 1889–1902 and later (D).

EASTON, WILLIAM: Mortehoe. Repairer, 1902 (D). The same name at Barnstaple later.

ECKHART, ANDREW: Buckfastleigh, Chapel St, 1878 (D).

EDGCOMBE (EDGCUMBE), JAMES: Plymouth, George St, 1883–9 (D).

EDGCUMBE, SAMUEL: Plymouth. Watch manufacturer, 11 Cornwall St, 1883–1902 and later (D).

EDGCUMBE, STEPHEN LUKE: Plymouth. Listed at Russell St, No 23, 1883–9; No 34, 1893–1902 (D).

EDWARDS, JOSEPH: Plymouth, Buckwell St, 1873–97 (D).

EDWARDS, THOMAS: Exeter, 5 High St, 1897 (D).

EDWARDS, WILLIAM: Plymouth Dock (Devonport). Clock and watchmaker and gunsmith, 44 Market St, 1822; 64 St Aubyn St, 1838; & Son, 63 St Aubyn St, 1844–50 (D).

EFFORD, CHARLES: Plymouth. Union St, Stonehouse, 1893–1902 and later (D).

EGBERT, WILLIAM H: Devonport, Tavistock St, 1873–89 (D).

ELFORD, JOSEPH: Axminster. Took out licence 18 May 1725 to marry Dorothy Lowring, a widow. In November 1727 agreed to erect a new clock and chimes at the parish church: 'The said Joseph Elford doth promise & agree to make or cause to be made a new clock, quarter clock & chimes to be by him plac'd in the towr of the said parish church of Axminster on or before our Lady Day next, and also to make a new hand to ye dyal – and to make ye chimes to goe to the tune of ye One Hundred & Fiftyeth Psalm & Britons Strike Home six times round every time it goes which chimes are to play every four houres'. He was to be paid £22 and further agreed to keep the clock and chimes in good order for ten years for 5s per annum (DRO). A clockmaker of same name can be traced later at Lymington, Hants, taking apprentices in 1738 and 1740.

ELLETT, CHARLES: Plymouth, 15 George St, 1856 (D).

ELLIOTT, JOHN: Plymouth. Watch and clockmaker; provided bond for marriage licence, 6 Nov 1732. Eight-day longcase clock, 'Jno. Elliott Plymouth', sold in May 1919 at Leigh House, Winsham, Somerset, auction

(Tapley-Soper). Another with brass-silvered dial with rocking figure of Father Time in the arch, in lacquered and stained case with chinoiserie decoration, was sold in 1978.

ELLIOTT, JOSEPH: Merton, 1883 (D).

ELLIOTT, MARY: Plymouth. Watchmaker; Jas. Pearce was apprenticed to her *c*1762, premium £20. Lacquered longcase clock, 'M. Elliott Plymouth', with brass and silvered dial sold at Devon auction in 1921 (Tapley-Soper).

ELLIOTT (ELLIOT), WALTER: Plymouth. Clock and watchmaker, St Andrew's parish. Took out licence 15 Feb 1747/8 to marry Mary Bidlake of Totnes (see Elliott, Mary). Apprentices included Edward Dilke (*c*1751) and Richard Heles (*c*1755), each with a premium of £30. Carried out repairs to the old clock and chimes at Ashburton *c*1754 and agreed 'to give security to maintain the same and keep it in good repair for the space of seven years barring accidents' (churchwarden's accounts). Brass arch-dial longcase clock with 5-pillar 8-day movement.

ELLIS, HENRY (1790–1871): Exeter. Clock and watchmaker, jeweller and silversmith; his Memoirs are featured in Chapter 5.

ELLIS & CO: Exeter. Mahogany 8-day longcase clock with white dial signed 'Ellis & Co Exeter'. Henry Ellis traded as H. Ellis & Co *c*1830 and while briefly in partnership with Robert Drake Gant; dissolved by mutual consent 11 June 1834, 'late firm of Ellis & Co . . . to be carried on as H. Ellis' (EFP).

ELLIS & SON: Exeter. Partnership between Henry Ellis and his son Henry Samuel Ellis, *c*1850. Dissolved by mutual consent 25 June 1854; 'the business to be carried on by H. S. Ellis and William Horton Ellis as Ellis Brothers' (EFP). 'Pendules de Voyage. Messrs. Ellis and Son are importing a further selection of carriage and other clocks' (*Western Times*, 10 Dec 1853).

ELLIS, HENRY SAMUEL (1825–78): Exeter. Partner in Ellis Brothers. Played active part in campaign to have Greenwich Mean Time adopted in Exeter and in 1854 wrote a paper, describing an experiment to link the striking of the Exeter Cathedral clock with an electric regulator at Exeter Guildhall. Was the city's mayor in 1869, and at the time of his death was chairman of the Culm Valley Railway and also of the Brixham line.*

ELLIS, WILLIAM HORTON (1832–1905): Exeter. Partner in Ellis Brothers; apprenticed 1848 to John Mallett jun, Barnstaple, watch manufacturer. Completed a carriage clock with duplex escapement that was displayed at Ellis & Co's stand at the Great Exhibition in 1851; was twice mayor of Exeter; and in 1891 with his wife, Phoebe Elizabeth, moved to Budleigh Salterton, where he died.*

ELLIS BROTHERS. Exeter. 200 High St; partnership between H. S. Ellis and W. H. Ellis, 1854–78. 'Messrs Ellis (Brothers), sole agents in Exeter for Brocot's clocks, have imported a large selection of new clocks by Brocot and other eminent makers. Also a selection of Bardou's field and opera glasses' (*Western Times*, 17 Dec 1859).

ELLIS, DEPREE & TUCKER. Exeter. Jewellers, watchmakers etc at 200 High St, from 1878; at 17 & 18 High St, 1885 (EFP); then continued as F. Templer Depree. Imposing bracket clock with eight-bell chime, inscribed 'Ellis, Depree & Tucker, Exeter', sold at auction in 1980 for £880.

ELLIS, THOMAS: Plymouth, Old Town St, 1866–83 (D).

ELLIS, WILLIAM JAMES: Plymouth, 80 Treville St, 1873 (D).

ELLORY, THOMAS: Modbury, 2 Galpin St, 1878–83 (D).

ELMS, JOSEPH: Plymouth Dock (Devonport). Clock, watch and chronometer maker, 55 Pembroke St, 1822; 11 Duke St, 1830; 20 Fore St, 1838 (D). 'On Saturday last as Mr Elms, silversmith, Duke St, Devonport, and Mr Blackford, of Devonport, were out on an excursion in the parish of St Budeaux, whilst his gun was loaded, a small branch, it is supposed, caught the trigger, and the gun unfortunately went off, lodging the whole of its contents in his right arm. Mr Elms was taken to Weston Mills, and medical aid procured, when it was found necessary to amputate the arm' (EFP, 5 Feb 1829).

EMDEN, GOMPERT MICHAEL: Plymouth Dock. 'For sale at Plymouth Dock at the house Gompert Michael Emden, silversmith, a bankrupt. A large quantity of valuable plate, watches, jewels, &c' (*Sherborne Mercury*, 30 Nov 1767). Later recorded as Emdin, Michael Gomport, clock and watchmaker, formerly of London, late of Plymouth, insolvent (*London Gazette*, 15 Sept 1781 – Buckley).

EMDON, MARK: Devonport, 48 Fore St, 1878 (D).

EPLETT, JOSEPH: Torrington. Watchmaker, South St, 1838 (D).

EUSTACE, THOMAS: Exeter. Jeweller; admitted as Exeter freeman by succession, 25 Sept 1773, and as member of Exeter Goldsmiths' Company the following year; bankrupt in 1789. Gilt-metal pair-cased verge watch No 9619 sold at Sotheby's in Mar 1976.

EVANS, DAVID: Exeter. The Evans family were prominent throughout the eighteenth century in Exeter as blacksmiths, locksmiths, whitesmiths, and as repairers of church clocks. The elder David Evans served his apprenticeship with Robert Barter and was admitted as an Exeter freeman in 1708. He made church clocks for Broadhembury 1733–4 and for St Mary Arches, Exeter, 1749; and kept Dawlish church clock in repair for 5s a year from Apr 1720 (agreement in churchwardens' account books). Buried at Allhallows on the Walls, 7 May 1754.

EVANS, DAVID (ii): Exeter. Son of the above; baptised at Allhallows on the Walls Aug 1713; buried there 5 Mar 1780. Made a clock for Sowton Church in 1758, and among Exeter church clocks he kept in repair were those of St Mary Arches and St Olave's.

EVANS (EVENS), EVAN: Totnes. Watch and clockmaker and silversmith; listed in Fore St from 1838; address given as No 15 in 1878 (D). Wall clock, 'Evan Evans Totness', in rosewood case with mother-of-pearl inlay, sold at auction in 1979.

EVANS, JAMES: Bovey Tracey, East St, 1873–89 (D).

EVANS, P: Apprenticed *c*1756 to Abraham Thorne, Tiverton, clockmaker, premium £7.

EVANS, SAMUEL: Brixham, Beach, 1838 (D).

EVANS, WILLIAM: Devonport, 41 Queen St, 1889 (D).

EVELEIGH, JOHN PEARSE: Ridgway, Plympton. Clockmaker; stood surety for several publicans (DRO – Victuallers lists, Plympton hundred, 1808). Longcase clock, 'John Eveleigh Ridgway' (Tapley-Soper). Death notice: 'At Ridgway, Mrs Grace Eveleigh, widow of the late Mr John Eveleigh, of that place, clock and watchmaker' (*Exeter Gazette*, 9 Aug 1828).

EVENS, ——: Totnes. Supplied Broadhempston Church clock.

EVENS, NICHOLAS: Totnes. Watchmaker; listed in 1823; Upper Main St, 1830; Fore St, 1844–56 (D).

EZEKIEL, ABRAHAM (*c*1726–99): Exeter. Gilt-metal pair-cased verge watch, 'A. Ezekiel Exon 5562', in Exeter Museum collection. Described in lease dated 18 May 1757 as silversmith, aged 31, St Kerrian parish; later moved to St Petrock's parish and occupied a house fronting against Fore St (DRO). One of the founders of the Exeter Synagogue *c*1763. *Exeter Flying Post*, 28 Nov 1799, carries a brief obituary: 'On Wednesday last died, at Portsmouth, Mr Abraham Ezekiel, for 50 years and upwards a respectable tradesman of this city'.

EZEKIEL, EZEKIEL ABRAHAM (1757–1806): Exeter. Engraver, optician, goldsmith, and watchmaker. Son of Abraham Ezekiel, he was apprenticed *c*1772 to Alexander Jenkins, Exeter, goldsmith, premium £45, and later acquired considerable reputation as a miniature painter and engraver of portraits, tradesmen's cards and such things as the title page of Dunsford's *History of Tiverton* and the ornamental headline used for a while as the title of the *Exeter Flying Post* newspaper. In Mar 1795, announcing his removal to a house in Fore St, seven doors below North St, he stated that 'the business of clock and watchmaking will be carried on at the same shop by his brother Henry Ezekiel'. Died 13 Dec 1806, leaving estate valued at 'under £600', and appointed his brother Henry, then of the island of Guernsey, goldsmith, and his brother-in-law, Benjamin Jonas, of Plymouth Dock, watchmaker, as trustees of his will. His bequests included £8 for the purchase of a clock for the Exeter Synagogue.

EZEKIEL, C. & A: Exeter. Successors to E. A. Ezekiel, 1806. At 179 Fore St; listed until 1830 (EPJ).

EZEKIEL, ELEAZER: Appeared before Insolvent Debtors Court at Exeter, 7 Mar 1827. Described as formerly of Ashburton, jeweller, watchmaker and keeper of a public library, and late of Newton Abbot, watchmaker.

EZEKIEL, HENRY: Exeter. Goldsmith and watchmaker; brother of E. A. Ezekiel (see above). Worked in Guernsey for a time, but died in Exeter, aged 63, on 10 Nov 1835, 'after a painful and lingering illness, borne with that fortitude which evinced a true faith in the resurrection of the dead according to the Jewish creed, to which he was a strong adherent' (EFP).

EZEKIEL, PHILIP: Plymouth Dock (Devonport). Watch and clockmaker and silversmith, 6 Duke St, 1812–30 (D).

FARINTON, STEPHEN: Torquay, 18 Abbey Rd, 1883 (D). See also Stephens, Farinton.

FAWKES (FAWKS), CHARLES: Plymouth Dock. Victualler, clock and watchmaker, James St, 1812; Old Half Moon, Pembroke St, 1822; 19 George St, 1823–4 (D).

FEEVINGS, JOHN: Torrington. Early nineteenth-century watch (Baillie).

FEHRENBACH, J: Tavistock. White-dial longcase clock with angling scene in arch, the date dial and false plate both stamped 'Finnimore', *c*1840.

FEHRENBAK & KETTERER: Plymouth, 6 George St, 1850 (D).

FENN, J. S: Plymouth, 7 Regent St, 1873 (D).

FERENBACH (FEHRENBACH), JOSEPH & CO: Exeter. German clocks; Alphington St, 1848; Fore St Hill, 1849–51; Alphington St, 1852–7 (EFP).

FERENBACH, XAVIER. Plymouth. 80 Treville St, 1878–83 (D).

FERRIS, WILLIAM: Exmouth. Watch repairer and trunk maker, Margaret St, 1866 (D).

FESANT, JOHN: 'pd John Fesant for amending the clock £1' (Modbury churchwarden's accounts, 1700).

FEURIER, LEON: Barnstaple. 'Practical watch and clockmaker, jeweller, &c, from Paris, at W. Alexandre's, surgeon dentist, 15 Cross St', 1878 (D).

FINNIMORE, WILLIAM: Exeter. Watch and clockmaker; tenant of house on the New Bridge (EFP, 8 Nov 1810). His wife Mary having died, he took out licence 7 Mar 1817 to marry Miss Martha Radford of St John, Exeter. Alphington St, 1821; Exe Island, 1849–51 (EPJ).

FLASHMAN, GEORGE: Exeter. Clock and watchmaker; took out licence 9 Oct 1762 to marry Dorothy Cannington of St Martin's, Exeter. Rented George Sanderson's premises 'over the porch' of St Mary Major Church, but in 1765 removed 'to a convenient house opposite the Guildhall . . . where he continues to make and sell all sorts of clocks and watches' (*Exeter Mercury*). His wife ran school for teaching needlework. Mahogany longcase clock showing 'High Water at Topsham Bar' sold at auction in north Devon in 1918 (Tapley-Soper).

FLASHMAN, JOSEPH: Modbury. Watchmaker and jeweller, Broad St, 1850–89; Brownston St, 1893–7 (D).

FLOOD, ——: Exeter. Name seen on newly restored white-dial longcase clock. A Peter Flood worked as a silversmith *c*1794–1801 (*Catalogue of Exeter Silver* in Exeter Museum).

FOLLAND, WILLIAM: Exeter. Watchmaker, Fore St, from 1789 (EPJ). Kept St Olave's parish clock in repair for 5s a year from 1790 and was churchwarden in 1811. Died 11 Aug 1824, aged 53 (*Exeter Gazette*). Watch movement, 'Wm Folland Exeter', 1795, in Horstmann Collection (Baillie).

FOLLAND, WILLIAM JOHN: Exeter. Recorded in 1851 Census at 160 Fore St, watchmaker, aged 35, married, born Exeter. Churchwarden's accounts of St Olave's parish record payments to him for winding the church clock *c*1858.

FOLLETT, ——: 'Paid to Follett of Sidmouth for mendinge the clocke £1' (East Budleigh churchwarden's accounts, 1666).

FOLLETT, RICHARD: Dartmouth. Watchmaker; bound in £20 for good behaviour and appearance at Sessions, 5 Mar 1663/4 (DRO).

FONTANA, ——: Exeter. Early nineteenth-century watch (Baillie). The *Exeter Pocket Journal* for 1832 lists Grassi, Fontana & Bulla, barometer and thermometer makers, Batholomew St West.

FONTANA, BAPTISTA (BAPTISTE): Exeter. Jeweller, ironmonger,

hardware dealer, Beaufort Place, Cowick St, St Thomas. Declining business (EFP, 2 Mar 1864).

FOOT, RICHARD: South Brent; early nineteenth century (Bellchambers).

FORD, CHARLES: Barnstaple. Clock and watchmaker and jeweller, 89 High St, 1866–83 (D). He supplied the clock for the Albert Memorial Tower, Barnstaple (1862); the stables clock at Arlington Court (*c*1864); and Chittlehampton (1878) and Braunton (1882) church clocks. Two large slate dials, 'C. Ford Barnstaple', at Barnstaple Museum. The business was taken over *c*1887 by Sly & Co.

FORD, JOHN jun: Exeter. Clock and watchmaker, 1831 (EPJ).

FORD, JOHN: Totnes. Watch and clockmaker, Fore St, 1838 (D).

FORD, ROBERT: Tavistock. Watch and clockmaker, West St, 1830 (D).

FORD, THOMAS: Dartmouth, Lower St, 1873 (D).

FORD, W. C: Plymouth, 11 George St, 1873 (D).

FORD, WILLIAM L: Plymouth, 54 York St, 1878 (D).

FORREST, JONATHAN: Apprenticed *c*1744 to Walter Mitchell (Michel), Plymouth, watchmaker, premium £44. Described as watchmaker, of Launceston, Cornwall, when he took out a licence on 6 Apr 1757 to marry Jane Catchcombe of Lifton.

FOSTER, H: Tiverton. Watchmaker, Fore St, 1830. Consulted Wasbrough Hale & Co, Bristol, about turret clock for Tiverton pannier market (reply in Tiverton Museum collection).

FOSTER, HENRY: South Molton, South St, 1844; East St, 1850 (D).

FOSTER, HENRY: Tiverton. Watch and clockmaker, Gold St, 1852 (D).

FOSTER, JAMES: Ashburton. Longcase clock, 'Jas. Foster Ashburton', second half of eighteenth century (Tapley-Soper).

FOSTER, T: North Tawton, 1866 (D).

FOSTER, THOMAS: Tiverton. Watch and clockmaker and jeweller, Fore St, 1823–56; also listed as auctioneer, appraiser, dealer in game, and agent to Universal Life Assurance Company (D). House and shop to let (EFP, 7 Dec 1826). He represented Castle ward on the borough council. White-dial longcase clocks; Tiverton Museum has 8-day example, also a watch No 7103, the case hallmarked London 1826.

FOX, J. F: Barnstaple, 99 High St (John Gaydon's former shop), 1902 (D).

FRANCIS, EDWARD: Winkleigh, 1902 (D).

FRANKE (FRANKS), ALEXANDER: Tiverton area. He kept the church clock at Halberton in repair from about 1619 to 1636 for 2s a year (churchwarden's accounts); and also carried out work on the clock and bells at St Peter's, Tiverton.

FRAZER, ALEXANDER: Kingsbridge. Clockmaker; took out licence on 20 June 1740 to marry Hannah Phillips of Kingsbridge.

FREEMAN, RICHARD: Exeter. Goldsmith, jeweller and engraver. 'Has opened a shop opposite the Guildhall, Exon, where he has laid in an entire new assortment of chas'd, gadroon, and double-polished plate, with a great variety of diamonds and other jewellery goods, and a great choice of gold, silver, and Pinchbeck watches' (*Exeter Evening Post*, May 29 and June 5, 1767). Two children baptised at St Martin's Church.

FRIEND, JOHN WALTER: Totnes. Clock and watchmaker; married Eliza Stone, daughter of the proprietor of the York Hotel, Sidmouth, at Sidmouth in 1830.

FRIEND, ROBERT: Totnes. Watch and clockmaker; aged 21 when he took out licence on 20 Dec 1832 to marry Elizabeth Bunker, also 21, at Totnes parish church. Listed at Fore St in 1838 (D). Died 18 Mar 1840, leaving estate valued at under £300. A copy of his will is in the Westcountry Studies Library, Exeter (Moger).

FRIEND, WALTER: Holsworthy. Clock and watchmaker, 1823 (D).

FRIEND, WALTER: Newton Abbot. Watchmaker and jeweller, East St, 1830–8 (D). Wrote a letter, dated 2 Nov 1829 and published in the *Exeter Flying Post*, on printing in colours by the compound-plate process.

FROST, CHARLES: Plymouth, 7 Regent St, 1883; 31 York St, 1893–7 (D).

FROST, JONATHAN: Exeter. Clock and watchmaker, New Bridge, 1789 (EPJ). Four silver watches and large pair of silver buckles stolen in shop break-in (EFP, 25 Sept 1783). Exeter Quarter Sessions jury list dated 29 Dec 1788 includes 'Jonathan Frost of St Olave, watchmaker'. Longcase clocks including one with brass-silvered dial showing 'High Water at Topsham' (formerly at Bradfield House – Tapley-Soper).

FROST, WILLIAM: Exeter. Clock and watchmaker; son of William Frost, of Moretonhampstead; baptised at parish church there 28 Nov 1790 (register). The *Exeter Flying Post*, 6 Feb 1834, reported that Lovelace's famous Exeter Clock had been 'put into perfect repair by W. Frost, clock and watchmaker, Paris St, in this city, himself a self-taught artist'. After carrying out further work on this clock he was sent by its owner, Mr C. Brutton, to exhibit it at the Great Exhibition in May 1851, but after being at the exhibition six weeks he was taken ill and had to return to Exeter. Died on Sept 28 that year, aged 62. Will in DRO mentions wife and six children: 'To my son William – All my clocks, watches, tools, stock in trade, shop fixtures and materials of every description . . . To my son Robert – the largest of my turning lathes and all its contents . . . also a watch when of age of twenty one years of the value of six pounds sterling to my son David'. Estate valued at 'under £200'.*

FROST, WILLIAM (ii): Exeter. Paris St, son of the above; moved to 82 Sidwell St *c*1854. Recorded in 1861 Census, aged 49, master watchmaker, born Exeter. The business listed under his name until 1872, then as Frost & Johns until 1875 (EPJ).*

FROST & JOHNS: Exeter, 82 Sidwell St, 1873–5 (EPJ).

FULFORD, ROBERT: Plymouth Dock. Clockmaker; married Mary Clark at Stoke Damerel parish church, 21 Nov 1785 (register). Longcase clock *c*1800 (Bellchambers).

GALE, JOHN BALKWILL: Topsham, Fore St, 1897–1902 and later (D).

GALE, WILLIAM CHARLES: Budleigh Salterton, Fore St, 1897 (D).

GALLEY, ARTHUR RUSSELL: Torquay, 26 Higher Union St, 1883–9 (D).

GANT, ROBERT DRAKE: From Plymouth. Was briefly with firm of Ellis & Co, 200 High St, Exeter; partnership dissolved by mutual consent, 11 June 1834.

GARD, EDWARD: Plymouth. 8 Townsend Hill, Mannamead, 1889 (D).

GARD, HENRY: Exeter. Clock and watchmaker, Cathedral Yard; apprenticed *c*1753 to Edward Upjohn, premium £3. Married Susanna Jenkins at Holy Trinity Church, 31 Mar 1766, and their family included a son, William, and several daughters. Sat many times on juries at Exeter Quarter Sessions. In May 1789 advertised for an apprentice – 'an ingenious sober lad not exceeding 14 years of age will be taken on reasonable terms' – and for a journeyman watchmaker (EFP). 'Last week died suddenly, Mr Henry Gard, of this city, watchmaker, who for more than 30 years had been eminent in his profession . . . in private life he recommended himself so as to be respected, as now lamented, by all' (EFP, 11 May 1797). The funeral took place at Exeter Cathedral; his widow died in 1800. Several longcase clocks noted, one signed in a semicircle around the top of the dial arch in which is depicted a harbour scene with shipping; also a brass-dial, mahogany-cased 'Act of Parliament' clock; and a watch, No 209, hallmarked 1790 and bearing its owner's name 'Willm. Cross, Exeter'.

GARD, WILLIAM: Exeter. Clock and watchmaker, Cathedral Yard; son of Henry and Susanna, baptised Exeter Cathedral, 20 Sept 1767 (register). Worked 14 years with his father and, after latter's death in May 1797, continued the business 'in all its branches, in the same shop, for the joint benefit of the family' (EFP). 'Lost, at or near Totness, in the Race Week, a plain metal watch, cap'd and jewel'd, maker James Masters, with a steel chain and gold seal, with arms and crest. Whoever will bring it to Mr Gard, Church Yard, Exeter, shall receive one guinea reward' (EFP, 30 Aug 1798). 'Friday last died, after a long and severe illness, Mr William Gard, late a respectable watchmaker of this city' (EFP, 28 Nov 1805). He carried out several repairs to the Exeter Cathedral clock, receiving a payment of £2 2s in 1797 and £18 18s in 1801. Bracket and longcase clocks; also watches.

GARLAND, RICHARD: Plymouth. Watch and clockmaker, 21 Market St, 1822; 41 High St, 1830–50 (D). Report of his wife's death (EFP, 10 Apr 1823). White-dial longcase clock with maker's bill for £6 10s pasted inside it (Tapley-Soper). Another, 8-day, with silvered dial and mahogany case seen at a Dartmoor inn.

GAWMAN, EDWIN: Devonport, 6 Chapel St, 1897 (D).

GAYDON, FREDERICK: Torrington. Watchmaker and silversmith; married Mary Jane, youngest daughter of John Lake, of the Exeter Inn, Torrington, on 17 Dec 1850 (EFP). Tendered successfully to keep Torrington Church clock in repair (EFP, 27 Jan 1853). Listed at High St, 1850–6 (D).

GAYDON, FREDERICK: Barnstaple, Litchdon St, 1889–93 (D).

GAYDON, JOHN: Barnstaple. Clock and watchmaker; flourishing 1855–95 at 99 High St. Supplied a clock for Barnstaple police station in 1855, also a clock for the Globe Hotel with two dials about 610mm (2ft) apart, one showing in the hotel bar, the other in the adjoining smoke room. He or his successors installed church clocks at Atherington, Bratton Fleming, Landkey, Mortehoe and Parracombe. The church clock at Swimbridge has an inscription: 'This clock was presented & erected in commemoration of the restoration of this church, 1880, by the brothers John Gaydon, High Street, Barnstaple, Henry Gaydon, Richmond, Surrey, Edwin Gaydon, New Brentford, Middlesex, Francis Gaydon, Upper Norwood, Surrey, watch and clock manufacturers, natives of Swimbridge' (Scott). Business listed under the executors of John Gaydon in 1897 (D).

GENDLE, THOMAS: Plymouth. Watchmaker and silversmith; listed in the *Universal British Directory*, 1798, and in subsequent directories – at Lower Broad St, 1812–22; Bilbury St, 1830–44. Died 1845 (EFP); 8-day longcase clock reported.

GENT, JOHN ROBINS: Barnstaple, Newport Rd, 1883–93; Post Office, 5 Summerland St, 1897–1902 and later (D).

GEORGE, SILAS W: Buckfastleigh, Fore St, 1893–1902 and later (D).

GERRY, FREDERICK WILLIAM: Newton Abbot, 44 Courtenay St, 1883–1902 and later (D).

GERRY, NICHOLAS: Torquay, 3 Braddons Hill Rd West, 1878 (D).

GIBBS, GEORGE: Budleigh Salterton. Watchmaker, jeweller and lapidary, 17 High St, 1866–1902 (D).

GIBBS, JAMES: Sidmouth, High St, 1850; New St, 1873–8 (D).

GIBBS, WILLIAM: Plymouth. Watchmaker; took out licence on 4 Oct 1802 to marry Elizabeth Taylor of St Andrew's, Plymouth.

GIBBS, Mrs: Plymouth. Watchmaker and silversmith, Frankfort Place, 1814 (D).

GIFFORD (GYFFORD), JOHN: He carried out work on the Exeter Cathedral clock between 1376 and 1395 (fabric accounts).

GILES, GEORGE: Plymouth, Deptford Place, 1889; 8 Armada St, 1897 and later (D).

GILL, JOHN: North Tawton. Watch and clockmaker, 1850 (D).

GILL, THOMAS: Brixham, Beach, 1850; the Quay, 1856 (D).

GILLARD, SAMUEL: Barnstaple. Watch and clockmaker and jeweller; a witness to will dated 24 Mar 1827 of Edward Snell, clockmaker, Barnstaple. Listed at High St in 1830 (D). White-dial clock, 'S. Gillard Barnstaple'.

GILLHAM, WILLIAM: Ottery St Mary, Butts Lane, 1844; Silver St, 1850–83 (D).

GILMORE, GEORGE: Plymouth. Importer, 5 Market Buildings, East St, 1897 (D).

GILPIN, ROBERT JOHN: Exeter, 134 Fore St, 1883–1902 and later (D).

GILPIN, WILLIAM R: Teignmouth, 24 Regent St, 1878 (D).

GODFREY, ——: Plymouth. Clock and watchmaker; listed in the *Universal British Directory*, 1798; at Market St, 1814 (D).

GODFREY, HENRY: Ottery St Mary, Mill St, 1873–1902 and later (D).

GODFREY, THOMAS: Plymouth. Watchmaker and silversmith, Lower Broad St, 1814 (D).

GODFREY, THOMAS DOVE: Plymouth. Watch and clockmaker and silversmith, 10 Frankfort Place, 1822 (D). 'Devon Assizes: Robert Amphlett, 18 (a marine, and who evinced at the bar the most daring hardihood) was charged with having broken the shop of Thos. Dove Godfrey, watchmaker, Whimple St, Plymouth, on the evening of 11th of May last, and stolen 2 gold, and 2 silver watches, of the value of £16, his property. It appeared that on this evening, Mr Godfrey was preparing to close for the night, when a smash was made at his window, he instantly darted in the direction of the sound, and seized a hand and arm which had been thrust through, saying "you blackguard, what did you break my window for?" A voice, that of the prisoner, replied "I did it to steal the watches". Guilty. Death Recorded' (EFP, 13 Aug 1829).

GODFREY, WILLIAM: Plymouth. Watch and clockmaker; listed at Lower Broad St, 1812; Old Town, 1822; 59 Market St, 1823–4; and at 17 Catherine St, Stoke, 1830 (D). Death notice: 'At Devonport, Mr Godfrey, silversmith & jeweller' (EFP, 2 May 1833). At the Devon August Assizes in 1829 William Dunn, 20, was found guilty of breaking into William Godfrey's house and stealing five silver watches. He was sentenced to death.

GOLD *see* GOULD

GOOD, EDWARD DARE: Seaton. He supplied the movement of Seaton Clock Tower, *c*1887; listed 1889–1902 (D).

GOOD, JACOB: Apprenticed *c*1711 to William Clement, Totnes, clockmaker, premium £16 2s 6d.

GOOD, SAMUEL: Seaton, Fore St, 1856–83 (D). 'After he retired from business at Seaton he used to go to sales, buy up old lantern clocks and convert them into grandfathers for which he would get as much as £20' (Tapley-Soper).

GOODFELLOW, E. H: Torquay, 2 Lucius St, 1902 (D).

GORFIN, HENRY: Plymouth Dock (Devonport). Clock and watchmaker, King St, 1822; 2 Marlborough St, 1850 (D).

GORFIN, WILLIAM: He and Henry Gorfin were among six Plymouth/Devonport watchmakers who signed a letter dated 20 June 1833, expressing willingness to have their stocks of foreign watches stamped (Clockmakers' Company archives).

GORMULLY, PHILIP: Devonport, 10 Duke St, 1838; 28 St Aubyn St, 1844; Fore St, 1850–6 (D).

GOSSIER (GOSSIERE), DANIEL: Plymouth Dock. Described as clock-maker, 'Hamoze Dock', when Thomas Browne was apprenticed to him *c*1721, premium £20; but as goldsmith, Stoke Damerel, when he took out licence 16 Apr 1735 to marry Blanch Wood. Longcase clock signed 'Dan Gossier, London', with five-pillar movement (*Horological Journal*, Nov 1957).

GOULD, GEORGE: George Nympton. Name noted (DNQ, vol 9). 'Was a maker' (Tapley-Soper).

GOULD, GEORGE: South Molton. Clockmaker; Ilfracombe Museum has 30-hour longcase clock, the dial signed 'George Gould Sth Molton 1740'. The clock looks twenty-five years later in style, so the date 1740 may not refer to its construction.

GOULD, GEORGE: South Molton. Probably the Geo. Gold who was apprenticed *c*1755 to Christopher Day, South Molton, clockmaker, premium £10 15s. Maker of 2.4m (8ft) mahogany longcase clock with complex astronomical dial (illustrated *Horological Journal*, July 1979); other longcase clocks reported, one with ship and cottage engraved in dial centre (Tapley-Soper); and verge watch. Listed as freeman and watchmaker in the *Universal British Directory*, 1798. A George Gould buried at South Molton on 12 June 1809 (register).

GOULD, THOMAS: South Molton. Watch and clockmaker, Barnstaple St, 1866–73; East St, 1878 (D).

GOULD, THOMAS: North Tawton, 1883 (D).

GOULD, WILLIAM: Bishops Nympton, 1850–6 (D).

GOULD, WILLIAM: 'pd William Gould for mending ye clock £1 5s' (East Budleigh churchwarden's accounts, 1710). A 'Mr Gould' is mentioned in the Feniton churchwarden's accounts; there, in 1688, he organised the setting-up of a clock dial and the following year was paid 10s 'for mending the Clock'. A new clock at Woodbury was made in 1697 by a 'Mr Gold'. There were Goulds living at Feniton at that time and the family seem to have had connection with St Edmund's parish, Exeter.

Plate 31 William Stumbels, of Totnes. Small ebonised bracket clock with silver spandrel mounts, calendar and mock pendulum apertures, and a particularly elegant nameplate. The dial in the arch is for regulating the pendulum

Plate 32 (*above left*) Arch dial of mahogany-cased tidal clock by William Stumbels. The time of high water is indicated in the arch. The regulator-type movement has shaped plates, a deadbeat escapement, maintaining power, and a very large pendulum disc with a flat rod. Mid-eighteenth century

Plate 33 (*above right*) 'William Stumbels Totness'. Arch dial of green lacquer longcase clock shown in Plate 45. The small silvered plate marked S and N frames the lever for silencing the strike

Plate 34 (*left*) A comparable dial from a longcase clock by Stumbels's apprentice George Paddon, of Kingsbridge. The lettering on the chapter-ring is similar to that on the Stumbels clock in Plate 25

Plate 35 (*opposite*) Adrian's Clock. A clock and carillon made under the direction of Maine Swete, *c*1725/30, as a gift for his brother Adrian, of Modbury. The walnut-veneered cabinet is said to have been made in Plymouth. Height 2.9m (9ft 8in), width 1.2m (3ft 5½in), depth 72cm (2ft 2½in). Victoria and Albert Museum, London

Plate 36 'Samuel Northcote Plymouth'. Table clock of *c*1765/70 with moonwork, showing the time of high water and the state of the tide at any time of the day or night. It was designed by James Ferguson the astronomer, who visited Plymouth and stayed for several months as a guest of Northcote's friend, Dr John Mudge

Plate 37 Samuel Northcote (ii), clock and watchmaker, of Plymouth. Portrait by his brother, James Northcote, RA. Plymouth Museum and Art Gallery

GOULDING, FRANK H: Plymouth, Bedford St, 1850–6; 49 George St, 1878 (D).

GOULDSWORTHY, EDWARD: Exeter. Watch and clockmaker; was apprenticed firstly to George Flashman, *c*1765, premium £14 14s, and then, the following year, to William Hornsey, premium £17 5s. Listed at Bell Hill, South St, 1789 (EPJ). Mahogany-cased bracket clock, 'Edward Gouldsworthy Exon', with silvered dial surmounted by a picture of a wild boar and lion whose paw, tail, tongue and eyes move with the beat of the pendulum.

GRAF: Exeter. See Stoneman & Graf.

GRAF, F: Tavistock, Brook St, 1866 (D).

GRANT BROTHERS: The Civet Cat establishment, 228 High St, Exeter, jewellers and watchmakers, 1856–83 (D). 'English, French, Dutch, and American clocks' (1866 advertisement). Also at 6 Victoria Parade, Torquay; wholesale only, 7 Little Nightrider St, St Paul's, London.

GRANT, JOHN D: Ilfracombe, Promenade, 1878 (D).

GRANVILLE & LINDE: Devonport. Watchmakers and jewellers, 42 Fore St, 1866–73 (D).

GRANVILLE, SAMUEL: Devonport, 40 Fore St, 1838; 48 Fore St, 1844–56 (D). His marriage was reported in *Exeter Flying Post* of 7 Nov 1839: 'On the 20th ult at the Registrar's Office, Devonport, Elizabeth, fourth daughter of Mr Gilbert Hutchings, of Moretonhampstead, to Mr Samuel Granville, watchmaker and silversmith, Devonport'.

GRASON, JAMES: Tiverton, Fore St. Took over J. W. T. Tucker's business; listed in 1878 as jeweller and silversmith (D). Mayor of Tiverton 1888–9. He had the contract for ordering and fixing the 1883 clock for St Peter's Church, Tiverton, and 'Grason Tiverton' is inscribed on the setting dial; a watch similarly inscribed was sold at auction in 1977.

GREEN, JAMES: Exeter. Watchmaker; a son, Henry, baptised at St Petrock's, 3 Jan 1702/3 (register).

GREEN, JOHN: Plymouth. Clockmaker; married Ann Leach at Stoke Damerel parish church, Plymouth, on 10 June 1776 (register).

GREEN, JOHN: Barnstaple. Early nineteenth-century mahogany, white-dial longcase clock.

GREEN, THOMAS: Torrington. Watchmaker; married Elizabeth Halfyard Boone at the parish church, 23 July 1820 (register). Listed at Potacre St, 1823 (D).

GREENLEAF, WILLIAM: Plymouth. Watchmaker and working jeweller, 44 Frankfort St, 1866 (D).

GREENWAY, JAMES: Tiverton. Apprenticed *c*1717 to Simon Thorne jun, watchmaker, Tiverton, premium £7. Among the bundles of the Exeter diocesan marriage licences is a curious entry: '16 Apr. 1725. Let no marriage licence be granted to James Greenway, junior, of Tiverton, clockmaker, and Elizabeth Ralley, alias Rawles, of ye same, spinster, before John Greenway, ye father, be first called, or Mr P. Cooke, his proctor'.

GREGORY, WILLIAM ARTHUR: Exeter. Chronometer, watch and clockmaker, jeweller and silversmith; his signature, 'W. A. Gregory 1854', is in the Exeter Cathedral clockroom. Recorded in 1861 Census, aged 21. 'The Practical Watchmaker W. Gregory (Ten years with Mr Upjohn, High St) . . . 1 Bampfylde St (opposite Castle St), Exeter' (advertisement, *Devon Weekly Times*, 23 Oct 1863). At 3 High St, 1869; at 2 High St, 1876–1902 and later (D). Clock formerly in Broadhembury Church tower has inscription: 'Repaired by Wm. Gregory Exeter 1900'.

GREGORY, W. T: Exeter. Watchmaker, East Gate St. Died pre-1858 (newspaper references).

GRIFFIN, HENRY: Devonport. Watch and clockmaker, 44 King St, 1830 (D).

GRIFFIN, JOHN: Plymouth. 35 Union St, East Stonehouse, 1883–9 (D).

GRIFFITH, WILLIAM: Combeinteignhead, 1883–9 (D).

GRIGG, HUMPHREY: Plymouth. Watchmaker, Drake St, 1830–50; 17 Camden St, 1856; 45 Tavistock St, 1866 (D).

GRIGNION, DANIEL (1684–1763): Topsham. Formerly of London, 'finisher to the late Mr Daniel Quare'. The Topsham parish registers record his burial on 10 Apr 1763 and that of Eleanor Grignion, widow, on 25 Mar 1766. Topsham longcase clock reported.

GRIMES, EDWARD: Bideford, 81 High St, 1883–1902 and later (D).

GRINKING, ROBERT: Exeter. Clock and watchmaker; married 1632.*

GRINKING, THOMAS: Exeter. Was paid in 1690 for 'keeping ye church clock' (Holy Trinity churchwarden's accounts).

GROVES, J: Hatherleigh, Market Place, 1883 (D).

GROVES, J: Exeter, Bridge End, St Thomas, 1893–1902 (D).

GROVES & LEYMAN: Exeter, Bridge End, St Thomas, 1889 (D).

GRUTE, WILLIAM: Totnes, Queen St, 1844; Warland, 1878 (D).

GUBB, THEOPHILUS: Topsham, High St, 1873–89 (D).

GUBB, T. & H: Topsham, High St, 1893 (D).

GUDRIDGE, PERCY: Devonport, Marlborough St, 1897 and later (D).

GUILLAUME, BENJAMIN: Plymouth, 36 Tavistock Rd, 1878–83 (D).

GUILLAUME, GUILLAUME (WILLIAM): Exeter. 'Practical watchmaker, from Geneva (late from Ellis Brothers), 25 Cathedral Yard . . . Repeating and musical work repaired' (*Western Times*, 8 Dec 1860). Also clock and chronometer maker, 21 Queen St, 1864; 91 Queen St, 1878; & Son, 1889. Recorded in 1861 Census at St Leonard Terrace, married, aged 24, born Amsterdam.

GUILLAUME, WILLIAM: Exeter, 88 Queen St, 1893–7 (D).

HADDON, C: Buckfastleigh, Fore St, 1883 (D).

HADDON, CHARLES: Dartmouth, Foss St, 1902 (D).

HADDON, JOHN CHARLES: Kingsbridge, Fore St, 1889–93 (D).

HADDY, WILLIAM: Plymouth. Watchmaker, jeweller and engraver, Exeter St, 1836; 45 Old Town St, 1856; 58 Bedford St, 1866; Frankfort St, 1873–8 (D).

HALEY, SAMUEL B: Plymouth. Watch and clockmaker, Stonehouse; listed in 1830; at 44 Union St, 1856 (D).

HALL, EDWARD: Plymouth Dock. Watchmaker, etc, Fore St, 1814 (D).

HALL, JOHN: Exeter. Dissolving partnership with Uglow (EFP, 25 Sept 1861).

HALL, JOHN JAMES: Retired to Exeter after career with the London and South Western Railway and achieved national recognition for his restorations of the old clocks of Ottery St Mary and Exeter Cathedral. A prolific contributor to scientific journals and a skilled technical artist. The Devon Record Office possesses more than forty volumes, recording his work as a writer, illustrator and lay preacher. Died 5 Jan 1941, aged 95, and his ashes were interred in the Sylke Chantry of Exeter Cathedral, beside the ancient clock he had so carefully restored some thirty years earlier. He is described on the memorial tablet as astronomer and horologist.

HALL, THOMAS: Ilfracombe, 43 Fore St, 1889 (D).

HALLETT, JOHN W. Chagford, 1893–7 (D).

HALLETT, WILLIAM & CO: Chagford, 1902 and later (D).

HALSE, JOHN: Plymouth. Watch and clockmaker and silversmith, 55 Old Town St, 1822 (D).

HALSE, WILLIAM HENRY: Plymouth, 13 Flora Cottages, 1873–1902 (D).

HALSTAFFE (HALSTAFF), PETER: Exeter. Locksmith; apprentice of William Hoppin (qv); admitted as Exeter freeman, 18 Sept 1637. Was paid on several occasions for repairs to St Kerrian's, Exeter, parish clock between 1652 and 1663; and also mended the St Mary Steps clock in 1661–2 (churchwardens' accounts). Listed among the poor of St Paul's parish in 1671 Hearth Tax.

HALSTAFFE (HALSTAFF), PETER (ii): Exeter. Gunsmith; son of Peter Halstaffe, locksmith; admitted as Exeter freeman by succession, 16 Dec 1695. High spirits landed him in the dock at Exeter Quarter Sessions in 1657 when he was one of seven fined 20s and ordered to be committed till payment for his part in a 5th of November riot; a Quarter Sessions case in the 1680s has a passing reference to an anvil in his workshop on which some silver coins were hammered. Several payments made to him (or his father) in 1671, 1672 and later for work on the St Petrock's, Exeter, clock, chimes and bells. A church clock at Drewsteignton was set up *c*1680 by 'Halstaff ye Clockmaker & his Men'.

HAMLIN, JOHN: Crediton, High St, 1856–73 (D).

HAMLYN, THOMAS jun: Ashburton. Clock and watchmaker; listed in the *Universal British Directory*, 1793.

HAMMOND, FREDERICK SAMUEL: South Molton, 7 West St, 1883; East St, 1889–1902 and later (D).

HARDING, CHARLES: Ashburton. Clock and watchmaker; listed in the *Universal British Directory*, 1793; address given as Bull Ring in 1823 (D). Died 14 Dec 1844 (EFP).

HARDING, CHARLES W: Sidmouth. Watchpaper: 'C. W. Harding, Watch & Clock Maker, Sidmouth, Plate & Jewellery repaired'. Listed in directories from 1823; at High St, 1830; Post Office St, 1838; Back St, 1844; Old Fore St, 1850–6. Death notice: 'Oct 3, in his 74th year Mr Charles Harding, for many years watchmaker of Sidmouth, much lamented' (EFP, 12 Oct 1864).

HARDING, JOHN: Ashburton. Watch and clockmaker, North St, 1830; East St, 1844–56 (D). Died at Ashburton 1 Jan 1860, aged 56 (EFP).

HARDING, JOHN (ii): Ashburton. Watchmaker; married Miss Ann Giles at Ashburton 22 May 1860 (EFP).

HARDING, RICHARD: Ashburton, East St, 1866 (D).

HARLEY, JOHN: Newton Abbot, 20 Courtenay St, 1897 (D).

HARNER, ——: Membury. Clock so inscribed.

HARNER, SYDNEY: Beer, 1873–97 (D). Supplied new dial for Beer Church clock.

HARNER, WILLIAM: Colyford, 1838–66 (D). Made or supplied Bishopsteignton Church clock, 1850.

HARRIS, BARTHOLOMEW SQUIRE: Ilfracombe, High St, 1883–1902 (D).

HARRIS, CHARLES: Chudleigh, Fore St, 1897–1902 (D).

HARRIS, FRANCIS: Ilfracombe. Watch and clockmaker, Market St, 1866 (D).

HARRIS, FRANCIS: Lynton, Lynbridge, 1873–8 (D).

HARRIS, ISRAEL: Exeter, 166 Fore St, 1850 (D).

HARRIS, JOHN: Ilfracombe, 41 Fore St, 1873 (D).

HARRIS, NATH: Apprenticed *c*1737 to John Tickell, Crediton, £5 premium. Son of John Harris of Crediton.

HARRIS, THOMAS: Bideford. Clockmaker; his wife was called Elizabeth; a child, Frances, was baptised at Bideford Independent Meeting on 23 May 1757 (register). Longcase clocks.

HARRIS, WILLIAM: Hatherleigh. Watch and clockmaker and shopkeeper; listed in the *Universal British Directory*, 1798. Longcase clocks.

HARRISON, ROBERT: Exeter. Working brazier and clockmaker, North St. Was given bail at Exeter Quarter Sessions on 25 Jan 1783: 'Robert Harrison of St Keryan Clockmaker in £20, James Hine of do. Brazier in £5, and John Shilston of do. Basketmaker in £5. For the appearance of the said Robert Harrison next Sessions to answer in Bastardy on the Information of Mary Bale widow' (Books of Recognizances 77). He was described as a widower when he married Ann Rositer at St Martin's on 25 Mar 1792; buried at St Kerrian's, Exeter, 8 Dec 1799 (registers). Mahogany longcase clocks with moonwork and sometimes tide indication.

HARRY, JOHN: Plymouth. 'A caution. The public are hereby cautioned not to trust Elizabeth Harry, wife of me John Harry, clock and watchmaker, and master of the East and West Country House on the Barbican, in the borough of Plymouth, and county of Devon, as I am determined to discharge no debt or debts she may contract after this public notice. As witness my hand this 4th day of August, 1792 JOHN HARRY' (*Exeter Gazette*). A John Harry was apprenticed *c*1769 to Thomas Olive, of Penryn, Cornwall, watchmaker, premium £17.

HARRY, WILLIAM: Apprenticed *c*1761 to Nathaniel Upjohn of Plymouth, watch and clockmaker, premium £68 5s.

HART, C: Cullompton, 3 Victoria Terrace, 1902 and later (D).

HART, EMANUEL: Plymouth. Watchmaker; listed in the *Universal British Directory*, 1798.

HART (HEART), HENRY: Plymouth. Watch No 9099 lost (*Daily Advertiser* 10 Mar 1778 – Buckley). Listed as silversmith in the *Universal British Directory*, 1798.

HART, MOSES: Plymouth. Watchmaker, Market St, 1822 (D).

HART, MOSES: Exeter. Watchmaker and clothesman, West St, 1828; Fore St Hill, 1830–47 (EPJ).

HART, SAMUEL: Plymouth. Watchmaker, jeweller and miniature painter; apprenticed to Abraham Daniel (qv) in 1779. Tenant of house for sale in Market St (EFP, 8 Oct 1795). Listed in the *Universal British Directory*, 1798. Wife died (EFP, 2 Mar 1809). Was the father of Solomon Hart, RA (1806–81), and began life as a worker in silver and gold at Bath; mentioned by Bromley (*Catalogue of Engraved British Portraits*, 1793) as a mezzotint engraver; studied painting under Northcote in London in 1785. Father and son went to London from Plymouth in 1820; the former taught Hebrew and the latter prepared drawings to become a student at the Royal Academy (*Dictionary of National Biography*).

HARVEY, GEORGE: Dartmouth, Higher St, 1866 (D).

HARVEY, ROBERT: Exeter. 'Married . . . March 5th at St Paul's, by the Rev. A. T. R. Vicary, Mr Robert Harvey, watchmaker and silversmith, to Miss Elizabeth Sclatter, both of this city' (EFP, 9 Mar 1837). St David's Hill, 1849–57 (EPJ).

HARVEY, WILLIAM: Plymouth, 4 Exeter St, 1883; 30 & 31 Treville St, 1889–1902 and later (D).

HARVIE, J. C: Totnes. Watchmaker; listed among agents for a State Lottery (Richardson, Goodluck & Co) in 1816 (EFP). Eight-day mahogany veneered longcase clock with shell inlay and silvered arch dial with rocking ship.

HARVIE, J. C: Plymouth. 115 Union St, East Stonehouse, 1838 (D).

HARVIE, WILLIAM C: Plymouth. Watchmaker, Old Town St, 1836 (D).

HAWKINGS, ——: Torrington. Clock and watchmaker; report of his wife's death (EFP, 11 Dec 1834).

HAWKINS, AMBROSE: Exeter. Maker of top quality clocks; died 1705.*

HAWKINS, HENRY: Plymouth. Watchmaker; listed in the *Universal British Directory*, 1798. A signatory to registration of Baptist meeting house at Catdown, 3 Jan 1803.

HAWKINS, JOHN: Exeter. 'John Hawkins Exon' on dial, *c*1680–1700. Mentioned in letter to Exeter library; 'Could not trace this maker' (Tapley-Soper).

HAWKINS, JOSEPH: Axminster. Watch and clockmaker; listed in the *Universal British Directory*, 1793.

HAYDON, JOHN: Exeter. Lantern maker, apprentice of John Turner, admitted as Exeter freeman 31 Aug 1579. Lived in St John's and kept the parish clock in order from *c*1585 to 1609 for 8s a year (churchwardens' accounts); died 1614.

HAYHURST, JAMES: Plymouth. 51 Union St, Stonehouse, 1893 (D).

HAYMAN, WILLIAM FRANCIS: Brixham, 1 Fore St, 1902 (D).

HAYS & PARSONS: Ashburton, North St, 1873 (D).

HAYS, ADRIAN: Ashburton. 'Watch and clockmaker, jeweller, &c. Clocks and watches regulated by the year. Market Place', 1878; North St, 1883–1902 and later (D).

HAYWARD, FREDERICK: Devonport. Watchmaker and jeweller, 48 Fore St, 1866 (D).

HAYWOOD, JOHN: Crediton. Watchmaker; on 7 Jan 1766 he patented (No 836) 'a lunar or callendar ring' (DNQ, vol 9).

HEALE, ABEL: Chulmleigh. Watch and clockmaker and grocer, New St, 1838–83 (D). White-dial longcase clocks.

HEALE, JAMES BROOKE: Crediton, High St, 1866–1902 (D). Watch-paper: 'J. B. Heale, watch & clockmaker, silversmith & jeweller, High St, Crediton. Plate & jewellery repaired. Wedding rings'.

HEALY, THOMAS: Devonport, 93 Fore St, 1838; 64 Duke St, 1856 (D).

HEARD, ——: Chawleigh. Thirty-hour white-dial longcase clock, nineteenth century.

HEARD, WILLIAM: Plymouth Dock. Watchmaker and gunsmith, Market St, 1822 (D).

HEARD, WILLIAM: Hartland. Watch and clockmaker, 1830 (D).

HEARN, ALBERT: Exeter. Watchmaker, Fore St, Heavitree, 1890 (D).

HEARN, H: Shebbear, 1897 (D).

HECTOR, WILLIAM: Crediton, High St, 1878–1902 (D).

HEINE & CO: Devonport. Clockmakers, 10 Tavistock St, 1866 (D).

HELE, W. W: West Teignmouth, Bank St, 1856 (D).

HELE, WILLIAM R: Teignmouth, Bank St, 1878 (D).

HELES, RICHARD: Apprenticed *c*1755 to Walter Elliot, Plymouth, watchmaker, premium £30.

HELLIER, JOHN: Mended South Molton Church clock, 1680 and 1688 (see *Transactions of the Devonshire Association*, vol 36, pp 248–9).

HELLYER, RICHARD: Plymouth Dock. Clockmaker and grocer, Pembroke St, 1822 (D).

HENDRICK, CHARLES SEPTIMUS: Colyton, 1873 (D).

HERBERT, SYDNEY T: Exeter, Sidwell St, 1883–1902 (D). He added minute hand to Brampford Speke Church clock (*Exeter Gazette*, 5 Aug 1881).

HERBERT, THOMAS: Exeter, Sidwell St, 1866–78 (D).

HETTISH & CO: Exeter. Sun St, 1861; Fore St Hill, 1862; then S. & F. Hettish a year or so later (EPJ).

HETTISH, A: Exeter. Mid nineteenth century 30-hour white-dial longcase clock.

HETTISH, GORDON: Exeter. Watch and clockmaker, jeweller etc, Magdalen St; bankrupt 1891 (Fig 11). Later worked at South St and then at 2 Catherine St.

HETTISH, S. & F. (SIMON & FRED): Exeter. Clockmakers, 121 Fore St Hill, *c*1864–78. They advertised American Waltham watches for sale in 1875.

HICKS, JOSEPH: Exeter. Watchmaker. St Kerrian parish (Quarter Sessions jury lists, Oct 1778 and 1780).

HICKS, WILLIAM: Clockmaker. Buried at Milton Abbot 6 Jan 1758.

IN BANKRUPTCY Re GORDON HETTISH

MAGDALEN STREET,

EXETER.

MESSRS.

HUSSEY & SON

Are instructed by Thomas Andrew, Esq. (Official Receiver), to SELL BY AUCTION,

On TUESDAY, August 4th, 1891,

At TWELVE O'CLOCK, on the Premises, the

STOCK-IN-TRADE

Fixtures, Household Furniture, &c.,

Of Mr. Gordon Hettish, Watch and Clock Maker, Jeweller, &c., comprising

STOCK-IN-TRADE, &c. Advertisement Clock outside shop, 4 Handsomely Carved Spring Cuckoo and Quail Clocks, Eight-Day Spring do., 4 Chain do., handsome Carved, 30 hour Clock with Brass Face and Oak Case, 2 do., 8 Clocks, Regulator by Hettish 6 American Timepieces, 52 Nickle Alarum Timepieces, Parkers Alarum Clock, Field Glass, Aneroid, 9 doz. Clock Main Springs, Clock and Watch Dials, Main, Waltham, and other Springs, Hands, Cylinders, Pinions, Glasses, Wheels, Keys, Movement Catches, Vurges, Escape Wheels, Balances, Side Clicks, several Boxes of Watch Screws, Ruby Pallettes, Watch Jewel Holes, Locket Glasses, Brooch Backs, Conflex Dial Glasses, 9 doz. Pairs Spectacles, 3 doz. Pairs Nippers, 3 doz. Cases for do., Model Steam Engine and Boiler, Small Lathe, lot old Brass, &c.

SHOP FIXTURES, &c.—Counter, large Show Glass, 2 small do., Mahogany Work Bench with Drawer, large do., Glass Window Case, Painted Material Case, [illegible] Watch Glass Case, Mahogany Desk, 3 Stools, 3 Cane-Seat Chairs, Umbrella Stand, Brass Scales and Weights, Gas Stoves, Gas Fittings, &c.

FURNITURE. Dining Table with Lap, Mahogany and Deal Tables and [illegible] Chess Table, Marble Top Cheffioneer, 6 and 1 Arm Leather Covered Chairs, Sofa, Gilt Chimney Glass, Gilt Clock, Carpet, Hearth Rugs, Linoleum, Fenders and Fire Irons, Chimney Ornaments, Iron Bedsteads and Palliasses, Wool Mattress, Feather Bed, Bedding, Set Mahogany Drawers, Marble Top Washstand and Ware, Dressing Table and Glass, Cane Seat Chairs, Plated Tea Pot and [illegible] Kitchen Table, Cupboard, China, Ware, and Sundry other Effects.

NINE FOWLS, HEN AND CHICKEN.

ON VIEW MORNING OF SALE FROM ELEVEN TO TWELVE O'CLOCK.

11, Queen Street, Exeter, July 23rd 1891. Agents to the Alliance Assurance Company.

WILLIAM POLLARD & Co., PRINTERS, EXETER.

Fig 11 Poster advertising clockmaker's sale

HILL, ALBERT: Plymouth, 53 Old Town St, 1873 (D).

HILL, BENJAMIN: Barnstaple, Joy St, 1873–8; & Sons, 1883–1902 and later (D).

HILL, HENRY: Exeter, 19 Union Terrace, St Sidwell, 1849; 9 Paragon Place, South St, 1851; 38 South St, 1855 (EPJ). 'A refractory prisoner: A few days since, Henry Hill, a watchmaker, who was recently sent to prison for neglecting his wife and family, attempted to commit suicide by hanging himself. During his imprisonment he has been very violent, and refuses to submit to the ordinary discipline of the gaol. There appears, however, to be some doubt as to his sanity' (*Western Times*, 8 Mar 1856).

HILL, HENRY H: Hatherleigh, Church Gate, 1889; Bridge St, 1893–1902 and later (D).

HILL, JOHN: Teignmouth, 46 Teign St, 1897; & Son, 1902 (D).

HILL, RICHARD: Plymouth, 11 Mutley Plain, 1902; Armada St, 1906 (D).

HILL, WILLIAM: Devonport, George St, 1873–8 (D).

HILL, WILLIAM jun: Cullompton, Fore St, 1850–6 (D).

HILL, WILLIAM W: Exmouth. Watchmaker and jeweller, Strand, 1850 (D).

HILLMAN, WILLIAM: Plymouth Dock, 1770 (Britten).

HILLSON, RICHARD: Plympton. Turret clock made for Devon country house in 1795 has inscription on setting dial: 'Richd Hillson Plympton MDCCXCV'. Another, at Antony parish church, Cornwall, was modified by Hillson. Its added setting dial is inscribed: 'Rd Hillson Plympton 1810'. Listed as clockmaker in the *Universal British Directory*, 1798.

HINE, JAMES (& CO): Exeter. A partnership between Hine and James Osmond Band at Bridgwater as braziers and ironmongers was dissolved 17 Oct 1772, 'Mr Band to continue'. Hine moved to Exeter where he seems to have maintained his connection with Band, for in 1779 was published an advertisement detailing the wide range of goods sold at Band and Hine's warehouse in North St, including church and other clocks. In a notice dated 8 Oct 1799, it was announced that 'Hine's wholesale copper, brass & plate iron manufactory, is removed from North St, to . . . St Sidwells, a few doors above the Old London Inn'. His son, W. J. Hine, took over the business in 1801. A mahogany longcase clock with arched silvered dial, signed 'James Hine & Co Exeter' and showing 'High Water at Topsham Bar', is illustrated in C. N. Ponsford, *Time in Exeter*. Hine died of dropsy, aged 66, and his burial was recorded in St Kerrian's on 16 Nov 1811; his wife Elizabeth died in 1795. At Exeter Quarter Sessions in Jan 1783, Henry Sobey was found guilty of stealing tools and other goods belonging to James Hine and was sentenced 'to be publickly whipt in the Market on Friday next and imprisoned three kallendar months and publickly whipt again on the expiration of his imprisonment and then discharged'.

HOARE, WILLIAM HENRY: Plymouth, King St, 1873–8 (D).

HOBBS, ELIAS: Ilfracombe. Clockmaker; '. . . Among other buildings which used to be at the Cove was a hut-like building, where for many years Elias Hobbs worked as a clockmaker. He died in 1845, aged 82 years, and was buried in the churchyard at the Holy Trinity' (from: *Around the Harbour* by John Longhurst, Ilfracombe Local History Group, 1978). Clock reported (Tapley-Soper).

HOBBS, ELIAS: Bampton. Listed in 1838 (D); died, aged 53, in 1861 while at Lynton, north Devon, which place he visited twice a year in pursuance of his business (EFP, 20 Mar 1861).

HOBBS, Mrs ELIAS: Bampton, 1873 (D).

HOBBS, ELIAS: Bampton, 4 Fore St, 1878–97 (D).

HOCK, PRIMUS: Exeter, 14 Market St, 1883 (D).

HOCKING, HENRY ROBERT: Plymouth, 18 Cecil St, 1873–93 (D).

HODGE, FREDERICK: Plymouth, 41 Clarence St, 1889–1902 (D).

HODGE, JOHN: Dartmouth. Watchmaker, St Saviour's parish; married Mary Cole at St Petrox Church, Dartmouth, 11 July 1788 (parish register). Listed in the *Universal British Directory*, 1793.

HODGE, JOHN: Plymouth Dock (Devonport). Clock and watchmaker; listed in the *Universal British Directory*, 1798. The *Exeter Gazette*, 18 July 1835, reported his death: 'July 8, at his house in South-hill Buildings, Stoke, Mr John Hodge, aged 77, formerly a respectable clock and watchmaker of Devonport'; 8-day longcase clock seen at Dartmouth hotel (Tapley-Soper).

HODGE, JOHN: Barnstaple. Watch and clockmaker, Boutport St, 1823–38 (D). His wife Martha died on 12 Nov 1846, aged 73 (EFP).

HOEFLER (HOFLER), FIDEL: Plymouth, 48 Southside St, 1878; 32 Exeter St, 1893; 11 High St, 1897 (D).

HOEFLER, R: Dawlish, 2 Lansdowne Place, 1897 (D).

HOEFLER, SEVERIN: Devonport, 10 Tavistock St, 1873–8; 23 Marlborough St, 1883; & Co, 1889–1902 and later (D).

HOIDGE, JOHN: He entered into agreements in 1782 and 1790 to repair and maintain Tavistock Church clock and chimes (churchwardens' account book).

HOLE, WILLIAM HENRY: Dartmouth, Duke St, 1873; New Quay, 1883–9; the Quay, 1893 (D).

HOLWILL (HOLOWILL), ROBERT: Sidmouth. Watch and clockmaker; married Ann Calcombe, of Sidmouth, at the parish church 25 Feb 1811; listed in 1823 (D). Eight-day longcase clock, silvered arch dial with moonwork, showing 'High Water at Sidmouth', the back of the moon dial scratched with name Billeter, mahogany case with black stringing (Loomes).

HONEY, JOHN B: Torquay. Listed at Lower Union St; at No 32 in 1866; at No 19 in 1873 (D).

HOOPER, HUMPHREY: Okehampton. Longcase clock: 'it bears upon the case the date 1748, dial inscribed "Hum. Hooper Oakhampton"' (DNQ, vol 8). Okehampton parish register of burials records that of a Mr Humphrey Hooper on 16 Oct 1760.

HOOPER, RICHARD: Tiverton. Watch and clockmaker, Bampton St, 1823; Towns End, 1830 (D).

HOPPEN, WILLIAM VOSPER: Plymouth, 55 George St, 1873–83 (D).

HOPPIN (HOPPINGE), CHARLES: He kept Woodbury Church clock in repair, *c*1592–1604 (churchwardens' accounts). This craftsman may be identified with a locksmith of St Sidwell's, Exeter, who died in Nov 1606 and whose estate was administered by his widow, Mary, and by Matthew Hoppin, of St Sidwell's, lockier. An inventory of his goods, including a great bible and two other books, came to £4 9s 6d (Moger).

HOPPIN (HOPPINGE), JOHN: 'Itm pd to John Hoppinge for mendinge of the Clocke 3s' (Woodbury churchwarden's accounts, 1588–9).

HOPPIN, MATTHEW: Exeter. Maker of the original Matthew the Miller clock, *c*1619–21, at St Mary Steps Church, Exeter.*

HOPPIN, WILLIAM: Exeter. Locksmith and church clock repairer; died 1643.*

HORNBROOK, AARON: Plymouth, 23 Southside St, 1856; 31 Treville St, 1873; 26 Bath St, 1878 (D).

HORNSEY, CHARLES: Exeter, 3 Northernhay St, 1893; then at Cowick St and Ironbridge (D).

HORNSEY, CHARLES: Cullompton. Watch and clockmaker and photographer, High St, 1878; also listed at Tiverton Lane the same year; Duck St, 1883–9 (D).

HORNSEY, WILLIAM: Exeter. Clock and watchmaker; a 1758 jury list records him in St Mary Major parish, but he later moved to Holy Trinity, where he was living at the time of his marriage to Miss Hannah Bidwell at St George's Church 12 May 1767. Mended the clock of Old Trinity Church in 1778–9 and was churchwarden there from Easter 1787 to Easter 1788. In the *Exeter Flying Post* of 5 Feb 1789, Thomas Holman, boot and shoe maker, announced that he had moved to 'the house lately occupied by Mr Hornsey, watchmaker, in Southgate St'. Hornsey's apprentices included John Lakeman, William Adams and Edward Gouldsworthy. A William Hornsey was buried 3 May 1801 at Allhallows on the Walls. His watch No 4 lost (*Bristol Journal* 15 Oct 1792 – Buckley). He made the gallery clock at the Old Meeting House, Sidmouth, 'The gift of M.S. 1767'; also longcase clocks, one showing 'High Water at Topsham Bar' illustrated in C. N. Ponsford, *Time in Exeter*. Another, 30-hour, was washed away in the Lynmouth flood disaster in Aug 1952, recovered from the shore and subsequently restored.

HOWARD, WILLIAM: Exeter. Brassfounder; extensively employed as mender of church clocks and added the minute-dial to the Exeter Cathedral clock in 1759. Son of a butcher, he was admitted as Exeter freeman by succession on 2 Sept 1734. Lived in St Edmund's parish where he had a foundry near the church on the old Exe Bridge; was churchwarden in 1762; was buried on 5 May 1768. Kept the Exeter Cathedral clock in order for 50s a year between 1759 and the time of his death. The business was continued by his widow, Elizabeth. She died in Sept 1781 and the stock in trade, which included clocks and jacks, was sold the following April. A brass chandelier, now at Hickleton Church, Yorkshire, is inscribed 'William Howard Fecit, Exon: AD MDCCXLVI', and is modelled upon those in Exeter Cathedral.

HOWE, Mrs C. & F: Exeter, 121 Fore St, 1889 (D).

HOWE, FREDERICK: Exeter, 123 Fore St, 1893; the business still listed there in 1956 (D).

HOWE, JOSEPH: Exeter, 40 Paul St, 1878; 121 Fore St, 1883 (D).

HUDGE, WILLIAM: 'Paid to William Hudge for mending the clock & chaimes 5 yeares agone £2 10s' (Ashburton churchwarden's accounts, 1706–7). He also mended the clock in 1720, 1721 and 1728.

HUET, JOHN W: Plymouth. Watch and clockmaker, 134 Union St, 1866 (D).

HUGHES, DAVID: Kingsbridge. Watchmaker and ironmonger; reported bankrupt, 'late of Kingsbridge' (EFP, 30 Mar 1815).

HUME, WILLIAM: Rockbeare wardens' accounts, 1784–1821, contain a vestry minute, 20 Dec 1804, recording agreement with William Hume 'to repair and put into compleat order the clock at Rockbeare Church, to find and put a new dial plate, three feet with gilt letters, to cover the upper square of the dial plate with lead to project three inches, and to do the whole in a workmanlike manner, finding every material for the sum of four guineas'.

HUMPHRY, HENRY: Brixham. Watchmaker, jeweller, silversmith and postmaster, Fore St, 1866–73 (D).

HUNN, HENRY H: Topsham. 'Watch and clockmaker, jeweller, and photographic artist, High St . . . every description of picture frames made to order' (advertisement in *Richards's Topsham Herald*, 24 Dec 1863). Listed in directories 1866–78.

HUNT, HARRY PARNELL: Barnstaple, 92 High St, 1866; 14 High St, 1873–89 (D).

HUNT, JOHN: Barnstaple, Joy St, 1844–50 (D).

HUNT, WILLIAM: Exeter. Clockmaker; submitted evidence at Exeter Quarter Sessions, 15 Jan 1697/8. Worked on Exeter church clocks at various times *c*1710–*c*1731 (churchwardens' accounts). Longcase clock with early type of chapter ring and crown and cherub spandrels, 'Wm Hunt Exon Fecit', *c*1700.

HUNTER, THOMAS JAMES: Exeter, 80 South St, 1890 (D).

HUNTLEY, WILLIAM: Barnstaple. A blind watch and clock repairer; he tendered to attend and keep the town clock in repair (EFP, 26 Oct 1826).*

HURCOMBE, HARRY: Lynton and Lynmouth, 1893 (D).

HURD, THOMAS: Honiton. Watchmaker, 1850 (D).

HURFORD, JOHN: Hatherleigh. Eight-day arch-dial longcase clock, the chapter ring with wavy 'Dutch' minute band, the dial centre with engraved shipping scene, and with a convex disc in the arch engraved with the words 'Tempus fugit', *c*1760.

HURLSTONE, JOHN: Exeter. 'To John Hurlstone for keeping ye clock and chimes and making a weele 11s' (St Sidwell's churchwarden's accounts, 1713). He was also paid in later years for work on the clock, up to about 1725.

HUTCHINGS (HUTCHINS), WILLIAM: Cullompton. The Cullompton churchwardens' accounts have these items: 1705 'pd in beer in bargaining to right the clock 8d. Itm pd Wm Hutchings for righting the clock £3'. Further payments were made to him in 1706–7, 1712, 1723 and 1730. Longcase clock in yew-wood case (Tapley-Soper).

HUTT, HENRY: Torquay. 7 Union Terrace, St Marychurch, 1889-93 (D).

HUTTON, JOHN: Apprenticed *c*1743 to Samuel Northcote, Plymouth, watchmaker, premium £40.

HUXTABLE, ——: Chittlehampton. White-dial oak longcase clock at North Devon Athenaeum, Barnstaple.

HUXTABLE, EDMUND: Chittlehampton area. Clockmaker, locksmith and agricultural dealer; his account book preserved in the Devon Record Office (2603M/B1) includes occasional clock entries *c*1776–88.

HUXTABLE, EDMUND: Bishop's Tawton churchwarden's accounts for year ending 8 May 1828 have these items: 'Mr Huxtable for putting the church clock in repair as per order of vestry £4 . . . for altering the striking part from the tenor to the treble bell 5s'. Vestry 4 May 1828: 'That Mr Huxtable be paid four pounds for repairing the clock, upon his undertaking to keep the clock in good repair for seven years at ten shillings a year. I do hereby agree that I will keep the church clock of this parish in good repair & that the sd clock shall keep good time and strike the hours for a term of seven years from this day for the sum of ten shillings a year the parish to find ropes for the weights – Edmund Huxtable'.

HUXTABLE, EDMUND: Newton Abbot. Watchmaker, 13 East St, 1866; 38 Queen St, 1873–83 (D).

HUXTABLE, JOHN: Sampford Courtenay. Thirty-hour white-dial elm longcase clock. 'Pd John Huxtable for new cutting the wheels &c of the church clock £2 5s' (South Tawton churchwarden's accounts, 1816–17).

HUXTABLE, Mrs SARAH JANE: Newton Abbot, 38 Queen St, 1889–1902 and later (D).

HUXTABLE, WILLIAM: South Molton, Broad St, 1856; Square, 1866; 28 Broad St, 1873 (D). White-dial longcase clocks.

HYMAN, ABRAHAM: Plymouth. Watch and clockmaker, 36 Treville St, 1830 (D).

HYMAN, H: Plymouth. Watchmaker and pawnbroker, Barbican, 1838 (D).

HYNE, CHARLES: Newton Abbot, Bridge St, 1850 (D).

ISACK, MABELL: Woodbury churchwardens' accounts record payments made to him in 1645 'for making clean of the clocke & mendinge of him & mendinge the loke of the tower dore 3s 8d'; and in 1649 for mending the clock, 6s 8d. Items in Chudleigh churchwardens' accounts include: 1651, 'for a new peise weighing 18lb for ye clock 4s 6d . . . to Mabell Isack to mend the clock £1 5s'; 1658, 'Mabell Isack hath now signed his undertaking to repaire the clock for 7 years having recd. 14s & 12d to be payd to him yearly by the churchwardens'.

ISLE, JOHN: Watchmaker. At Exeter Quarter Sessions, 5 Sept 1823, Jane Harrison, single woman, informed against him for assault.

IVERSON BROS: Kingsbridge, 35 Fore St, 1902 and later (D).

JACKSON, J: Tavistock. Thirty-hour one-handed longcase clock.

JACKSON, JOHN: Tavistock. Ironmonger and watchmaker, West St, 1838–50 (D). Died 5 Dec 1850; will in DRO mentions house and premises in West St 'in which I now reside'; estate valued at under £4,000. See also Jessop & Jackson.

JACKSON, JOHN & EASTCOTT, WILLIAM: Tavistock. Watch and clockmakers and ironmongers, West St, 1850 (D).

JACKSON, J: Plymouth, 32 Frederick St East, 1889 (D).

JACKSON, JOSEPH: Plymouth, 25 Edgcumbe St, 1893 (D).

JACOB, J: Totnes. White-dial longcase clock *c*1800 with moonwork in arch and showing 'High Water at Torbay', inlaid mahogany case with flat-topped arched hood. Eight-day bracket clock inscribed 'Jacob & Sons Totnes'.

JACOBS, ALEXANDER: Torquay. Described as silversmith when he signed petition in 1836 'for the making of the Torbay Road' along Torquay seafront; listed as Madrepore manufacturer, 14 Strand, in 1850 (D). Eight-day white-dial longcase clock with swans depicted in the arch, good quality case (Tapley-Soper).

JACOBS, JANE: Totnes. A notice in the *Exeter Gazette*, 15 Aug 1840, reads: 'Totnes. Jane Jacobs, widow of the late Lewis Jacobs, goldsmith, jeweller, and watchmaker, returns her sincere thanks for the very liberal patronage evinced towards her late husband by his friends and the public; and respectfully informs them that the business heretofore conducted by him and his predecessors (now of about seventy years standing,) will in future be carried on by herself, assisted by her brother, Arthur Jackson; and she solicits a continuance of their favours. She also takes this opportunity of stating, in

Plate 38 (*above left*) Thomas Mudge's own striking clock, the last that he made. It was in his possession when he died in November 1794. Mudge made his three famous marine timekeepers while living at Plymouth

Plate 39 (*above right*) The under-dial work of Mudge's striking clock

Plate 40 Replica hallmarked 1795 of the first detached lever escapement watch, made by Thomas Mudge in 1770 for Queen Charlotte. The original is at Windsor Castle

Plate 41 (*above left*) Henry Ellis (1790–1871), the Exeter watchmaker whose memoirs provide a vivid insight into the jewellery and watch and clockmaking trade in the provinces

Plate 42 (*above right*) Henry Ellis's regulator, made in 1821 for his shop in Exeter High Street. It was sold in 1893 by a successor in the business and is still in existence. Painting by George Townsend

Plate 43 Regulator made by John Perryman, of Barnstaple, at the end of his apprenticeship *c*1834. He took it with him to Australia, where in 1853 he established a business in Adelaide. The case was made by his brother Joseph. Western Australia Museum, Perth

consequence of frequent enquiries, that this establishment has not, nor ever had, any connection with any other house in the country, or elsewhere. The same workmen employed as before'.

JACOBS, LEWIS: Totnes. Watch and clockmaker, goldsmith and jeweller, Lower Main St, 1830 (D). Infant son died (EFP, 8 Mar 1838). Inlaid oak longcase clock (Tapley-Soper). Bracket clock with brass inlay, 'Jacobs Totnes', striking and repeating on a bell. The business was continued after his death *c*1840, by his widow Jane.

JACOBS, MORRIS: Exeter, 1825–34 (EPJ). 'Tailor, draper & hatter; silversmith & watchmaker; plate of every description; gentlemen's clothes & children's dresses made in the most elegant style of fashion at reduced prices; watches warranted. 22 Butcher Row and 169 Fore St, Exeter' (1827–8 billhead, DRO).

JAGO, H: Plymouth. 50a Edgcumbe St, East Stonehouse, 1889 (D).

JAGO, WILLIAM HENRY: Paignton, Winner St, 1893–1902 and later (D).

JAMES, GEORGE: The name is thus recorded in *Clocks & Clockmakers of Tiverton*, but should probably read George Eames; he carried out work on the Old Blundell's School clock *c*1865–73.

JAMES, NICHOLAS: Plymouth. Watchmaker, 29 Old Town St, 1844; 21 Bilbury St, 1866 (D).

JARMAN, ALLAN: Tiverton, Bampton St, 1883–1914 and later (D).

JARMAN, HENRY: Tiverton, Bampton St, 1878 (D).

JARVIS, HENRY: Devonport, 26a Marlborough St, 1889 (D).

JEFFARD, ——: Newton Bushel (Newton Abbot). Watchmaker, 1795 (Baillie).

JEFFERY, ALFRED BYRON: Plymouth, 69 Treville St, 1883; 20 Treville St, 1889–1902 (D).

JEFFERY, F: Devonport. 18 Tavistock Road, Stoke, 1893–7 (D).

JEFFERY, WILLIAM ABBOTT: Plymouth. Watchmaker, jeweller and chronometer manufacturer, 21 Whimple St, 1866; 1 St Andrew's Place, 1873–89 (D).

JEFFERY, WILLIAM HENRY: Torquay. Watchmaker, 76 Lower Union St, 1866 (D).

JENKIN, P. H: Bideford, 12 Allhalland St, 1889-1902 and later (D).

JERGER, JOHN: Devonport, 42 Fore St, 1883 (D).

JERWOOD, JOHN: Kentisbeare churchwardens paid him £2 in 1722 'for righting the clock & making two wheels' and 6s in 1725 'for righting the clock'.

JERWOOD, THOMAS: He kept Kentisbeare Church clock in repair for several years, commencing about 1727; he also made a weathercock for the church in 1732 at a cost of £6 (churchwardens' accounts).

JESSOP & JACKSON: Tavistock. Watchmakers and silversmiths, King St, 1823; Fore St, 1830 (D). Longcase clock reported (Loomes).

JESSOP, PETER: Tavistock. Watchmaker; took out a licence 22 July 1808 to marry Grace Coad of Tavistock.

JESSOP, THOMAS: Tavistock. Entries in Tavistock parish accounts include: 1764 – Thomas Jessop will keep the clock and chimes in order for £2 3s, 'only abuse excepted'; 13 Mar 1796 – 'Mr Jessops bill for keeping the clock in repair £4 10s'; 1798 – 'To Thos Jessop his salary for ye chimes'. Eight-day lacquer longcase clock with moonwork, signed 'Thomas Jessop Tavistock' in capitals on strip around top of dial arch, *c*1755/65 (Plate 47).

JESSOP, THOMAS & PETER: Tavistock. Watchmakers and silversmiths, Higher Market St, 1838 (D). They were brothers. Thomas died 10 Jan 1842, will in DRO; Peter named as executor; estate valued at under £100.

JEWEL (JEWELL), LEONARD: He received a salary of £2 6s in the early 1750s for keeping Tavistock Church clock; he also received payments for repairs (churchwardens' account book).

JOHN CLOCKMAKER: 1536–7: '£9 payd unto John Clockmaker makyn of the clock and the chyme' (Ashburton churchwarden's accounts).

JOHN (clockmaker): He kept Dawlish church clock in repair for 2s a year *c*1589–93, receiving the money annually on midsummer day (churchwardens' accounts). This John was possibly John Crowder (qv).

JOHNS, RICHARD: Exeter. Clockmaker; admitted as Exeter freeman 17 Mar 1628, on payment of £1. Was 'servant' to the clockmaker John Savage, who left him 20s in his will and whose daughter, Margaret, he married. 'To Richard Johns for mending of the Clocke, takeing him downe and setting of him up, 11s' (St Petrock's, Exeter, churchwarden's accounts, 1648).

JOHNS, SAMUEL: Exeter, 186 Sidwell St, 1876; Portland St, 1883; Paris St, 1902 (D); *see also* Frost & Johns.

JOLE, JOHN: Exeter. Watchmaker, 253 High St. Shop door wrenched open and eleven watches and a silver watch-case stolen (EFP, 23 Feb 1826).

JONAS, ——: Exeter. Five watches stolen from shop of Mr Jonas, of Bartholomew St, silversmith &c (EFP, 23 Nov 1848).

JONAS, B: Exeter. Tool and clock dealer and watchmaker, Bartholomew Yard, 1849–51 (EPJ).

JONAS, BENJAMIN: Plymouth Dock. Watchmaker and silversmith, James St, 1812–14 (D). Married Anne, sister of E. A. Ezekiel, the Exeter engraver, and was named as a trustee in the latter's will dated 18 Sept 1806.

JONAS, BENJAMIN: Teignmouth. Watch and clockmaker, 8 Wellington Row, 1823; Old Market St, 1830 (D).

JONAS, HENRY: Torquay. Watchmaker, Torwood Lane, 1823 (D).

JONAS, JONAS: Newton Abbot. Listed at East St, 1823 (D). 'The creditors of Jonas Jonas, of Newton Abbot, in the county of Devon, watchmaker, who hath lately taken the benefit of the Insolvent Act, are requested to meet at the Globe Tavern, in the city of Exeter, on Thursday the 27th day of November instant, at 12 o'clock in the forenoon precisely, in order to choose an assignee of his estate and effects' (EFP, 20 Nov 1817).

JONAS, SAMUEL: Exeter. Silversmith, Friernhay Lane, 1796 (EPJ). 'All persons having any demands on the late Mr Samuel Jonas, of St Thomas, near this city, goldsmith, deceased, are desired to apply to Mr Emanual Levy, silversmith, Fore St' (EFP, 29 Sept 1803). An eighteenth-century watch signed by Jonas is numbered 887.

JONES, ARTHUR and THOMAS: Barnstaple. Lake, 1902 and later (D).

JONES, EDMUND: Barnstaple. Lake, 1889–97 (D).

JONES, ELI: Axminster. Watchmaker, 1850 (D).

JONES, JAMES: Sidmouth, Church St, 1902 and later (D).

JONES, JOHN WILLIAM: Barnstaple, 78 High St, 1838–50 (D).

JONES, THOMAS HENRY: Barnstaple. 27 Victoria St, Newport, 1897 (D).

JONES, WILLIAM: Tiverton, Angel Hill, 1830 (D).

JOSEPH, ——: Tavistock. Listed in the *Universal British Directory*, *c*1795.

JOSEPH, A. & SONS: Plymouth. A reward was offered following the loss of a box of watches being sent to them by mail coach from Baily & Upjohn, London (EFP, 8 Dec 1791).

JOSEPH, B. & B: Exeter. Watchmakers, jewellers and silversmiths, 137 Fore St Hill. Partnership to be dissolved and the stock sold off . . . Benedict Joseph to carry on the pawnbroking business (*Exeter Gazette*, 6 Dec 1828).

JOSIAS: Exeter. On 30 Nov 1648 it was 'ordered that Mr Receiver give Josias ye keeper of ye clock at ye Great Church for keeping it and cleansing ye gutters by ye walks in ye greate Churchyard 10s to be allowed on his acc' (Act Book of the Mayor and Chamber). This was during the Cromwellian period when Exeter Cathedral was in civic hands.

JOYCE, JOHN: Barnstaple. Watchmaker; married Margaret George at the parish church 13 Aug 1757 (parish register).

KALLENDAR (KALENDER), ORLANDO: Exeter. Lived in the Close (1773 jury list), later moving to St Stephen's. Son of Thomas Kallendar, hellier, admitted as Exeter freeman by succession, 26 Jan 1784; married Mary Alley at St Stephen's Church, 7 Mar 1784 (register). His watch No 4326 reported lost in 1781 (*Oxford Journal* – Buckley).

KALLENSEE, CHARLES HENRY: Devonport. Clock and watchmaker, 71 Fore St, 1850–6 (D). A witness to the will of John Gay Towson, of Devonport, retired watchmaker, dated 2 Sept 1856 (DRO).

KAMMERER, SERAPHIN: Plymouth, 118 Exeter St, 1878 (D).

KAUFMAN, BERNARD: Devonport, 102 Fore St, 1873–83 (D).

KELLOND, GEORGE ALBERT: Torquay, 21 Torwood St, 1893 (D).

KENDRICK, SAMUEL: Brixham. Son of a Brixham innkeeper, he was apprenticed *c*1802 to John Tucker in Exeter, accompanying him to Tiverton where he completed his apprenticeship. Went to Spain, working for the army, and set up in business in Brixham on his return. Died, aged 32, and was buried at Brixham 22 Sept 1819 (parish register). The Torquay Natural History Museum has a silver pair-cased watch, 'S. Kendrick Brixham'. White-dial longcase clock with rocking ship in arch, neat mahogany case with black stringing in trunk door and around the arch top (see Chapter 5).

KENNELLY, SAMUEL: Lynmouth, 1883 (D).

KENNETON, JAMES: Axminster. Watchmaker; married Ann Tomkins at the parish church 15 Oct 1754 (parish register).

KENNETT, ARTHUR W: Plymouth. Repairer, 1893–1902 and later (D).

KENT, SOLOMAN: Plymouth. Watchmaker, Bilbury St, 1866 (D).

KENT, WILLIAM: Barnstaple. Longcase clock *c*1700 reported; no further details.

KENT, WILLIAM: Plymouth, Frankfort St, 1850–56 (D).

KERFUTT, WALTER: Described in 1727 as watchmaker of Exeter. Provided two sureties at the city Quarter Sessions in 1734 and was on a jury list in 1740. Paid £20 in 1743 for repairing the Exeter Cathedral clock; died the following year and was buried in St Mary Major parish 6 Dec 1744. Arch-dial longcase clock with moonwork, 'Walter Kerfutt Exon', seen in hall of Devon stately home (Tapley-Soper).

KERRIDGE, FREDERICK (FRANK): Tiverton. Watch and clockmaker, jeweller and silversmith, 15 Bampton St, 1897–1914 (D). Moved to Fore St, continuing into the 1920s.

KETTERER, M. & CO: Plymouth, 39 Frankfort St, 1856 (D).

KEYS, JOSEPH: Exeter. Clock and watchmaker; listed at West St in 1822 (EPJ) and recorded there in the 1841 Census together with Ann Keys. He later moved to James St, which was his address when he died in Aug 1865, aged 76 (St Mary Major burial register). Worked on Exeter church clocks, making a new brass wheel for that of St Mary Steps in 1832. The escape wheel of the Exeter Cathedral old clock is inscribed 'JK 1841'. A 30-hour painted-dial, mahogany longcase clock by this maker, dated 15 Apr 1825, is discussed and illustrated in *The White Dial Clock* by Brian Loomes.

KEYS, WILLIAM: Exeter. Apprenticed *c*1766 to Richard Upjohn, watchmaker, premium £30. Recorded as clockmaker, Holy Trinity parish, 1803 Militia List.

KINGDON, SAMUEL (1745–97): Exeter. 'Wanted immediately, five or six journeymen clock or jack-makers. Apply to Mr Samuel Kingdon, iron-monger, Exeter' (EFP, Jan 1790). Died 31 Oct 1797, having 'conducted an extensive business through a long period with honour and integrity' (*Gentleman's Magazine*). Eight-day mahogany longcase clock, the dial freely engraved with the four seasons, some swans, a house and an eagle with '*Tempus Fugit*' in the arch, sold at auction in 1978.

KISTLER, A: Barnstaple, Boutport St, 1838 (D).

KNEEBONE, ARNOLD: Plymouth, 14 Ebrington St, 1902 (D).

KNOTT, FREDERICK: Braunton, Caen St, 1893–7; East St, 1902 (D).

KRAMM, CHARLES: Torquay, 12 Lucius St, 1897 (D).

KRAMM, JOHN: Torquay, Fleet St, 1878–93; & Son, 1897–1902 and later (D).

KRESSINGER, CHRISTIAN: Plymouth. Clockmaker and repairer, 17 York St, 1856; 44 Frankfort St, 1873–1902 and later (D).

KYMM, —— (1576): 'Paid Kymm for mending of ye clock & wheles thereof for making of a new spryng and for wyer for ye same 4s 8d . . . and for one rope for ye great pease of ye clock this year 4s 4d' (Crediton Church governors' accounts).

LAITY, EDWARD: Chudleigh, Fore St, 1873–93 (D).

LAKE, HENRY: Exeter. Watchmaker, silversmith, manufacturing jeweller, malachite manufacturer, fancy hair worker etc; married Elizabeth Ellett at St Thomas 2 May 1832. Listed in the High St: at No 266, 1834–9; No 214, 1840–56; and No 43, 1857–75 (EPJ).

LAKE, JOHN ELLETT: Exeter. Son of the above, born 4 Mar 1833. He continued the business at 43 High St from about 1875. By 1902 the business had become John Ellett Lake & Son. The Westcountry Studies Library, Exeter, has a photograph of the shopfront.

LAKEMAN, GEORGE: Plymouth, 27 King St, 1878–1902 (D).

LAKEMAN, JOHN: Exeter. Watch and clockmaker, Cathedral Close. Apprenticed *c*1761 to William Hornsey, premium £10 10s. He and Hannah King, both of the Close, were married at Exeter Cathedral 27 Feb 1766. Three children of the marriage baptised there, and cathedral register records John Lakeman's burial on 9 Oct 1785. Longcase clock reported.

LAKEMAN, R: Plymouth. Watchmaker; married Julia, 'daughter of Mr R. Weeks, RN' (EFP, 14 June 1832).

LAKEMAN, ROBERT: Plymouth, Exeter St, 1856–78 (D).

LAKEMAN, WILLIAM: Exeter. Watchmaker, Cathedral Close. His name appears on Exeter Quarter Sessions jury lists Apr 1774, and Jan and Oct 1776.

LAMBY (LAMBEY), JOHN: Exeter. Clockmaker and jackmaker, St Edmund's parish. Took out licence 12 Apr 1700 to marry Dinah Legge of Exeter, widow. At Exeter Quarter Sessions in July 1705 it was stated that 'John Lamby hath for some time absconded himselfe & left his habitation'. The court heard he had made no provision for the maintenance and instruction of his apprentice, Joseph Purchase, who was then 'discharged of & from his sd. master'. Lamby is incorrectly recorded as Lamley in some lists of makers.

LAMPORT, HENRY: Plymouth Dock. Watchmaker and jeweller; listed at Pembroke St in the *Picture of Plymouth*, 1812. His death was reported in the *Exeter Gazette*, 30 Aug 1828: 'On Tuesday, in Oxford-place, Mr Henry Lamport, jeweller, of Plymouth, deservedly regretted by his relatives and friends'.

LAMPORT, HENRY (ii): Plymouth. Signed letter dated 20 June 1833, stating willingness to have his stock of foreign watches stamped (Clockmakers' Company archives).

LAMPORT, WILLIAM: Plymouth. Gold and silversmith, watchmaker, etc, agent to Bish's Lottery Office, Market Place, 1814 (D).

LANCASTER, THOMAS: Plymouth Dock (Devonport). Watch and clockmaker; son of William Lancaster, who died in 1820 and whose business he continued. Listed at Queen Street, 1822–50 (D).

LANCASTER, WILLIAM: Plymouth Dock. Watch and clockmaker, Queen St. Married Mary Pinhey at Stoke Damerel parish church 20 Nov 1771 (register). The watchmaker Richard Lear Pinhey was a witness, and he also witnessed Lancaster's will, dated 23 Feb 1815 (DRO). Gold watch lost out of the Western Mercury coach, 'Wm. Lancaster Plymouth Dock' (EFP, 15 Jan 1795). Listed in the *Universal British Directory*, 1798. Died 12 Feb 1820, leaving estate valued at under £800. He left his working tools to his son, Thomas Lancaster, watchmaker, Plymouth Dock. His widow Mary was named as executrix and the will also mentions sons James and William and a daughter, Mary. Lancaster's name is to be found on good quality mahogany longcase and bracket clocks, also watches. He was a customer of Thwaites of Clerkenwell, and other London suppliers.

LANCEY, W. J. O: Ottery St Mary, Broad St, 1850 (D).

LAND (LANDS), THOMAS: Dartmouth, Tiverton and Honiton. Described as a watchmaker of St Saviour's parish, Dartmouth, in 1705 when, on July 24, he married Leah Hewiss (or Harris) at Honiton parish church. He later worked at Tiverton, having the care of St Peter's Church clock and chimes in 1718. A yew-cased longcase clock with square dial is signed 'Thos. Lands Tiverton'; a bracket clock also noted (Baillie), and a newspaper advertisement of 1776 refers to a lost watch – described as old-fashioned – by Land, Tiverton. Entries in Gittisham churchwarden's accounts for 1725–6 indicate that Land had by then moved on to Honiton: 'Mr Land for new making ye clock £2 12s 6d; taking abroad ye clock & cleaning 8d; boiling ye clock & brasses 1s; carrying ye clock to Honiton & fetching 1s 4d; one day to help up ye clock with Mr Land 1s'. One of his Honiton longcase clocks is said to have a mahogany case. Honiton burials include: 'Land, Leah wife of Mr Tho' 18 Nov 1737.

LANE, JOHN: Crediton. Was apprenticed on 3 June 1815 for seven years with James Bucknell & Son, of Crediton, premium £5; described as son of John Lane, late of Crediton, baker (indenture in DRO). Subsequently listed as watchmaker at West Town, 1823; High St, 1838–50; and North St, 1856 (D).

LANG (LANE), JOHN: Apprenticed *c*1756 to John Nathaniel Tickle, of Crediton, clockmaker and gunsmith, premium £3.

LANG, JOHN: Moretonhampstead. Clockmaker and gunsmith; listed as John Long in the *Universal British Directory*, 1798. Brass and white dial longcase clocks reported, including 30-hour example in narrow mahogany case topped by three large gilded wooden finials, the dial centre engraved in a floral pattern; another has an engraving of a ship and castle. The Moretonhampstead parish registers record the marriage of a John Lang to Mary Hutchings on 2 May 1771. The clocks are signed with the place-name 'Moreton'.

LANG, JOSEPH: Moretonhampstead. A clock reputedly signed 'Jos. Lang', but possibly a confusion with Jno. for John (see above).

LANG, RICHARD: Plymouth, 21 Drake St, 1883–1902 and later; Henry Lang listed at the same address in 1893 (D).

LANGBRIDGE, JAMES: Torrington. Watchmaker, High St, 1823–30 (D).

LANGBRIDGE, JOHN: Dolton. Good quality square-dial 8-day longcase clock, 'John Langbridge Doulton', *c*1730 (Loomes). Dolton parish registers record the marriage of John Langbridge and Margery Howard 13 Dec 1729, and the burial of John Langbridge sen 22 Dec 1775.

LANGMEAD, GEORGE: Dartmouth, Lower St, 1844 (D). On 28 Apr 1847 was examined in bankruptcy at Exeter and stated to be aged 27. A white-dial longcase clock reported.

LANGWORTHY, WILLIAM: Exeter. Watchmaker, North St, 1789 (EPJ).

LARMUTH, THOMAS: Silver verge watch by this maker, hallmarked London 1823–4, in Exeter Museum collection. According to a note by the donor, this watch was presented by Larmuth to his son, James William Larmuth, whose name it bears. The note adds that Thomas Larmuth served his apprenticeship with Upjohn in Exeter.

LASHMORE, CHARLES E: Plymouth, 53 Old Town St, 1878 (D).

LASKEY, SAMUEL: Moretonhampstead. Clockmaker; longcase clocks including a 30-hour one with musical train, playing every four hours on six bells; another, *c*1770–80, in pine case (Loomes). The surname Laskey appears in the Moretonhampstead parish registers; the clocks are signed with the place-name 'Moreton'.

LATY, EDWARD: Colyton, 1866 (D).

LAVENDER, JOHN: Crediton. Clock and watchmaker; listed in the *Universal British Directory*, 1793. Was a widower when on 9 Oct 1793 he took out licence to marry Sarah Pasmore, widow, of Crediton.

LAZARUS, ISAAC: Exeter. Watch and clockmaker, 12 Northernhay St, 1850 (D).

LAZARUS, J: Exmouth, Chapel St, 1838 (D).

LAZARUS, MOSES: Exeter. Watch manufacturer; recorded in Holy Trinity parish, aged 25, in 1841 Census. Holloway St, 1849–50; 5 Lansdowne Terrace, 1851–4 (EPJ).

LEAR, JAMES: Exeter. Watch and clockmaker and church clock repairer, Synagogue Place, 1825; Fore St Hill, 1827; Cathedral Yard, 1832; Mary's Yard, 1834; Gandy St, 1836; North St, 1837; Bear St, 1842; Deanery Place, 1851–60 (EPJ). Two of his daughters were married at Exeter Cathedral in 1858. He kept St Thomas Church clock in repair between 1858 and 1864.

LEAR, JOHN: Dawlish, Brunswick Place, 1883; & Son, 1889–1902 and later (D).

LEAR, JOHN: Torquay, 7 Sandhill Rd, 1893 (D).

LEAR, RICHARD: Plymouth. Longcase clock with modified Ferguson-type tidal dial with subsidiary dials and sea in the arch, *c*1770 (Bellchambers: *Devonshire Clockmakers*, p24); another, in mahogany case, showing 'High Water at Plymouth Dock' (see also Richard Lear Pinhey).

LEAR, SAMUEL: Torquay, 1 Strand, 1844; Park Lane, 1856 (D).

LEDDRA, JOHN: Devonport. Watch and clockmaker and jeweller, King St, 1850–66 (D).

LEDUM, JOHN: Plymouth. His watch No 433 reported lost in 1785 (Buckley).

LEE, WILLIAM: Okehampton. Watchmaker, 1844–50 (D).

LEE, WILLIAM: Bideford. High St, 1856–73; Mill St, 1878–89 (D).

LEGG, JOHN: Axminster. Clockmaker; a church clock he made for Gittisham, near Honiton, is now at Bristol Museum. 'Paid Mr John Legg for a new clock £18 10s' (Gittisham churchwarden's accounts, 1729–30). On 2 Aug 1776 he and William Oliver of Axminster, tailor, and Robert Bryant, barber, entered into a bond to administer the estate of Catherine Oliver, of Axminster, deceased (Moger).

LEMON, HUBERT. Appledore, 3 Meeting St, 1902 and later (D).

LEMPERT, SAUL: Plymouth. 51 Union St, Stonehouse, 1897–1902 and later (D).

LETCHER, HENRY: Ivybridge, 20 Fore St, 1889–97 (D).

LEVI & CO: Devonport. Watch and clockmakers, 71 Fore St, 1830 (D).

LEVI, LION: Plymouth. Silver masonic watch hallmarked 1810.

LEVY, A: Plymouth, 45 Bedford St, 1856 (D).

LEVY, BARNETT: Plymouth, 5 Treville St, 1830 (D).

LEVY, EMANUEL: Exeter. Watch and clockmaker, goldsmith and silversmith, at Fore St in 1803. Witness to the will of E. A. Ezekiel who died in 1806. Listed at Cowick St in 1831 (EPJ).

LEVY, JONAS: Exeter, 52 Alphington St, 1878 (D).

LEVY, JOSEPH & SON: Exeter, 137 Fore St Hill. Pawnbrokers, jewellers, watchmakers and general dealers (*Exeter Gazette*, 21 Jan 1826).

LEVY, MARKS (MARKES, MARCUS): Plymouth. Goldsmith, jeweller, watch and clockmaker, Treville St, 1836; 50 Bedford St, 1838–50 (D). Surname also given as Levi.

LEWIS, HENRY: Exeter, Trinity St, 1870 (D).

LEWIS, WILLIAM CHARLES: Plymouth, 23 King St, 1893–7 (D).

LEWIS, WILLIAM THOMAS: Plymouth. 52 Adelaide St, East Stonehouse, 1873; Union St, 1878–1902 and later (D).

LEYMAN, FRANK (1856–1952): Exeter. Watch and clock repairer and jeweller, 6 (now 44) Fore St, Heavitree. Served apprenticeship with John Parrish, commencing on his own in 1887; the business has been continued to the present day by his son John; account book in DRO (see also Groves & Leyman).

LIBBEY, DANIEL: Stonehouse, Plymouth. Watch and clockmaker, Chapel St, 1822; 4 Quay, 1823–4 (D).

LIBBEY, DANIEL HARSON PEARCE: Plymouth Dock (Devonport). Watch and clockmaker and jeweller, Tavistock Lane, 1822; Fore St, 1830–56 (D).

LIBBY, J: Plymouth Dock. Watchmaker, Market Lane, 1814 (D).

LIDSTONE, GEORGE. Dartmouth. Watchmaker, jeweller and nautical instrument maker, New Rd, 1838; Foss St, 1844–66; Duke St, 1873–89 (D).

LIDSTONE, GEORGE HENRY: Tavistock, 78 West St, 1866–89 (D). Advertisement in Kelly's 1866 *Devonshire Directory*: 'G. H. Lidstone, Watch and Clockmaker, Jeweller, &c. A good assortment of new and second-hand watches, jewellery, spectacles, electro-plate, &c on sale. Agent for Keyzor & Bendon's opera, marine & race-course glasses, telescopes &c. West Street, Tavistock'.

LIDSTONE, JOHN: Dartmouth, Lower St, 1878–1902 and later (D).

LIDSTONE, JOHN FREDERICK: Paignton, 40 Victoria St, 1897; 2 Dartmouth Place, 1902 (D).

LIDSTONE, ROBERT: Dartmouth. Watch and clockmaker, Lower St, 1823 (D).

LININGTON, WILLIAM RICHARD: Winkleigh, 1883 (D).

LINTON, GEORGE: Exeter, 4 Summerland St, 1850; Sidwell St, 1850–89 (D). 'Died . . . Apr 2, at Michael Hill, Bristol, Mr J. Linton (one of the Waterloo heroes), father of Mr G. Linton, watchmaker, of this city, aged 82' (EFP, 9 Apr 1862).

LISLE, WILLIAM: Exeter. Clock and watchmaker, 27 New Bridge St, 1849–66 (D).

LISLE, WILLIAM RICHARD: Exeter, 27 New Bridge St, 1873; 179 Fore St, 1889–1902 and later. Died 1938, aged 90. St Thomas Church clock is inscribed, 'W. R. Lisle Exeter 1883'; watch with 'W. R. Lisle Exeter' on dial.

LOCK, CHARLES: Plymouth Dock. Watchmaker, Pembroke St, 1814 (D).

LOCK, GEORGE: Barnstaple, Castle St, 1889–93; 52 Boutport St, 1897–1902 and later (D).

LOMBARDINE, FRANCIS: Totnes, Fore St, 1844 (D).

LOMBARDINI, F: Torquay, 6 Lower Union St, 1856 (D).

LOMBARDINI, FRANCIS: Newton Abbot, 44 Courtenay St, 1866–78 (D).

LONG & CO: Cullompton, Cockpit Hill, 1889 (D).

LONG, JOHN: Tiverton and Bradninch. Son of stonemason, he was born at Tiverton 10 Mar 1821; died at Bradninch 23 Oct 1887. Established a business in Bampton St, Tiverton in 1842. Longcase clocks noted, one dated

1847 illustrated in C. N. Ponsford *et al*, *Clocks & Clockmakers of Tiverton.* Tiverton Museum has several receipted bills signed by him for clock and watch repairs. Moved to Bradninch in 1863 and on a trade card described himself as 'goldsmith, jeweller, watch and clock manufacturer, inventor, patentee & sole manufacturer of the new lock to prevent the loss of brooches &c . . . patronized by the late J. Heathcoat Esq., MP for Tiverton, the original inventor of the lace machine . . . and by the Baroness Burdett Coutts, of London'. A correspondent noted that 'in his workshops there was old-fashioned machinery for cutting gear wheels etc' (Tapley-Soper). His name seen on English and foreign wall clocks. Married Emma Hurst, alias Cousins, daughter of John Hurst of Westminster, and they had an extensive family. Business listed as J. Long & Co in 1902.

LONG, ROBERT T: Barnstaple, Litchdon St, 1866 (D).

LONG, SAMUEL: Tiverton, 1 Wellbrook St, 1894–1902 (D).

LONG, SPECCOTT THOMAS: Uffculme. Watchmaker; brother of John Long (qv); born at Tiverton 12 June 1829; died 1892. Established business at Uffculme in 1859. Married Susan Puddicombe, who outlived him many years and died aged 92; four children. The business was continued by his sons James and Walter (see C. N. Ponsford *et al*, *Clocks & Clockmakers of Tiverton*, pp 48–9).

LONG & SONS: The Watch and Clock House, Uffculme, founded 1859; 'grandfather, chime & other repairs a speciality' (advertisement); still in business in 1921.

LONG, THOMAS: Bradninch, 1889 (D).

LONG, THOMAS: Tiverton, Wellbrook St, 1893–7 (D).

LONG, WILLIAM: Tiverton, Bridge St, 1850 (D).

LOOSEMORE, JOHN: Teignmouth, Regent St, 1889–1902 and later (D).

LOTT, GEORGE: Honiton. Eight-day longcase clock, square brass dial with chapter ring and water jets engraved in centre, simple oak and mahogany case, *c*1770.

LOVEGUARD, Mrs FLORA: Devonport, 42 Marlborough St, 1902 and later (D).

LOVELACE, JACOB: Exeter. Married 1712, died 1755.*

LUCKHAM, HENRY: Kingsbridge, 49 Fore St, 1897–1902 and later (D).

LUGG BROTHERS: Okehampton, 1873–8 (D). Also at North Tawton.

LUGG, THOMAS JAMES: Okehampton, Fore St, 1883–9; West St, 1893; the Arcade, 1897–1902 (D).

LUKE, WILLIAM: Exeter. Eighteenth-century red-lacquer longcase clock, 'Wm Luke Exon'. The Luke family were prominent as ironmongers and grocers.

LUSCOMBE, RICHARD: Totnes. Ashburton churchwardens' accounts have these entries: 1729, 'paid Mr Harris for makeing ye bond for Mr Luscomb to ye parish to keep ye clock & chaims in repair for 20 yeares 3s 4d'; 1730, 'paid Mr Luscomb of Totnes for keeping the clock & chaims in repair £1'. Subsequent entries identify this craftsman as Richard Luscomb, or Luscombe. Thirty-hour clock with 190mm (7½in) dial, exquisite hour hand and finely turned pillars, trains side by side, signed 'Rich. Luscombe Totnes fecit'; also 8-day oak longcase with arched dial (Bellchambers). Luscombe was maker of a brass chandelier inscribed 'for the ringers use forever', now in the south chapel of Totnes parish church (illustrated in *The Connoisseur Year Book*, 1958).

LUSCOMBE, ROBERT: Probably Totnes. 'Paid Robert Luscombe for looking after the clock &c £1' (Ashburton churchwarden's accounts, 1740).

LUXMORE, HENRY: Devonport, 19 Princes St, 1878–93 (D).

LUXTON, JOHN HAWKINS: Plymouth, 53 Old Town St, 1883 (D).

LYDDON, WILLIAM: South Molton. Watchpaper: 'Lyddon, Watch & Clockmaker, South Molton, Sells Plate, Wedding Rings &c.' Listed at Market Place, 1823; Broad St, 1830–44; a draper also (D). 'A daring robbery was committed on Saturday night in the house of Mr W. Lyddon, watchmaker, South Molton, when 21 watches, 3 silver spoons, money to the amount of £3 17s and a quantity of old silver coin, was stolen from him. Suspicion fell on John Wotten, a young man who had been doing some carpenter's work for Mr L., and on enquiry the constables of that place found he had left the town; with their usual promptitude, they commenced an immediate pursuit, and afterwards discovered and secured him on the Tiverton Road, coming to this city [Exeter], with the whole of the stolen property about him. He was immediately brought to this city, and committed to the County Gaol, to take his trial at the next Assize. It is supposed he concealed himself on the premises before dark, and had lain there until the family had retired to rest, when he effected his purpose' (EFP, 23 Dec 1824). Aged 18, he was subsequently sentenced to death at Devon Assizes.

LYON, FRANCIS: Bideford. Silversmith and watchmaker; listed in the *Universal British Directory*, 1793. Longcase clock reported.

LYON, FRANCIS: Plymouth. Watchmaker, Pike St, 1822 (D).

LYON, JUDAH: Plymouth. Watch and clockmaker and silversmith, Bedford St, 1830–6; 7 Union St, 1850–6 (D).

LYONS & SONS: Exeter. Watch and clockmakers, jewellers, 3 George St, 1890 (D). J. Lyons listed as jeweller, George St, in 1870 (EPJ).

MABIN, WILLIAM S: Torquay, 62 Lower Union St, 1883–93; 1 Union St, 1897–1902 and later (D).

MACEY, ROBERT: Plymouth. Watchmaker, silversmith, repairer of mathematical instruments and milliner, listed in the *Universal British Directory*, 1798; at Market St, 1812; Briton-side, 1822; South Side St, 1823–44 (D).

McKENZIE, JOSEPH: South Molton. Watchmaker, East St, 1838 (D).

McNAY, A. & SON: Exeter. Watchmakers, jewellers and wholesale watch-glass manufacturers, 107 Fore St, *c*1873–5.

MAISTERS (MASTERS), JOHN: Apprenticed *c*1756 to James Pike, Newton Abbot, premium £17.

MAJOR, ORLEY: Beer, 1883–1902 and later (D).

MALLETT, JOHN: Barnstaple. Watch and clockmaker and silversmith, 4 High St; business established in 1811. Described as widower, of Barnstaple, when he married Susanna Mary Upjohn at Ealing, Middlesex, 4 June 1818; she was the sister of William John Upjohn, watch manufacturer, of 11 St John's Square, Clerkenwell. Mallett supplied a new clock (replaced 1913) for Barnstaple parish church in 1820. In the same year his apprentice Joseph Webb ran away (EFP, 27 July 1820). His wife died, aged 46, on 1 July 1835 (EFP); and he retired *c*1850 with his two daughters to Marine Cottage, Instow. Mahogany 8-day longcase timepiece with circular white-painted dial; other longcase clocks reported, also watches (see Chapter 5).

MALLETT, JOHN (ii): Barnstaple. Son of the above; in 1848 returned from London after working in the watch trade, married, and succeeded his father in the business, establishing himself also as a watch manufacturer. William Horton Ellis was apprenticed to him in 1848, premium £150. Some time after 1856 the business at 4 High Street was succeeded by that of Short & Bailey. The Mallett family later became prominent as antique dealers, first at Bath and then in London.

MALLETT, STEPHEN: From Devon. Apprenticed in 1689 to John Trubshaw, London.

MANLEY, J: Watchmaker. Died at Topsham 16 July 1862 (EFP).

MANLEY, JOHN jun: Dartmouth. George III longcase clock.

MANLEY, WILLIAM: Topsham, Fore St, 1850 (D).

MANNER, JOHN: Exeter, *c*1780 (Baillie).

MANNING, G. & CO: Exeter. Office, 11 Colleton Buildings, The Friars, 1889 (D).

MANNING, JOHN: High Bickington, 1878 (D).

MANNING, JOHN: Barnstaple. 80 Boutport St, 1883–97 (D). Wall clock.

MANNS, JOHN: Barnstaple, 15 Joy St, 1902 (D).

MARCH, H: Devonport, 24 Marlborough St, 1856 (D).

MARGRIE, S: Topsham. Early nineteenth-century longcase clock with automata in the arch.

MARKES, CHARLES: Plymouth, 22 Whimple St, 1844–50 (D).

MARKES (MARKE, MARKS), THOMAS: He looked after the clock of Old Trinity Church, Exeter, in the 1540s and 1550s for 2s a year (churchwardens' accounts).

MARKEYS, JOHN: He was paid 12d for mending the clock at St John's, Exeter, *c*1539 (churchwarden's accounts).

MARKS, C. A: Buckfastleigh, Fore St, 1889 (D).

MARKS, CHARLES: Exeter, 31 North St, 1897 (D).

MARKS (MARKES), RICHARD: South Tawton. 'Pd Richd Marks new makeing the clock as agreed at a vestery £4' (South Tawton churchwarden's accounts, 1741). Several generations of the Marks family were blacksmiths at South Tawton.

MARKS, SIDNEY: Exeter, 9 High St, 1883 (D).

MARKS & ELSEY: Exeter, 9 High St and 108 Fore St, 1889 (D).

MARSH, E. & J: Axminster, Victoria Place, 1838 (D).

MARSH, HENRY: Apprenticed *c*1761 to John Tucker sen, Tiverton, watchmaker, premium £18.

MARSHALL, JOHN: Of Plymouth, apprenticed *c*1715 to John Richards, of ——llompton (Cullompton, presumably), clockmaker, premium £30.

MARSHALL, JOHN: Chulmleigh. Clockmaker; died, aged 72, on 24 May 1817 and was buried in Chulmleigh churchyard. His will, dated 22 Sept 1814, mentions his son Robert, daughter Betty, and leasehold property in East St, Chulmleigh; 'to my beloved wife Mary Marshall the house which I now reside in called Wackrells House'; effects sworn at under £2,000 (DRO). Eighteenth-century brass-dial 30-hour longcase clock (Loomes); another, white dial, with rocking ship in good plain oak case.

MARSHALL, ROBERT: Chulmleigh. Son of the above; described as clockmaker in his father's will made in 1814; listed also as draper and gunsmith in 1823 (D). Took out licence on 8 Oct 1825 to marry Mary Clarke of Chulmleigh; died 25 Dec 1844, leaving estate valued at under £3,000. Will in DRO.

MARTIN, EDWARD BRAKER: Totnes, 23 Fore St, 1883 (D).

MARTIN (MARTYN), JOHN: Cullompton. Watch and clockmaker; took out licence on 10 Aug 1775 to marry Jane Salter, otherwise Frost, of Cullompton, a minor. Was paid £2 5s in 1778 for repairing Cullompton Church clock and chimes, and £1 2s 6d for cleaning the clock in 1783 (churchwardens' accounts). Several brass-dial longcase clocks reported.

MASNE, THOMAS: Cullompton. Longcase clock, 'Thomas Masne Columpton', with painted dial decorated with flowers, reported by correspondent.

MASON, CHARLES: South Brent, 1883–93 (D).

MASTERS, ELIZABETH: Dartmouth. Eighteenth-century longcase clock (Bellchambers). A Betsy Masters married James Traies (qv) at Dartmouth, 21 Nov 1790.

MASTERS, JOHN: Dartmouth, *c*1810 (Guildhall Index, London).

MASTERS, PHILIP D. F: Axminster, Lyme St, 1878 (D).

MATHER, EDWARD: Exeter, 1761; and Leiden, Holland (Nanne Ottema: *Geschiedenis van de Uurwerkmakerskunst in Friesland*, p94).

MATTHEWS, EDWARD: Exeter, 46 Paul St, 1883 (D). Longcase clock, 'E. Matthews Exeter'.

MAUNDER, JOHN: Apprenticed as gunsmith to John Tickell, of Crediton, clockmaker and gunsmith, *c*1737, son of Sarah, widow, premium £3.

MAUNDER, JOHN (ii): Crediton. Listed at High St, 1844; East Town, 1850; and Charlotte St, 1856–83 (D). At one stage went bankrupt and his stock was advertised for sale (EFP, 7 Jan 1858).

MAUNDER, JOHN: Exeter. 100 Regent St, St Thomas, 1889–97 (D).

MAUNDER, MICHAEL: Plymouth. Union St, Stonehouse, 1873–83; 16 Octagon St, 1889 (D).

MAUNDER, WILLIAM: Tiverton. Watchmaker; recorded at the address of William Eames, watchmaker, Tiverton, in the 1861 Census, aged 22, born Crediton.

MAUNDER, WILLIAM DAIL: Tiverton. Watch and clockmaker and jeweller, Fore St, 1866; Gold St, 1873 (D).

MAY, STEPHEN: Bovey Tracey, Fore St, 1889–1902 and later (D).

MAYER, JOSHUA HEATH: Ottery St Mary, Silver St, 1893–7 (D).

MAYNARD, GEORGE: Exmouth, 54 Albion St, 1866; 28 Strand, 1873–83; Rolle St, 1889–97 (D).

MAYNARD, J: Teignmouth. Longcase clock *c*1800.

MAYNARD, WILLIAM: Exmouth. Silversmith, jeweller and watchmaker, Parade, 1844; Staples Buildings, 1850; Strand, 1856–66 (D).

MAYNARD, WILLIAM THOMAS: Exeter. Watchmaker and working jeweller, North St, 1833–40 (EPJ). Suspicious watch brought to his shop (EFP, 13 Feb 1834); wife Elizabeth died, aged 42, 11 Feb 1837 (*Exeter Gazette*).

MAYNE, THOMAS: Plymouth. Watch and clockmaker and silversmith, 111 Union St, Stonehouse, 1830–50 (D).

MAYNE, WILLIAM: Cullompton. Fore St, next to the White Hart Hotel, 1873–8 (D). The business listed under Mrs Jane Mayne, 1883–9 (D).

MEDLAND, R: Bow, 1856 (D).

MENDS, BENJAMIN: Plymouth. Mahogany arch-dial longcase clock of *c*1770–80 illustrated in *Antiquarian Horology*, Summer 1978, p865.

MENDS, JOHN: Plymouth. Watchmaker; took out licence 7 May 1767 to marry Mary Clements of Stoke Damerel.

MERRYFIELD, JOHN: Hatherleigh. Thirty-hour brass-dial longcase clock in oak and elm case.

METHERELL, JOSIAH: Plymouth, Russell St, 1873–1902 (D).

MICHELL, JOHN: Chardstock. 'An excellent maker of lantern clocks about 1700, and judging from the number of specimens still existing, he must have had a considerable connection. His frets were good and bore a distinctive character . . . Michell was succeeded by the family of Drayton, of which several generations successively carried on the business' (Britten). Lantern clock signed 'John Michell Chardstock fecit' (Tapley-Soper); a top quality walnut longcase clock is illustrated in Plate 1. For more details see *Dorset Clocks and Clockmakers* (Tribe & Whatmoor).

MICHELL, RICHARD: Highweek, Newton Abbot, 2 Miller View, 1893 (D).

MICHELL (MITCHELL), WALTER: Plymouth. Clock and watchmaker, Market Street. Jonathan Forrest was apprenticed to him *c*1744, premium £44. Took out licence 12 Apr 1746 to marry Ann Turtliff, of St Andrew's, Plymouth; and they were subsequently left a guinea in the will of Stephen Turtliff, of Moretonhampstead, proved 22 Dec 1749. Lantern clock at Buckland Abbey; many longcase clocks (Bellchambers).

MIDGLEY, RICHARD: Exeter, 1771. Longcase clock (Baillie).

MILES, ALBERT: Moretonhampstead, Ford St, 1883; Square, 1889–97 (D).

MILES, Mrs C: Tavistock, 37 Brook St, 1878 (D).

MILES, CHRISTIAN: Tavistock, 11 West St, 1883 (D).

MILES, JOHN: Okehampton, Fore St, 1883–9 (D).

MILES, JOHN: Tavistock. Watchmaker, Elbow Lane (also Exeter St), 1838; West St, 1850; Brook St, 1866; 28 West St, 1873 (D).

MILES, Mrs MARY JANE VIGARS: Tavistock, 11 West St, 1902 and later (D).

MILES, SAMUEL: Tavistock, 49 Brook St, 1873 (D).

MILES, THOMAS: Chagford. Watch and clockmaker, 1850–73 (D).

MILES, THOMAS: Moretonhampstead, Ford St, 1878 (D).

MILES, THOMAS: Bovey Tracey, Fore St, 1878; East St, 1883 (D).

MILES, THOMAS O: Tavistock. Succeeded John Miles at 28 West St; listed there in 1878; then at 11 West St, 1889–97 (D).

MILES, WILLIAM: Bovey Tracey, Town Cross, 1873 (D).

MILL, JOHN O: Holsworthy. Listed in 1873; at Higher Square, 1878; and at Stanhope Square, 1893–1902 (D).

MILL, WILLIAM JEWELL: Bideford. Watch and clockmaker, The Quay, 1830; High St, 1838–56 (D).

MILLER, JACOB: Exeter, 113 Fore St, 1866 (D).

MILLER, RICHARD: Lifton, 1866–89 (D).

MILLER, RICHARD: Exeter. Recorded in St Thomas in 1851 Census, clock and watchmaker, aged 17, born Cadbury; Cowick St, 1854–5 (EPJ).

MILLMAN, ARTHUR JOHN GAY: Plymouth. Edgcumbe St, Stonehouse, 1893–1902 (D).

MILLS, JOHN: Cullompton. Clock and watchmaker, 1838 (D).

MILLS (MILL), JOHN: Ottery St Mary, Cornhill, 1844 (D). Tendered unsuccessfully in 1846 to repair Thorverton Church clock and chimes.

MILLS, SAMUEL G: South Molton, Barnstaple St, 1850 (D).

MILLWARD, WILLIAM: Chudleigh. Name noted on 254mm (10in) dial plate from early eighteenth-century longcase clock.

MINCHINTON (MINCHINGTON), WILLIAM: Brixham, 62 Fore St, 1883; 1 Fore St, 1889–97 (D).

MINIFEE (MINIFEY), JOHN: Tiverton. Blacksmith; paid in 1614–15 for making a plate for St Peter's Church clock, Tiverton. Died *c*1647, leaving a widow Joan (Moger).

MITCHELL *see* MICHELL

MITCHELL, BARNET: Plymouth. Wholesale watchmaker, Ebrington St, 1836 (D).

MITCHELL, HEZEKIAH: He was paid for work on Rewe Church clock in 1762 and 1765 (churchwardens' accounts).

MITCHELL, JAMES: Plymouth, Exeter St, 1844 (D).

MITCHELL, RICHARD JOHN: Plymouth. Watchmaker and jeweller, 1 Saltash St, 1844; 3 Treville St, 1850–66 (D).

MITCHELL, THOMAS: Plymouth. Watchmaker, Cambridge St, 1836 (D).

MOGRIDGE, HENRY: Crediton, Charlotte St, 1878–89; Blagdon Place, 1893 (D).

MOLE, WILLIAM: Exeter, West St, 1857 (D).

MONK, RICHARD R: Plymouth, 26 Frankfort St, 1878 (D).

MOORE, HENRY: Sidmouth, Church St, 1883; Old Fore St, 1889–1902 and later (D).

MORCOMBE, JOHN: Barnstaple. Clockmaker; made the Hartland town clock 1622–3. His name as repairer is mentioned in the Hartland town accounts between 1614 and 1658. 'Itm to Morcombe the Clockmaker for mendinge the clock 12s' (Braunton churchwarden's accounts, 1643). It is possible the above references concern more than one maker of the same name.

MORGAN, JOHN: Exeter, 83 South St, 1893 and later (D).

MORGAN, ROBERT WILLIAM: Exeter, Beaufort Place, St Thomas, 1873; & Son, 'practical watchmakers, gold & silversmiths, opposite St John's Bow, 151 Fore St' (*Western Times*, 19 Dec 1876); 83 South St, 1883–9 (D).

MORLE, JOHN: Exeter, 166 Sidwell St, 1889–93 (D).

MORRISON, JAMES: Barnstaple, 4 High St, 1873–89 (D). The business listed as Morrison & Chapman at the same address, 1893–1902 and later.

MORSHEAD, ROBERT: Plymouth. Clockmaker; sought licence 20 Apr 1691 to marry Priscilla Hardey of Plymouth.

MORTIMER, J: Crediton, 1823 (D).

MORTIMER, WILLIAM: Plymouth. Mahogany longcase clock, detached columns to arched hood, painted dial with rocking ship in arch, 2.2m (7ft 3in), sold at Christie's for £580 (*Antiquarian Horology*, Summer 1977, p350).

MORTIMORE, HARRY: Exmouth, 3 Albion St, 1893–1902 and later (D).

MORTIMORE, ROBERT: Dartmouth, New Quay, 1823–30 (D). 'Died – At Dartmouth, on the 13th inst., after a long illness, Mr Robert Mortimore, watchmaker, of that town' (EFP, 21 Apr 1831). White-dial longcase clock.

MORTIMORE, ROBERT M: Dartmouth, Duke St, 1838–66 (D).

MOSER, B: Devonport, 10 Tavistock St, 1856 (D).

MOSER (MOSSER) & WEHRLE: Devonport. German clocks; 13 High St, 1838; 1 King St, 1844; 10 Tavistock St, 1850 (D).

MOSS, JOHN: Exeter. Watchmaker, North St, 1819 (EPJ).

MOXEY, MILFORD KEMP: Exeter, 5 Eastgate Arcade, 1883; & Co, 1897–1902 (D).

MUDFORD, FREDERICK: Tiverton. Recorded in 1871 Census, aged 25, clock and watchmaker, born Tiverton, unmarried, son of William Mudford, basketmaker.

MUDGE, THOMAS (1715–94): London and Plymouth.*

MUNFORD, EDWARD HODDER: Tiverton. Watchmaker, gold and

silversmith, Angel Hill, 1878; Fore St, 1889–1906 (D). Was succeeded by T. Leonard Crow.

MURCH, JEROM: Honiton. Clock and watchmaker; John Skinner was apprenticed to him *c*1763, premium £6 16s 6d. He or a man of the same name was buried at Honiton on 6 Sept 1802 (parish register). Longcase clocks, including 30-hour brass-dial example (Loomes).

MURCH, JOHN: Exeter. Watchmaker; son of William Murch; admitted as Exeter freeman by succession, 22 Aug 1698.

MURCH, JOHN: Apprenticed *c*1754 to Francis Pile, Honiton, premium £15. Was a witness to Pile's will, dated 22 June 1763 (DRO). The *Universal British Directory* lists a John Murch as a silversmith at Honiton in 1798.

MURCH, JOHN: Honiton. 'Wanted, an assistant in the clock and watch line, or watch line only. Apply to Mr John Murch, clock and watchmaker, Honiton' (EFP, 3 Jan 1805). A disused church clock at Dalwood, east Devon, is inscribed 'Murch Honiton'. Mahogany longcase clock, 'Jno. Murch Honiton'.

MURCH, JOHN AND MATTHEW: Honiton. Watch and clockmakers, High St, 1823–30 (D).

MURCH, MATTHEW: Honiton. Watch and clockmaker, High St. Listed in partnership with John Murch, 1823–30; then alone, 1838–56; also described as jeweller and silversmith, stamp distributor, and agent to the Sun Life and Fire Office (D). He supplied church clocks for Uplyme (1846), Sowton (1846), and St Paul's, Honiton (1851). An old clock formerly at Broadhembury Church has a brass plate 'Repair'd by Mattw. Murch Honiton 1853'. A report in the *Exeter Flying Post*, 24 Aug 1848, reads: 'Honiton. The Rival clockmakers: Mr Murch, of this town, has erected, during the past week, a splendid dial in front of his house, showing railway time, and Mr Tovy has also one in front of his house with mean time attached. As competition always tends to do good – in no instance has it been seen to greater advantage – we congratulate the clockmakers upon the alteration, and no doubt a regular time will now be kept by the inhabitants in general'. He married Sarah Trump at Honiton 1 Jan 1821 (register).

MURRAY, ROBERT: Plymouth, 71 Old Town St, 1878 (D).

MUTTON, ROBERT: Exeter, Cowick St, 1857 (D).

MUTTON, SAMUEL: Plymouth. Discharged as a debtor; described as late of Edgcumbe St, East Stonehouse, watchmaker and jeweller (EFP, 17 July 1828). See next entry; probably the same man.

MUTTON, SAMUEL jun: Plymouth Dock. Watchmaker, jeweller, goldsmith, etc, Catherine St, 1814 (D). Samuel Mutton sen was landlord at that time of the Star and Garter, Fore St.

MYERS, T: Plymouth, 40 Greenbank Av, 1902 (D).

NATHAN, ——: Plymouth. Watchmaker, etc, Market St, 1814 (D).

NEAME, WILLIAM JOHN: Plymouth. Edgcumbe St, East Stonehouse, 1838–44 (D).

NEEDS, THOMAS: Exeter. Clockmaker, Sidwell St, 1837–8 (EPJ).

NETHERWAY, ALBERT: Torquay. 2 Goschen Terrace, Cockington, 1893–1902 (D).

NEUGARD, JOHN: Plymouth, 5 Sussex Place, Union Rd, 1856; 104 Union St, 1878–83 (D).

NEWCOMBE, HENRY: Exeter, High St, 1890 (D).

NEWCOMEN, THOMAS (1663–1729): Dartmouth. The steam-engine pioneer. He is said to have repaired clocks, including the one at Dartmouth parish church.

NEWEY, JOSHUA: Plymouth, 51 Exeter St, 1889 (D).

NEWMAN, GEORGE: Crediton. Clockmaker and whitesmith. Brass-dial longcase clocks; a dial from one of them at Tiverton Museum. His apprentices included William Esworthy, whitesmith, *c*1749; James Bucknell, clockmaker, *c*1756; and James Northleigh, smith, *c*1758. Carried out repairs to Newton St Cyres Church clock in 1757 and 1758 (churchwardens' accounts). 1764–5: 'To Mr George Newman for repairing the church clock & new placeing it as per contract £2. To carriage of the clock to & from Crediton 2s' (North Tawton churchwarden's accounts). He was paid 5s on 18 Jan 1780 after submitting proposals for erecting a new set of chimes at Crediton (governors' accounts).

NEWMAN, GEORGE: Topsham. Watchmaker; married Elizabeth Melhuish Butler at the parish church 30 Apr 1775. Kept Topsham Church clock in repair from 1776 to 1785 (churchwardens' accounts).

NEWTON, GEORGE: Exeter, 55 & 56 Cowick St, 1890 (D).

NICHOLS, JAMES: Exeter; '22 Paris St, practical watch and clockmaker, manufacturing and working jeweller, engraver, electro-plater, and gilder' (advertisement, *Devon Evening Express*, 14 May 1878). Listed until 1902; also dealer in watch materials (D).

NICHOLLS, JOHN: Ridgway, Plympton. Watchmaker, 1830–73 (D).

NICHOLSON, BENJAMIN: Plymouth, 13 Bedford St, 1836–44 (D).

NICKS, JOHN: Exmouth. Watchmaker and music seller, Parade, 1823–50 (D). Several painted-dial longcase clocks reported, including 8-day example in mahogany case.

NIX, B. W: Exeter. '. . . (late Stone), goldsmith, silversmith, jeweller, watch and clockmaker, 36 High St' (advertisement in *Devon Evening Express*, 1878). In 1887 he advertised 'highly finished watches and clocks for testimonials'.

NORTHCOTE, JAMES, RA (1746–1831): Historical and portrait painter; son of Samuel Northcote, Plymouth, with whom he served a full seven-year apprenticeship as a clock and watchmaker.*

NORTHCOTE, SAMUEL: Plymouth. Clock and watchmaker and optician; died in his eighty-third year on 13 Nov 1791.*

NORTHCOTE, SAMUEL (ii): Plymouth. Clock and watchmaker; son of the above. Died 1813, aged 71.*

NORTHWOOD, ERNEST: Bideford, 69 Meddon St, 1897–1902 and later (D).

NOTTLE (NATTLE), GEORGE: Holsworthy. Clock and watchmaker; listed in the *Universal British Directory*, *c*1795.

NOTTLE, JOHN: Holsworthy. Gunsmith and clockmaker; John Oatway was apprenticed to him *c*1760, premium £10.

NOTTLE, JOHN: Okehampton. Clockmaker; took out licence 11 Dec 1766 to marry Honour Hawish of Okehampton. A John Nottle, of Okehampton, was also married at Stratton, Cornwall, 23 Nov 1773, to Martha Wallis. Buried at Okehampton 10 Feb 1793. Listed in the *Universal British Directory*, 1798 (out of date information?). North Tawton churchwardens' accounts have these items: 1773–4, 'To Mr John Nottle to repairs to ye clock £1 19s'; 1774–5, 'To Mr Nottle a new weight for the clock a new lifting piece and doing other things to the sd clock 10s 6d'.

OATWAY, ALFRED: Appledore, 1883–93 (D).

OATWAY, ALFRED: Bideford, High St, 1866–97 (D).

OATWAY, JOHN: Torrington. Apprenticed *c*1760 to John Nottle of Holsworthy, gunsmith and clockmaker, premium £10. 'Lost, between Buckland Town and Woodford Bridge, about a week since, a silver watch, maker's name, John Oatway, Torrington, No. 295. Whoever has found it, and will bring the same to Mr John Oatway, as above, shall receive one guinea reward. Torrington. Jan. 25, 1791' (EFP). His watch No 7336, hallmarked 1770, with painted polychrome scene in dial centre, was sold at Sotheby's in 1975 for £132. He agreed to provide an 8-day town clock for Hartland in 1797, but the order was not executed. Longcase clocks reported.

OLIVER, THOMAS: Plymouth. St Andrew's parish. Described as watchmaker and widower when he took out licence 18 Feb 1806 to marry Mary Cudlipp of Charles parish, Plymouth.

OLIVER, W: Plymouth. Watchmaker, Parade, 1814 (D).

ORGAN, WILLIAM: Ilfracombe. Watch and clockmaker and silversmith, 1830 (D). 'A daring robbery was committed at Ilfracombe one evening last week, at the shop of Mr Organ, watchmaker, when some villain had the temerity to thrust his hand through the glass, and steal four silver watches of considerable value' (EFP, 28 Feb 1828).

ORGAN, WILLIAM: Bideford, Mill St, 1838–56 (D). 'Mr Organ, a watchmaker, of Bideford, unfortunately had one of his legs fractured on Saturday se'nnight, when going from Barnstaple to the former place in Pridham's stage cart. Mr Martin, a surgeon, of Bideford, happened to be near and rendered such timely assistance that Mr Organ is doing well' (EFP, 4 June 1835). His wife died in 1854 (EFP).

OSBORN, W. G: Ashburton, West St, 1889–1902 and later (D).

OSBORN, WILLIAM: Kingsbridge. Watchmaker and jeweller, Fore St, 1850–66 (D). White-dial longcase clocks.

OSBORNE, HENRY: Devonport, 52 Cornwall St, 1873 (D).

OSBOURN, HENRY J: Devonport, 10 Albany St, 1889 (D).

OSBOURN, JOHN: Devonport, 10 Boot Lane, 1883 (D).

OSMOND, JOHN: Pilton, Barnstaple. Longcase clock, 'Osmond Pilton', brass dial, silvered chapter ring, in massively carved oak case. John Cook was apprenticed to him *c*1736.

OWEN, ALFRED WESTLAKE: Plymouth, 32 Tavistock Rd, 1897–1902 (D).

OWEN, PETER CHARLTON: Crediton. Watch and clockmaker, opposite the Angel Inn. Served his apprenticeship with Thomas Upjohn in Exeter, but was living in Crediton at the time of his marriage to Mary Woodrow at St Stephen's Church, Exeter, 7 Jan 1782 (register). Advertised for 'a journeyman clockmaker, who can produce a good character from his late master; if he understands the watch branch the more agreeable . . . also wanted an apprentice' (EFP, 29 Jan and 5 Feb 1784). Described as 'the late', Oct 1787. Longcase clock reported.

OYENS, PETER jun: Plymouth Dock. Watchmaker; married Elizabeth Coe, widow, at Stoke Damerel parish church, 10 Nov 1784 (register). F. J. Britten mentions a token issued by Oyens, No 60, Fore St, Dock. 'Bankrupts . . . Peter Oyens, of Plymouth Dock, Devon, watchmaker, to surrender Feb 15, & March 11, at ten, at Guildhall, London' (EFP, 2 Feb 1797). Longcase clocks; one in mahogany case has figure of Father Time in arch with skeleton following behind.

PACE, HENRY: Exeter. 'To watchmakers, manufacturers, and the public in general. Henry Pace, watch manufacturer (removing to London), offers his well made assortment of warranted gold and silver watches, at very reduced prices, at 274 High St, Exeter, from Monday the 17th instant, till Saturday the 29th, inclusive. An assortment of gentlemen's gold guards, London-made seals and keys, equally low' (EFP, 27 Feb 1840).

PACK, ISAAC AND JACOB: Tiverton. The watch and clockmaking business established by the brothers in Leat St in 1875 was transferred to Bridge St *c*1910. Tiverton Museum has Victorian 'boneshaker' that was ridden by Jacob Pack on his rounds, together with a photograph showing him standing beside it (illustration in C. N. Ponsford *et al*, *Clocks & Clockmakers of Tiverton*).

PADDON, GEORGE: Kingsbridge. Apprenticed *c*1735 to William Stumbels, of Totnes, clock and watchmaker, premium £10. He later set up in business at Kingsbridge, John Randle being apprenticed to him *c*1752 for a fee of £10 10s. A bracket clock, 'George Paddon Kingsbridge', illustrated in J. K. Bellchambers's *Devonshire Clockmakers*. It has an alarm, rise and fall pendulum regulation, and can be made to repeat the hours and quarters on two bells. Longcase clocks. Buried at Kingsbridge in 1771.

PADDON, GEORGE (*c*1741–1822): Modbury. Was paid annually for looking after Modbury Church clock from 1767 to 1783 and occasionally thereafter (churchwardens' accounts). Listed as clockmaker in the *Universal British Directory*, 1793. Died, aged 81, and was buried at Modbury 4 Mar 1822 (parish register). Thirty-hour and other longcase clocks reported.

PADDON, WILLIAM: Modbury. Watch and clockmaker, Brownston St, 1823; Church St, 1850 (D). Payments to him are recorded in Modbury churchwardens' accounts for care of the parish clock from about 1830. Died 6 June 1863, aged 71 (EFP).

PAGE, KEEN & PAGE: Plymouth. Goldsmiths, jewellers, watch and clockmakers, George St, 1866–1902 and later (D). James Andrew Page listed earlier as silversmith and jeweller.

PALMER, CHARLES: Exmouth, 31 Rolle St, 1902 and later (D).

PALMER, THOMAS: Exeter. Recorded in 1851 Census, watchmaker and finisher, aged 40, born Kennington, Surrey; together with wife Rebecca, 34, and seven children.

PARE, ——: Carried out work on the Exeter Cathedral clock in 1679.

PARKER, HARRY: Torquay, 3 Braddon Terrace, 1883–9 (D).

PARKER, HENRY M: Torquay, 2 Lower Union St, 1878 (D).

PARKER, HENRY MAY: Tiverton, 2 Church St, 1883 (D).

PARKER, HENRY MAY: Paignton, Winner St, 1889–93; 15 Victoria St, 1902 and later (D).

PARKER, PHINEAS (PHINEHAS): Barnstaple. Clockmaker; buried at the parish church on 10 Apr 1695 (register); mentioned in property deeds commencing 26 Mar 1679 (DRO). His daughters were called Pascha, Elizabeth and Mary.

PARKHOUSE, WILLIAM: Appledore, Bude St, 1878 (D).

PARKHOUSE, WILLIAM H: Bideford, 25 Mill St, 1889 (D).

PARKIN, EDWARD A: Exeter, 128 Sidwell St, 1897–1902 (D).

PARNELL, JOHN: Tavistock, 78 West St, 1893–1902 (D).

PARRISH, JOHN: Exeter, Paris St, 1873–97; & Son, 9 Eaton Place, Heavitree Rd, 1902 (D). His apprentices included Frank Leyman and George Conibear. The business was being continued in 1939 at 21 Heavitree Road by Arthur Symes Parrish. The clock from the old city workhouse can still be seen affixed to the premises.

PARROTT, E: Cullompton. Watchmaker; report of theft of several watches from his shop (EFP, 14 Sept 1826).

PARSONS, WILLIAM: Plymouth. Watch and clockmaker, 47 Frankfort St, 1830 (D).

PARTRIDGE, JAMES T: Kingsbridge, Fore St, 1866–73 (D).

PASMORE, S: Plymouth, 43 Union St, 1850 (D).

PASSMORE, HERCULES: Exeter, Butcher Row, 1830 (D).

PASSMORE, JOHN: Morchard Bishop, 1883 (D).

PASSMORE, RICHARD: Morchard Bishop. Longcase clocks, one with 30-hour, post-framed movement, 'Rd Passmore Morchard Bishop'.

PASSMORE, RICHARD: Barnstaple. Watch and clockmaker and working jeweller, High St, 1823–44 (D). Died 23 Dec 1844, leaving a widow, Susannah, and a brother, Thomas. Will in DRO. Numerous longcase clocks.

PASSMORE, RICHARD: Torrington, High St, 1844 (D).

PASSMORE, RICHARD: Sidmouth, High St, 1878–97; & Son, 1902 and later (D).

PASSMORE, WILLIAM: Probably north Devon. Longcase clocks, including 30-hour brass-dial example signed 'Wm Passmore 1780'.

PASSMORE, WILLIAM HENRY: Chulmleigh, Fore St, 1873–1902 and later (D). He supplied Chulmleigh Church clock, 1891.

PATCH, ALFRED: Budleigh Salterton. Watchmaker, jeweller and bible repository, High St, 1866; 4 South Promenade, 1883–97 (D).

PATCH, ROBERT: Budleigh Salterton, 1850 (D).

PATTERSON, THOMAS: Plymouth. Watch and clockmaker and jeweller, Old Town St, 1866–73 (D).

PAUL, THOMAS: Sidmouth, Prospect Cottage, 1889–93 (D).

PAYNE, J: Tiverton. Possibly John Ellis Payne. Watchmaker and repairer, 3 Angel Terrace. Advertisement in *Tiverton & District Trades Directory*, 1894–5: 'Note the big clock in window . . . People in country waited on and work done on their premises'.

PAYNE, JOHN ELLIS: Tiverton, Leat St, 1893; West Exe North, 1897 (D).

PEARCE *see also* PEARSE

PEARCE, CHRISTOPHER: Torquay. Fore St, St Marychurch. Listed in 1878 and 1902 (D).

PEARCE, H. J: Exeter, 31 New Bridge St, 1897 (D).

PEARCE, JAMES: Apprenticed *c*1762 to Mary Elliott, Plymouth, watch-maker, premium £20.

PEARCE, JOSEPH WILLIAM: Plymouth, Cecil St, 1889; 26 King St, 1902 and later (D).

PEARCE (PEARSE), SAMUEL: Honiton. Clock and watchmaker; listed in the *Universal British Directory*, 1798. Was apprenticed *c*1750 to Francis Pile,

Honiton, premium £15. The *Exeter Flying Post* has two references to watches with their owners' names on the dials: 'Lost at Tor-Key . . . a silver watch, with the name Robert Troke on the dial-plate, maker's name Samuel Pearce, Honiton' (notice dated 4 Sept 1781); and, among watches stolen from the Sidmouth shop of John Blackmore, clock and watchmaker, 'a silver watch with John Woolcott marked on the face, and Samuel Pearce, Honiton, on a cap which covers the work' (23 Apr 1795). A Samuel Pearse died aged 80 and was buried at Honiton on 28 Jan 1814 (register).

PEARCE, WILLIAM: Plymouth. Apprenticed *c*1750 to Henry Raworth, Plymouth, clockmaker, premium £10. Described as watchmaker, of Charles, Plymouth, when he took out licence 10 July 1761 to marry Lydia Summers of St Mary Major, Exeter. Listed as wholesale dealer in plate and watches, Market St, in Bailey's *Western Directory*, 1783; also listed in the *Universal British Directory*, 1798.

PEARCE, WILLIAM: Plymouth. Watchmaker, How's Lane, 1822 (D).

PEARCE, WILLIAM PHILLIPS: Devonport, York Place, 1897 (D).

PEARSE *see also* PEARCE

PEARSE & BOUNSALL: Tavistock, Barley Market St, 1844 (D).

PEARSE, FREDERICK: Devonport. William St, Morice Town, 1883–93 (D).

PEARSE, JAMES: Ashburton. Thirty-hour brass-dial longcase clock *c*1775.

PEARSE, JOHN: Tavistock. Ironmonger and clock and watchmaker who patented, 27 July 1822, improvements in spring jacks and their connection with roasting apparatus (No 4693); in two subsequent patents for wheeled carriages he is described as ironmonger only (DNQ, vol 9). Listed at Barley Market St, 1823–30 (D).

PEARSE, JOHN: Bideford, Grenville St, 1838 (D).

PEARSE, SAMUEL: Axminster. 'Wanted an apprentice to a watch and clockmaker . . . Apply, letters post paid, to S. Pearse, Axminster. A premium will be expected' (EFP, 13 Feb 1806). Listed at Lyme St, 1830 (D). His wife died 6 Apr 1837 (EFP).

PEARSE, WILLIAM: Axminster. Listed at Castle Hill in 1844 (D). Described in 1847 as watchmaker, auctioneer and general dealer (DRO).

PEARSE, WILLIAM: Tavistock, Barley Market St, 1850 (D).

PENBERTHY, FREDERICK: Devonport, Marlborough St, 1893–1902 and later (D).

PENGELLY, JOHN: Barnstaple. Listed in High St, at No 6 in 1844; at No 35 in 1850; and at No 7 in 1866 (D). Verge watch movement, 'Jno Pengelly Barnstaple 8929', at Exeter Museum.

PENGELLY, Mrs M. A: Barnstaple. Watchmaker and silversmith, 7 High St, 1873–8 (D).

PENNY, EDWARD: Buckfastleigh, Fore St, 1893–1902 and later (D).

PENNY, P. L: Plymouth Dock. 'Friday died, at Plymouth Dock, Mr P. L. Penny, of that place, formerly clock and watchmaker' (EFP, 4 Jan 1816).

PEPPERELL, ALBERT: Plymouth, 80 Treville St, 1893–1902 and later (D).

PERCY, WILLIAM: Tavistock. Watchmaker, Higher Brook St, 1838 (D).

PEREAM, GEORGE: Axminster. Clockmaker and gunsmith; married Hester Peream of Ottery St Mary, at Exeter Cathedral on 6 Jan 1742/3 (register).

PEREAM, JOHN: Ottery St Mary. Clockmaker and gunsmith. Will dated 19 Nov 1754, proved 21 May 1756.*

PERRIEN, JOHN HENRY: Okehampton. Watchmaker, Fore St, 1866–73 (D).

PERRY, WILLIAM: Hockworthy. Watch and clockmaker and beer retailer, 1856–66 (D).

PERRYMAN, JOHN: Ashburton. He had the care of the church clock in 1822–3 (Ashburton churchwarden's accounts). Listed at North St, 1830; Lad St, 1844; and Back Lane, 1850 (D).

PERRYMAN, JOHN (1813–71): Barnstaple. Watch and clockmaker and silversmith from *c*1838 to Oct 1852 when he disposed of his business at 18 High St to John Mallett. He then went to Australia, taking his shop regulator with him (now in Western Australia Museum, Perth), and established a business in King William St, Adelaide (Plate 43).

PERYAM, CLEMENT: 'To Clement Peryam for worke donn about ye clock' (East Budleigh churchwarden's accounts, 1673).

PERYMAN (PERIMAN), RICHARD: Repaired Sidmouth Church clock in 1714 and 1732 (churchwarden's accounts); probably a blacksmith.

PETERS, WILLIAM JAMES: Torquay, 14 Florence Terrace, 1878; Lower Union St, 1883–93; Union St, 1897–1902 and later.

PETHERICK, J: Plymouth. 93½ Union St, Stonehouse, 1856 (D).

PFAFF & BROWN: Exeter, Market St, 1870–2 (EFP). The partners were probably John Pfaff and Anthony Bern, also German born.

PFAFF, JOSEPH & JOHN: Exeter. Clockmakers, Market St. Recorded in the 1851 Census as Joseph, aged 33, and John, 28, brothers, both born in Germany and unmarried. Listed in partnership until *c*1866; then John Pfaff listed alone, 1868 (EPJ).

PHILIPS, WILLIAM: 1768: 'Paid William Philips for taking an estimation of the clock and chimes 1s' (Colyton churchwarden's accounts).

PHILLIPS, ——: Plymouth. Advertisement for lost watch, described as 'old fashioned' (*Public Advertiser*, 17 Nov 1775 – Buckley).

PHILLIPS, RICHARD: Was paid £12 12s in 1808 for repairing the clock at St Mary Steps Church, Exeter (churchwarden's accounts).

PHILP, RICHARD: Okehampton. 'Wanted, an apprentice to the clock and watch business, a lad about fourteen. Apply by letter, post-paid, to Richard Philp, clock and watchmaker, Okehampton' (EFP, 2 Dec 1819). Listed at Fore St, 1823–56 (D). Later bankrupt; watches, jewellery, and stock in trade offered for sale (EFP, 20 Jan 1859).

PICKFORD, W: Newton Abbot. White-dial longcase clock *c*1800 (Loomes).

PIKE, JAMES: Newton Abbot. Clock and watchmaker 'of Wolborough'. Took out licence 23 Jan 1756 to marry Ann Tickle, of same, spinster; died aged 70 and was buried in the parish 7 Apr 1798. At Chudleigh, 15 Nov 1756, agreed with the churchwardens 'to rectifie & enlarge all the wheels and other work . . . And sufficiently repair the newest clock now in ye tower of the parish church of Chudleigh. And likewise to put into good and sufficient repair the chimes thereunto belonging'. Entered into another agreement at Chudleigh in 1766, and in 1773 made the present tower clock at Powderham Castle. John Maisters was apprenticed to him *c*1756 for £17 premium, and John Brooking, *c*1759, for £31 10s. Kenton overseers' accounts for 1777 have entry: 'Mr Pike having neglected the parish clock, on that account he is not to be paid any money'. A silver pair-case verge watch hallmarked 1781 sold at auction in Torquay in 1979; numerous brass-dial longcase clocks, some with picturesque engraved scenes.

PIKE, JAMES jun: Newton Abbot. Clock and watchmaker, Wolborough. Took out a licence 13 Aug 1792 to marry Charlotte Vaughan, a minor; but in 1803 was in financial and other difficulties:
'The humble petition of James Pike, of Newton Abbot, Devon. The above James Pike, married in the year 1792, Miss Vaughan, daughter of the late

General Vaughan, by whom he now has a family of four infant children, all of whom, with the mother, he has by his own industry, and the late General's bounty, supported with decency; but on the General's death, the bounty flowing from that quarter entirely ceased. Since which he has met with a severe misfortune in breaking his arm; his family has been afflicted with sickness; and being himself incapable of working to his business, he is reduced to such distress, as to be not only unable to maintain his wife and children as formerly, but is even deprived of the means of procuring for them and himself the necessaries of life. He therefore, in behalf of his wife and her infant family, humbly solicits the humane assistance of those who can feel for their distressed situation. The object of this attempt is to raise a small sum of money from the friends of humanity, in order to put him in business, instead of the family's becoming a burden on his parish; and the humble petitioners will ever be bound to pray for those who can and will, in the smallest degree, relieve them. References can be given to persons of the utmost respectability for the truth of the above statement, and of the unhappy situation to which the family is reduced. The smallest donations will be received by Messrs Trewman & Son, printers of this paper. Already received: Miss Bampfylde 10s 6d, Mrs Granger 10s 6d, Mrs C. Hoare 10s 6d, Miss Pennyman 10s 6d, Mrs Russel 7s' (EFP, 5 May 1803).

PIKE, JOHN: North Tawton, 1856–78 (D).

PIKE, WILLIAM: Totnes. 'Received of Mr Geo. Taylor Excusing Taking a parish Apprentice for the House Mr Pike Clockmaker lives in £2' (1761 Poor Rate accounts). Buried at Totnes 29 Sept 1793 (register). Good quality bracket clock with mahogany bell-top case and silvered dial signed in the arch 'William Pike TOTNES'. Longcase clocks; Totnes Museum has 8-day example with shipping scene on dial, mahogany case; also a brass dial with raised chapter ring 'Willm Pike Totnes'. White-dial clock with name Wilson on back of dial; also 'small Parliamentary clock'.

PIKE, WILLIAM: Barnstaple. Clock and watchmaker; listed in the *Universal British Directory*, 1793. Longcase clocks suggest his working period extended back to the mid-eighteenth century; they include examples in lacquer cases with bull's eyes in the trunk doors. An engraved brass dial from one of Pike's longcase clocks, at Barnstaple Museum, features a church with steeple, a house, and a man holding flatfish, with a pack pony, and shouting 'Soles & Place'.

PILE (PYLE), FRANCIS jun: Honiton. Clock and watchmaker; active as a church clock repairer in east Devon 1731–63, working at various times on the clocks at Awliscombe, East Budleigh, Gittisham, Otterton and Sidmouth. His apprentices included Samuel Pearce for a £15 premium received in 1750, John Murch (£15 in 1754) and John Chubb (£20 in 1759). Following the death of his first wife, he married Anne Pinney at Honiton 11 Aug 1755, and was buried there 3 Aug 1763. Will dated 22 June 1763 in DRO; witnessed by Jno. Murch, Jno. Chubb and William Seaman; mentions Anne 'my present wife'; John, 'my eldest son'; 'my daughter Elizabeth'; three or more houses in or near New St to be let; and five little tenements or houses by the Prebyterian Meeting House. He directed his executors to sell his clocks, watches, materials and working tools, 'being first appraised by William Upjohn of ye city of Exon, watchmaker, or some other proper person'. He made the gallery clock at the United Reformed Church, Lyme Regis; also longcase clocks, including a good quality 8-day one in walnut case (Bellchambers MS) and 30-hour examples with a single hand. Several watches reported: No. 3837 with silver champlevé dial and maker's name 'Pile Honiton' is hallmarked 1762. 'Lost. On Sunday last, between Ide and Dunsford, in Devon, a silver watch, maker's name Francis Pyle, of Honiton, marked on the inner case RS 1752, in the dial plate, Pyle, Honiton. Whoever hath found the same, and will make it known to the printer of this paper, so as the watch may be had again, shall receive half-a-guinea reward. Exeter May 10, 1775' (EFP).

PILLAR, HENRY ALEX: Budleigh Salterton, 14 High St, 1902 (D).

PILSON, ABRAHAM: Plymouth. Watch *c*1700 (Britten).

PIM, JOHN RATCLIFFE: Described as watchmaker, of Bideford, when he married Elizabeth White, daughter of a Barnstaple maltster, at Kentisbury on 13 July 1835. Died at Barnstaple, aged 33, Sept 1847 (EFP).

PINHEY, RICHARD LEAR: Plymouth Dock. Was a witness when Mary Pinhey married the clockmaker William Lancaster at Stoke Damerel parish church, 20 Nov 1771 (register); was also a witness to Lancaster's will, dated 23 Feb 1815 (DRO). He and several others registered a meeting house at 21 Chapel St, Plymouth Dock, for Independents, 27 Mar 1813. Several good quality 8-day longcase clocks noted; one sold at auction in 1977 has a mahogany case and an engraved arched brass dial depicting cherubs, representing the weather, and showing chicken in a country landscape, above which there is a rocking movement and an inscription, 'Beware of the Hawk'.

PINKSTONE, Mrs CHARLOTTE MARIA: Tiverton. Watchmaker and

Plates 44 & 45 Two elaborately cased clocks. The one on the left is by George Boutcher, of Broadclyst, *c*1740. It can be made to play one of six tunes every three hours and has a walnut case. The other has a green lacquer case with gilt brackets under the hood. The maker was William Stumbels, of Totnes, *c*1735

Plate 46 John Tickell, of Crediton. Large dial clock made for the old Unitarian chapel at Crediton in 1727. Eight-day movement with one-second pendulum, sounding a single stroke at the hour. Black and gilt Japanned case; the dial 1.3m (4ft 4in) across, the trunk over 1.5m (5ft) long

Plate 47 Thomas Jessop, Tavistock. Hood and dial of eight-day longcase clock with moonwork, signed around the top of the arch. In 1764 Jessop agreed to keep Tavistock Church clock and chimes in order for £2 3s, 'only abuse excepted'

grocer. Bampton St, 1878–89 (D). Widow of Thomas D. Pinkstone, whose business she continued.

PINKSTONE, GEORGE: Tiverton, Bampton St. Recorded in 1861 Census, aged 49, watchmaker and innkeeper, born Tiverton. Listed in directories from 1838. Longcase clock reported.

PINKSTONE, RICHARD: Son of Thomas D. Pinkstone, of Tiverton, he emigrated to Philadelphia in 1875 and established a clock and watchmaking business which celebrated its centenary in 1975 when his son (also named Richard) retired. In his last year the son was held up three times by gunmen.

PINKSTONE, THOMAS D: Tiverton, Bampton St, 1850–66 (D). Recorded in the 1861 Census, aged 37, watchmaker and grocer, born Tiverton; together with wife, Charlotte M., 33, and sons Richard, 10, Thomas W., 8, and Samuel, 4.

PINKSTONE, WILLIAM R: Totnes, 23 Fore St, 1889; 32 High St, 1893–1902 (D).

PINN, RICHARD: Salcombe Regis. Took out licence 8 Feb 1763 to marry Anne West of Sidbury. This suggests a possible connection with the Sidbury clockmaker James West. 'To Mr Pinn for cleaning the clock 6s . . . to a journey to Salcomb about the clock 1s 6d' (East Budleigh churchwarden's accounts, 1761); he received a further payment in 1766. Brass-dialled longcase clocks signed with place-name 'Salcombe'.

PINN, RICHARD: Sidmouth. Lost watch reported (*Public Advertiser*, 3 Nov 1786 – Buckley).

PINN, RICHARD: Exmouth. 'On Tuesday last was married at Exmouth, Mr Richard Pinn, watch and clockmaker, of that place, to Miss Jane Bayley, daughter of Captain John Bayley, of Chester' (EFP, 28 Sept 1797). Shop broken open and robbed of fourteen gold and silver watches (*Exeter Gazette,* 28 Mar 1811. *Memorials of Exmouth*: 1827 'Paid Mr Pinn one year's salary for keeping in repair and winding up Chapel clock £6'. Listed at Fore St in 1830 (D). Died 20 July 1834, having made a will nine days earlier, appointing his wife as sole executrix and mentioning his children Henry, Mary and Eliza. Estate valued at under £100 (DRO). Longcase clocks include one with superbly engraved silvered arch dial in fine inlaid mahogany case.

PIPER, EDWARD: Exeter, 201 High St, 1836–7 (EPJ). Was apprenticed to Henry Ellis *c*1815 and remained with him some seventeen years before eventually being dismissed in 1832. He then foolishly opened a shop next door to his old master, 'but in a short period his effects were sold off for the benefit of his creditors'. After this he removed with his wife, Matilda, and family to London where he worked as a journeyman watchmaker (see Chapter 5).

PIPER, F. J: Ottery St Mary, Silver St, 1902 (D).

PLEACE, WILLIAM H: Plymouth, 43 Cecil St, 1873–8 (D).

PLUMLEIGH (PLUMLEY), THOMAS: Exeter. Watchmaker; married Anne Payne at St John's Church, 31 May 1697. His name noted on documents variously dated between 1707 and Apr 1727 (DRO). Insolvent (*London Gazette*, 8 June 1725 – Buckley). A man of this name was apprenticed in London on 4 July 1687 to Thomas Taylor for seven years (Clockmakers' Company register of apprentices).

PLUMLEY, WILLIAM: Exeter. Watchmaker; provided a £5 surety at Exeter Quarter Sessions, Oct 1702.

POLING (POULING), ——: Exeter. Smith; he made a new clock for St Edmund's Church, Exeter, *c*1731, for which he was paid £10; he received a further £1 11s 6d 'for new placing ye clock ironwork & frame' (churchwarden's accounts).

POLING, ABRAHAM: Exeter. Locksmith; apprentice of John Poling, admitted as Exeter freeman 6 May 1734. Carried out work on Exeter church clocks and was possibly the maker of the 1731 St Edmund's clock (see previous entry).

POLING, ANTONY: Carried out work on the Exeter Cathedral clock 1689–90.

POLING (PAULING, PAWLYN), JOHN: Exeter. Was paid 12s in 1670 for new making the clock at Old Trinity Church (churchwarden's accounts).

POLING (POLIN), JOHN: Exeter. Son of John Poling, smith; admitted as Exeter freeman by succession, 6 May 1734. Received a year's salary for looking after the St Edmund's Church clock in 1746 (churchwarden's accounts).

POLLARD, ——: Crediton. 'Paid Mr Pollard of Crediton in part towards the new town clock £26' (Sheepwash Church accounts). Ashreigney Church clock has brass plate: 'Pollard Crediton fecit 1813'.

POLLARD, JOHN: Crediton. Clockmaker. He was over 21 when he took out licence 13 Mar 1756 to marry Martha Commins of St Leonard's, Exeter. In 1787 he agreed to make for £30 a new set of chimes for Chudleigh parish church (churchwarden's accounts). 'On Tuesday died at Crediton at the advanced age of 85, Mrs Pollard, widow of the late Mr J. Pollard of that place' (*Exeter Gazette*, 24 Apr 1819). Numerous longcase clocks, including brass-dial 30-hour examples. Clockwork roasting jack with brass front engraved 'Pollard Crediton' in kitchen at Georgian House Museum, Bristol.

POLLARD, JOHN: Plymouth Dock. Clock and watchmaker; son of John Pollard, Crediton, baptised 28 Nov 1756. Described as 'of Stoke Damerel' when he married Mary Ewings at Exeter Cathedral on 31 July 1781 (witnesses Jonas Comins, Thomas Comins, John Pollard sen). Became a widower and married, secondly, Phillis Leah (or Leach) at Stoke Damerel parish church on 14 Sept 1797 (register). Listed in the *Universal British Directory*, 1798; and at Market St, 1812–14. He supplied Stoke Damerel Church clock, which has inscription, 'John Pollard Plymouth Dock 1811'. Longcase clocks.

POLLARD, JOHN: Crediton. Ironmonger and clockmaker; eldest son of Jonas Pollard; listed at High St, 1838 (D). Bankrupt 1848; died at Oxford 16 Aug 1854, aged 37 (EFP).

POLLARD, JONAS: Crediton. Clockmaker and ironmonger; son of John and Martha. Married Dorothy Burridge at Crediton 6 Apr 1806 and died after a long and severe illness on 30 Apr 1835, aged 62. In his will he makes provision for thirteen children; estate valued at under £1,500 (DRO). 'The inhabitants of the West Town of Crediton are much indebted to the public spirit of Mr Jonas Pollard, clockmaker and ironmonger of that place – who has, at his own expense, fixed up a neat clock with the dial against the front of his house, facing the market place, which is a great accommodation to the public' (EFP, 16 Dec 1824). Longcase and turret clocks.

POLLARD, T. & J: Exeter. They made, among others, the church clocks at Shillingford St George (1812–13), Talaton (1817), Dunsford (1830), and Broadclyst (*c*1830). The Talaton clock was originally at Haldon House; another country house clock by them is dated 1829. It would appear business was partnership between Thomas Pollard, Exeter, and Jonas Pollard, Crediton. The latter wrote letter mentioning the Dunsford clock and another made for Lord Exmouth when consulted in 1830 about erecting a clock at Tiverton pannier market; 'Mr Pollard works to low prices' commented Wasbrough Hale & Co of Bristol in letter to one of his rivals for the contract, H. Foster, watchmaker, Fore St, Tiverton (correspondence at Tiverton Museum). The order went to J. W. T. Tucker (qv).

POLLARD, THOMAS: Exeter. Watch and clockmaker; probably the son of the Crediton clockmaker John Pollard; baptised 22 June 1763. Listed at Alphington St, St Thomas, *c*1795–1844 (D). Died 6 Mar 1846. Will in DRO: estate valued at 'under £900'. He made a clock (since replaced) for Alphington parish church in 1797 and was extensively employed on church and other turret clocks. Brass-dialled and other longcase clocks noted (see also T. & J. Pollard).

POLLARD, THOMAS COMINS: Crediton. Watch and clockmaker, jeweller and printer, High St, 1856–73 (D). Son of Jonas and Dorothy; baptised at Crediton 31 Dec 1819. A Thomas C. Pollard was employed in E. J. Dent's turret-clock manufactory, Somerset Wharf, Strand, in 1845; the church clock at Combeinteignhead is inscribed: 'T. C. Pollard, Manufacturer, London, 1852'.

POPE, WILLIAM: Plymouth. Watchmaker, 3 King St East, 1866–73 (D).

PORTER, HENRY: Sidmouth, Fore St, 1850–73 (D).

POTTER, JOHN: South Molton. Watchmaker. Peter Allen was apprenticed to him *c*1758, premium £45.

POTTER, WILLIAM: Bideford, East-the-Water, 1897 (D).

PRATT, ALBERT J: Torquay, 39 Union St, 1897–1902 and later (D).

PRATT, WILLIAM & SON: Torquay, 13 Fleet St and St Marychurch, 1878 (D).

PREDDY, WILLIAM: Totnes, Fore St, 1850 (D).

PRESCOTT, JOHN: Beer, 1889 (D).

PRICKMAN, JOHN: Tiverton. Watch and clockmaker, Bampton St, 1830 (D).

PRIDHAM, LEWIS (alias GREENSLADE): Sandford. His first marriage was in 1687; died 1749.*

PRIDHAM, WILLIAM: Torquay. Watchmaker, 29 Torwood St, 1866–78 (D).

PROCTOR, WILLIAM: Devonport, 1873–1902 (D).

PROUT, JOSEPH: Newton Abbot, Highweek St, 1873–83 (D).

PUGSLEY, MARK: Morchard Bishop, 1883–1902 and later (D).

PULMAN, PHILIP: Axminster. Clock and watchmaker, jeweller, and silversmith, Lyme St, 1823–57 (D). Son of Philip and Mary Pulman; born 8 May 1791 and baptised at Colyton in July that year (parish register). Father a blacksmith and yeoman.

PULMAN, WILLIAM: Axminster. Watch and clockmaker, Lyme St, 1866 (D).

PURCHASE, JOSEPH: Exeter. Apprenticed 1700.*

PYE, JOHN ELLIOTT: Exeter. Watch and clockmaker, silversmith etc;

took over John Tucker's shop at 188 High St in July 1805 (EFP); married Harriet Branscomb ('late letter carrier') at St Stephen's Church 25 Oct 1810. 'Paid Pye for keeping & repairing the clock £6 10s 6d' (St Petrock's, Exeter, churchwarden's accounts, 1812–13). Longcase clock, 'J. E. Pye Exeter', seen at Devon hotel.

PYKE, H: Maker of Kingsteignton Church clock, 1811.

PYLE *see* PILE

PYNE & SONS: Exeter, 148 & 149 Fore St and 230 High St, 1873 (D).

QUICK, Mrs MARY JANE: Torrington, South St, 1878 (D).

QUINLAN, JOHN PATRICK: Plymouth, Bedford St, 1873 (D).

RAMSEY, EDWARD: Devonport, 54 Fore St, 1844–50 (D). An 8-day clock, 'E. Ramsey Devonport', dated 1850, is described and illustrated in Deryck Roberts's book *The Bracket Clock* (David & Charles, 1982).

RAMSEY, FREDERICK A: Ridgway, Plympton, 1873–1902 (D).

RAMSEY, JONATHAN: Devonport. 51 Fore St, Stoke, 1830 (D).

RANDLE, JOHN: Apprenticed *c*1752 to George Paddon of Kingsbridge, clockmaker, premium £10 10s.

RAW, HENRY: Ilfracombe, 43 Fore St, 1893–1902 (D).

RAWLE, FREDERICK: Torquay, 5 St Aubyn's Terrace, Higher Bronshill Rd, 1897–1902; at 121 Windsor Rd later (D).

RAWLING, ALFRED: Devonport. 15 College Rd, Ford, Stoke, 1902 (D).

RAWLINSON & DEACON: Plymouth. 25 Edgcumbe St, East Stonehouse, 1897 (D).

RAWLINSON, JOHN: Plymouth. 25 Edgcumbe St, East Stonehouse, 1902 (D).

RAWORTH, CHARLES: Plymouth. Some lists of clockmakers include this name, but it is probably a mistake for Raworth, of Charles parish (see next entry).

RAWORTH, HENRY: Plymouth. Clockmaker, Charles parish. William Pearce was apprenticed to him *c*1750, premium £10. Took out licence 10 Sept 1761 to marry Elizabeth Hutchings, widow. The St Andrew's, Plymouth, churchwardens' accounts record several payments made to him between 1780–1 and 1784–5 for repairing the church clock and chimes. Listed as watch and clockmaker in the *Universal British Directory*, 1798.

Bracket clock in bell-top case, with arched silvered dial, signed in arch, 'H. Raworth Plymouth'.

RAWORTH, SAMUEL: Plymouth. Good quality longcase clocks *c*1760–70; one, in rich mahogany case, has rocking see-saw in arch; another has moonwork. A silvered-dial bracket clock by this maker was advertised for sale in 1979; it has quarter chiming on six bells, with well-engraved backplate, pendulum rise and fall and strike/silent regulation, and mahogany case, *c*1770. Repeating watch (Baillie). Lacquer-cased clock.

RAYMONT, WILLIAM E: Brixham, Fore St, 1897–1902 and later (D).

REDDAWAY, ROBERT: Exeter. Recorded in 1861 Census in Combe St, aged 26, watchmaker, born Exeter.

REDSTALL, JOHN: Tiverton. Some lists of clockmakers include this name, but there was probably no such person (see Debnam, John).

REED, B: Plymouth. Watchmaker and silversmith; listed in the *Universal British Directory*, 1798. 'Paid Mr Reed balance due to him in 1804 for repairing & cleaning the clock, chimes etc. etc. £9 12s' (St Andrew's, Plymouth, churchwarden's accounts, 1806–7). 'Lost . . . having been left in the window of the little house in the yard of the Golden Lion Inn, Barnstaple, a gold watch, maker's name, Robert Reid, No 1794. Within is a watchpaper – REED, Watchmaker, Plymouth. Two guineas reward' (EFP, 20 Dec 1798). Longcase clock said to be signed 'Benjamin Reed Plymouth'.

REED (READ, REID), GEORGE: Barnstaple. Watchmaker and musical instrument maker; a break-in at his house reported (EFP, 20 Mar 1773). Listed in Bailey's *Western Directory*, 1783, and in the *Universal British Directory*, 1793.

REED, ROBERT: Totnes, 68 Fore St, 1889; 84 High St, 1893–1902; & Son, 1906 (D).

REED, WILLIAM: Bradninch. Watchmaker and jeweller, fancy goods dealer, dealer in furniture and cycle agent; Fore St, 1897 (D).

REED, WILLIAM: Crediton, 134 High St, 1902 (D).

REEPE, JOSEPH: Plymouth, *c*1650–70. Lantern clocks (Loomes); see also Rupe.

REEVES, JOHN: Crediton. Watch and clockmaker; on 6 Oct 1853 at Devon Quarter Sessions he was sentenced to three months' hard labour for stealing a watch and tools belonging to George Webber, watchmaker, by whom he had been employed for three years (DRO).

REID, FREDERICK TREEBY: Exeter, 5 Rockfield Rd, Howell Rd, 1889; 177 Sidwell St, 1893–1902 and later (D). He was the owner and constructor of one of Exeter's very first motor cars.

RENDELL (RENDLE), HENRY: Tiverton. 'On Thursday last was married by banns, at Tiverton, Devon, Mr Henry Rendle, of that place, (but late of Wiveliscombe, Somerset) Clockmaker, to Miss Susannah Owens, eldest daughter of John Owens, Esqur; (one of the principal corporators in that antient and loyal corporation) with a fortune of £1,000 besides a house furnished very handsomely with the best of furniture: . . . but what renders her price more, she is an agreeable young lady, and indued with every accomplishment requisite to make the marriage state happy' (*Sherborne Mercury* 13 July 1767). He rented premises (now part of Midland Bank) on south side of Fore St from 1770 to 1814; and kept several church clocks in the Tiverton area in repair. Small black-cased bracket clock noted; also longcase clocks, including one showing 'High Water at Topsham Bar' and playing a tune every three hours; 30-hour dial at Totnes Museum.

RENDELL (RENDALL), HENRY jun: Tiverton. Watchmaker, the *Exeter Flying Post*, 10 Oct 1793, records that at the annual meeting of the Humane Society in Tiverton he received a reward of a guinea and a half 'for taking from the mill leat before the cotton factory . . . a young woman named Mary Duckham, apparently dead, having been in the water nearly 20 minutes, who was almost miraculously restored to life and health, by the medical assistance and proper use of the apparatus belonging to the Humane Society'.

RENDELL, JAMES: Described in 1778 as watchmaker of Tiverton, son-in-law of John Owens (Harrowby Tiverton MSS). 'Tiverton . . . On Monday last was married, by special licence, in St Peter's Church, Mr James Rendell, of Wiveliscombe, Somerset, watchmaker, to Miss Sally Owens, second daughter of Mr Owens, sergemaker of this town, an agreeable young lady endued with every accomplishment requisite to make the marriage state happy' (EFP, 19 May 1775).

RENDELL, JAMES: Barnstaple. Ironmonger, clock and watchmaker; listed in the *Universal British Directory*, 1793.

RENDELL, JOHN O: Tiverton. Successor to Henry Rendell *c*1815, Fore St. Charged £2 12s in 1827 for cleaning, repairing and winding the clock at the Old Market House, Tiverton (receipt at Tiverton Museum). He also kept the town's church clocks in repair.

REYNOLDS, Mrs ELIZABETH: Plymouth. Watchmaker and jeweller. Listed at Old Town St; at No 6 in 1866; at No 5 in 1873–89 (D).

REYNOLDS, JAMES: Rockbeare. Oak longcase clock (Tapley-Soper) – possibly 1820 or earlier.

REYNOLDS, JAMES: Ridgway, Plympton. 1889–1902 and later (D).

REYNOLDS, JOHN: Barnstaple. Early English lantern clock, 'Johannes Reynolds de Barnstaple', mentioned in letter to the editor of *Antiquarian Horology* (Dec 1967, p327).

REYNOLDS, JOHN: Holsworthy. Watch and clockmaker, 1844 (D).

REYNOLDS, T. Plymouth, 22 Old Town St, 1856 (D).

REYNOLDS, WILLIAM: Plymouth. Watchmaker, 149 King St, 1866 (D).

REYNOLDS, WILLIAM: Ridgway, Plympton, 1873–83 (D).

RICHARDS, FRANCIS HENRY: Exmouth, 114 St Andrews Road, 1902 (D).

RICHARDS, JOHN: Tiverton, Cullompton and Honiton. Eight-day oak longcase clock signed 'John Richards, Tiverton', illustrated in *Country Life* (10 July 1975, p67). John Marshall, of Plymouth, was apprenticed *c*1715 to John Richards, of ——llompton, Devon, clockmaker. Pair-cased verge watch with silver champlevé dial and calendar aperture by John Richards, Honiton, sold at Sotheby's in July 1980 for £520. East Budleigh churchwardens' accounts have these items: 1718 – 'paid to Mr John Richards about ye clock £3'; 1722 – 'paid to Mr John Richards for ritting ye clocke 7s 6d'.

RICHARDS, JOHN: Apprenticed *c*1751 to William Stumbels & Co, Totnes, premium £21.

RICHARDS, WILLIAM: Tavistock, 10 Duke St, 1873–83; 39 Brook St, 1889–93 (D).

RICHARDS, WILLIAM JENKIN: Exeter, 177a Cowick St, 1883 (D).

RICKARD, HERCULES (*c*1712–73): Exeter. Clockmaker; son of John Rickard (Reckard), blacksmith; baptised at St Kerrian's and can be traced in the 1750s at St Thomas, where he kept the church clock in repair and was buried on 30 May 1773. His elder brother, John, was a locksmith, and a younger John Rickard was apprenticed to him in the 1760s as a watchmaker. Walnut and other longcase clocks; 30-hour example signed 'Hers. Rickard Exon' illustrated in C. N. Ponsford, *Time in Exeter*; another at Totnes Museum.

RICKARD, HERCULES (ii): Exeter. Watch and clockmaker, silversmith, jeweller etc; son of John Rickard, watchmaker, and Sarah; baptised at St Mary Steps 21 Dec 1787. Married Mary Sampson at St Mary Arches Church, 15 July 1812. 'Has just commenced business . . . at 246 High St' (EFP, 13

Jan 1814); 'moved from 246 to 268, nearly opposite the Grammar School, High St . . . Duplex, lever, repeating, and other watches carefully repaired' (EFP, 20 Apr 1815). His wife died, aged 40, in 1828, leaving him with five young children. Was churchwarden of St Lawrence's, 1830–2, and in 1841 moved to 6 Paris St, 'nearly opposite the side entrance to the Old London Inn'. The *Exeter Pocket Journal* lists him until 1854.

RICKARD, JOHN: Exeter. Watchmaker; apprenticed *c*1767 to Hercules Rickard, premium of £12 12s. Listed at Westgate in 1789 (EPJ); described as infirm in 1803 Militia List. Died, aged 73, on 23 Dec 1821 and was buried at St Thomas. Will in DRO mentions wife Sarah, sons John and Hercules, and three infant grandchildren, 'sons of my late deceased son James Rickard'; estate sworn in at 'under £600'. Exeter Museum has verge watch movement No 11686; also small faded mahogany, 8-day longcase clock, 'J. Rickard Exeter', with square silvered dial and a figure of Britannia inlaid in middle of trunk door, *c*1800.

RICKARD, JOHN (ii): Exeter. Watch and clockmaker; son of John and Sarah; baptised at St Mary Steps 16 May 1784. Notice in *Exeter Flying Post*; 'John Rickard, Jun., Watch and ClockMaker No. 4 Westgate-Quarter. A respectable youth wanted as an apprentice, Exeter, Aug. 14, 1815'. Married Elizabeth Barrell at Heavitree 2 Nov 1812 and, secondly, Damaris Woosley, widow, at Alphington 10 Nov 1825. Recorded in 1841 Census in Bridge St.

RICKARD, JOHN (iii): Exeter. 'Notice of removal. In consequence of the premises in St Martin's Lane being required for the enlargement of the Royal Clarence Hotel, John Rickard, chronometer, watch and clockmaker, has removed to 229 High St, Exeter, adjoining the corner of Gandy St. J.R. avails himself of the opportunity to return thanks for the kind patronage which he has received during the past twelve years, and to state that nothing on his part shall be wanting to justify a continuance of public confidence' (EFP, 3 Feb 1859). Not listed after 1861 (EPJ).

RICKARD, WILLIAM: Bideford. Watch and clockmaker, Churchyard, 1830 (D).

RIDGWAY, HARRY WHITE: Newton Abbot, Wolborough St, 1902 and later (D).

RIEDLINGER, LUDWIG: Plymouth, 2 Richmond St, 1878; 4 Duke St, 1889; & Sons, 1897 and later (D).

RIHLL, ALBERT JOHN: Moretonhampstead, New St, 1873; George Square, 1878; the Square, 1883–1902 (D).

RIHLL, WILLIAM: Exeter and Moretonhampstead. Entry in Moretonhampstead parish registers records the marriage of William Rihll of 4 Paris St, Exeter, clock and watchmaker, son of George, gentleman, and Mary Ann Dayment, daughter of William, cordwainer, on 31 Aug 1837. No 4 Paris St was the address of the watchmaker William Frost, who was himself from Moretonhampstead. Listed as watchmaker and shopkeeper at Moretonhampstead, 1838–66 (D), premises at New St. 'Paid Wm Rihll for repairing the tower clock 3s' (Drewsteignton churchwarden's accounts, 1846–7). Thirty-hour painted-dial longcase clock; others reported from as far afield as Australia.

ROBERTS, JOHN: Dartmouth. Watchmaker; listed in the *Universal British Directory*, 1793.

ROBERTS, R: Buckfastleigh, Station Rd, 1889 (D).

ROBERTS, ROBERT: Totnes. Watchmaker, Grove, 1866; Fore St, 1873; 1 High St, 1878–83 (D).

ROBERTS, WILLIAM: Plymouth. 35 Union St, East Stonehouse, 1873–8 (D).

ROBERTSON, JOSEPH: Maker of the present clock movement at St Mary Steps Church, Exeter, in 1725. He also carried out work on the church clocks at Sidmouth (1724) and at St Sidwell's, Exeter (1726 and 1728) (churchwardens' accounts).

RODD, WILLIAM: Plymouth, 3 Russell St, 1889; 66 York St, 1897–1902 (D).

RODGERS, F. J: Paignton, 32 Victoria St, 1902 and later (D).

RODGERS, F. R: Paignton, 1 Victoria St, 1897 (D).

RODGERS, WILLIAM HENRY: Torquay, 22 Higher Union St, 1873–93; 54 Union St, 1897–1902 and later (D).

ROGER THE CLOCKMAKER: In 1424–5 horsemen were sent from Exeter to seek him at Barnstaple. He rode back with them and carried out repairs to the Exeter Cathedral clock (fabric accounts).

ROGERS, ALBERT I: Bideford, Mill St, 1893–1902 and later (D).

ROGERS, JOHN: Dartmouth. Watch and clockmaker, Lower St, 1830 (D).

ROHRER, ANDREW: Plymouth, 39 Frankfort St, 1866; at 55 Union St also, 1883; & Son, 1893; & Sons, 39 Frankfort St, 54 & 55 Union St, and 2 Old Town St, 1897 and later (D).

ROOF, CHARLES: Exeter. Clockmaker and repairer, 16 Preston St, 1883 (D).

ROSE, GEORGE: Exeter, South St, 1883; & Co, 1897 (D).

ROSE, RICHARD: Exeter. Hairdresser and watchmaker, South St, 1857–78 (D).

ROSS, DANIEL: Exeter. 'Chronometer, watch, and clockmaker, 31 High St . . . has commenced business . . . and trusts the practical knowledge he has acquired of the trade from one of the first chronometer and watchmakers in London, will secure for him that patronage it will ever be his study to deserve. Watches and clocks on the most improved and scientific principles made and accurately repaired' (*Exeter Gazette*, 2 Sept 1837). 'Ship chronometers rated for finding the longitude by one of Troughton's transit instruments' (1848 advertisement). Moved to 230 High St in 1855 and to Bedford St in 1859; on the closure of the business in 1864 Ross's stock of clocks and watches was acquired by Messrs Ellis, 200 High St (EFP). He was recorded in 1861 Census at Bedford St, aged 49, watchmaker, born Exeter, 'blind about 10 years'. His wife, Mary Anne, died 10 July 1863 (EFP). Ross successfully tendered in 1844 to supply the clock at Exminster Hospital; he also supplied one for Old Blundell's School, Tiverton, and another for St Mark's Church, Dawlish (now demolished). His clocks and watches are invariably of a high standard; they include rosewood-cased bracket and travelling (four-glass) clocks. The name 'Ross Exeter' is on a handsome pair of burrwood library timepieces, one of which is illustrated in C. N. Ponsford, *Time in Exeter*. Exeter Museum has lever escapement, silver watch.

ROSSE, JAMES: Exeter. Watchmaker; his name appears as a surety on marriage allegation documents of 1687 and 1707.

ROSSITER, E: East Teignmouth, 4 Den Place, 1883 (D).

ROSSITER, GEORGE: Teignmouth, Bank St, 1902 and later (D).

ROSSITER, Mrs S: East Teignmouth, 4 Den Place, 1889; 15 Bank St, 1893–7 (D).

ROUCKLIEFFE (ROUCKLEIFFE), JOHN: Totnes. Apprenticed *c*1740 to William Stumbels, Totnes, premium £20. Silver pair-case watch with champlevé dial and wavy minute band illustrated in *Transactions of the Devonshire Association*, 1935. A 'Mr Rouckliefe' was assessed at 2d for his house in the 1762 Totnes Poor Rate. At Bridgwater, Somerset, later. Longcase clocks and watches.

ROUTLEIGH, GEORGE: Lydford. Watch and clockmaker; died, aged 57, on 14 Nov 1802 and was buried in Lydford churchyard, where his tomb chest with its famous and often quoted epitaph can still be seen. Worked at Launceston in Cornwall; 30-hour brass square dial and other longcase clocks (Miles Brown).*

ROW, WILLIAM HARRIS: Devonport, 34 Fore St, 1844 (D).

ROWE, HENRY: Devonport, 2 Fore St (Temperance Hall), 1889 (D).

ROWE, JOHN PASSMORE: Torrington. Clock and watchmaker, South St, 1866; Corn Market St, 1873–1902 and later (D).

ROWE, THOMAS SQUIRE: Okehampton. Watchmaker, West St, 1838 (D).

RUDALL BROTHERS: Plymouth, 42 Beaumont Rd, 1902 and later (D).

RUDD, JOHN WILLIAM: Plymouth. Watchmaker, Tavistock St, 1866–1902 (D).

RUFF, KONRAD: Exeter. Recorded in 1861 Census in South St, clockmaker, aged 46, born Germany. Possibly an employee of Anselm Spiegelhalder (qv).

RUNDELL, JOSEPH: Plymouth. A watch by him lost (*Public Advertiser*, 21 Mar 1788 – Buckley).

RUNDER, FRANCIS: Exeter. Jeweller, watch and clockmaker, Bedford St, 1830–7; 11 High St, 1838–48 (EPJ).

RUPE, ——: Plymouth. Watch *c*1720 at Liverpool Museum (Buckley).

RUSSELL, GEORGE P: Okehampton, 9 Fore St, 1878 (D).

RUTT, RICHARD: Plymouth Dock. Chronometer maker etc, St Aubyn St, listed in the *Picture of Plymouth*, 1812.

RYDER, JOHN W. & RICHARD J: Devonport. 6 William St, Morice Town, 1838–44 (D).

SALMON, WILLIAM: South Brent, 1897 (D).

SALTER, ISAAC: Cullompton. Was paid £1 2s 6d for mending Cullompton Church clock and chimes in 1737; also paid in 1749 (churchwardens' accounts).

SALTER, ROBERT: Topsham, Fore St, 1844–50; Lower Shapter St, 1856–89 (D). Recorded in 1851 Census in Shapter St, aged 34, watchmaker, born Topsham.

SALTER, THOMAS TOUZEAU: Honiton. Watchmaker; married

Elizabeth Boalch at Colyton 6 Nov 1833 (EFP). Listed at New St, 1838 (D).

SAMPSON, JOHN: Plymouth Dock. Eight-day longcase clock *c*1780 (Bellchambers).

SAMPSON, THOMAS: Plymouth Dock. Longcase clock *c*1790; no further details.

SANDERS, WILLIAM: Barnstaple, 33 Boutport St, 1878 (D).

SANDERS, WILLIAM: Dartmouth, St George's Square, 1902 and later (D).

SANDERSON, GEORGE: Exeter. Watch and clockmaker and jeweller; married 1747; died in Clerkenwell, London, 10 Oct 1764.*

SANDFORD, FREDERIC: Plymouth. Watchmaker and working jeweller, Bedford St, 1836 (D).

SARE (SAR), JOHN: Stokeinteignhead. Clockmaker; mended Exeter Cathedral clock several times 1452–6 (fabric accounts).

SAUNDERS, EDWARD: Winkleigh, 1883–9 (D).

SAUNDERS, E. JENKIN: North Tawton, 1883 (D).

SAUNDERS, JOHN PALMER: Crediton and Honiton. Recorded as retired watchmaker at Topsham in 1851 Census, aged 60, born Crediton. Was living at Crediton when he took out licence 11 Jan 1812 to marry Charlotte Knight. Listed as watch and clockmaker in High St, Honiton, in 1830 (D).

SAUNDERS, T. G: Plymouth, 51 Southside St, 1889 (D).

SAUNDERS, WILLIAM: Torrington, South St, 1850 (D).

SAVAGE, ABRAHAM: Exeter. Clockmaker; son of Peter Savage (qv). He (or a namesake) was apprenticed in London for eight years to Thomas Alcock on 12 May 1648 and then 'turned over' to Henry Child. Can be traced in St Stephen's parish, Exeter, in 1660 (Poll Tax). The lease of a tenement and garden on north side of Combe Rewe (later called Rocks Lane) in the parish of St Mary Major was assigned to him on 20 Mar 1670 (DRO).

SAVAGE (SAVIDGE), JOHN: Exeter. Made a clock for Plymouth Guildhall; died 1627.*

SAVAGE, JOHN (ii): Exeter. Clock and watchmaker; son of Peter Savage (qv), admitted as Exeter freeman by succession, 5 July 1658. Married Ruth Pince at St Mary Arches Church 24 Mar 1656 (register); recorded in St Stephen's parish in 1660 (Poll Tax) and in St Lawrence in 1671 (Hearth Tax).

SAVAGE, PETER: Exeter. Clock and watchmaker; died 1657.*

SAVAGE, THOMAS: Exeter. Clockmaker; son of John Savage the elder (qv). Married Agnes Barrons at St David's Church 2 June 1640 (register). In 1660 he and a fellow parishioner collected the Poll Tax in the parish of St Lawrence, where he had his shop; he later moved to St Stephen's, and was buried on 2 June 1679 (Holy Trinity parish register).

SAVAGE, THOMAS (ii): Exeter. Clock and watchmaker; son of the above. He (or a namesake) was apprenticed in London for seven years to Joseph Quash on 1 Mar 1659 (Clockmakers' Company register of apprentices). Recorded in St Stephen's parish, Exeter, in 1671 (Hearth Tax); and in Oct 1672 was stated to be aged 28 and a native of Exeter (Consistory Court records). Married Elizabeth Etheridge (died *c*1679) and several children were baptised at St Stephen's Church: Elizabeth (1669), Susanna (1674) and Thomas (1676). Quarter Sessions records suggest that he held Jacobite sympathies. In May 1691 was one of several men remanded to prison, and at the Sessions on 9 May 1692 he was again gaoled: 'Edward Paynter of this City, innkeeper, Thomas Savage of ye said City, watchmaker, & Anthony Turner of ye same, taylor, their fidelity to ye present government of this kingdome being suspected, were confined prisoners to ye White Swann Inn within ye said City [Exeter]'. It is not clear what happened after that, but it is possible that Thomas resumed his trade in London.

SAYER, MATTHEW: Exeter. Clock and watchmaker; married Mary Osmond at St Paul's Church in Mar 1734 and lived in the precinct of the Close. Kept the Exeter Cathedral clock in repair from 1746 to 1758 and also mended the St Stephen's Church clock several times. The Exeter Cathedral register records his burial on 7 Nov 1765. John Caine was apprenticed to him *c*1755. Exeter Museum has a watch by him hallmarked London 1757–8; longcase clock illustrated in C. N. Ponsford, *Time in Exeter.*

SAXBY, HARRY: Devonport. Repairer, 3 Ordnance St, 1897 (D).

SAXBY, HENRY jun: Bideford, Grenville St, 1889 (D).

SCADDING, JAMES: Sidmouth. Clockmaker; married Florence Follett of Sidmouth, at St Pancras, Exeter, 30 Aug 1728 (register transcript). Mentioned in Sidmouth documents of 1731–2 and 1762 (DRO).

SCHERZINGER, FRANK: East Budleigh, 1878 (D).

SCHERZINGER, JOSEPH: Axminster. Watch and clockmaker; listed in 1856; at South St in 1857; at Lyme St, 1866 (D).

SCHERZINGER, SELES: Budleigh Salterton, High St, 1873 (D).

SCHERZINGER, SILAS: Ottery St Mary. Clockmaker, Cornhill, 1866 (D).

SCOBELL, JOHN SPURWAY: Colyton and Ottery St Mary. Son of Joseph and Mary Scobell; born 14 July 1799; baptised Colyton 11 Aug 1799 (parish register). Listed at Broad St, Ottery St Mary, in 1830 (D). Several longcase clocks reported, including 30-hour example with 280mm (11in) square white dial signed 'Ino S. Scobell Colyton', the dial back stamped 'S. Wilkes Birmingham' and the movement, 'ISS 1837'.

SCOTT, FREDERICK G: Exeter, 124 Sidwell St, 1897; 5 York Rd, 1902 (D).

SEARL (SEARLE), GEORGE: Chudleigh. Clock and watchmaker; listed in the *Universal British Directory*, 1793. Longcase clock, 'Geo. Searle Chudleigh'.

SEARLE, GEORGE E: Chudleigh, 1883–1902 and later (D).

SEARLE, G. ELLIOT: Plymouth. Edgcumbe St, Stonehouse, 1850; 111 Union St, 1856–73; & Son, 21 Bedford St, 1878; & Sons, 1902 (D).

SELLICK, JOHN: Barnstaple, Pages Lane, 1878 (D).

SELLICK, JOSEPH G: Barnstaple, Holland St, 1856–78 (D).

SELLICK, RICHARD: Barnstaple, Pages Lane, 1893 (D).

SERMON, EDWARD: Torquay, 16 Victoria Parade, 1883; 10 Victoria Parade, 1889–1902 and later (D).

SEYMOUR & BOND: Okehampton. Ironmongers and watchmakers, Fore St, 1866–73 (D).

SHAPLAND, H: Barnstaple, Trinity St and High St, 1873 (D).

SHAPLAND, HENRY: Barnstaple, 74½ High St, 1873 (D).

SHARLAND, Mrs EMMA: Tiverton. Watchmaker and jeweller, Fore St, 1878 (D).

SHARLAND, HERBERT JOHN: Tiverton. '. . . watchmaker, jeweller and general factor, respectfully announces his having commenced business in the above branches in temporary premises, next door to Mr Clapp, saddler, Bampton St' (*Western Times*, 17 Nov 1868). At Fore St later; died 1877.

SHARLAND, JOHN: Tiverton. Born 5 Mar 1812, died 12 May 1903. Watchmaker and silversmith, perfumer and fancy goods dealer, Gold St. Established his business in the mid 1830s and travelled extensively in the West of England and as far afield as Birmingham and the Channel Islands. Was an active Liberal and from 1835 to 1865 was a member of Lord

Palmerston's local committee (Palmerston was one of Tiverton's two MPs at that time); he was also a freemason and borough councillor. Preserved in the United Reformed (former Congregational) Church, St Peter St, is a flute which Sharland played in church. Towards the end of his life he wrote a book entitled *Recollections of the Great Lord Palmerston and Old Times in Devon*. Several longcase clocks reported, one with rocking ship in arch; watch No 25241 hallmarked 1838. The business became John Sharland & Co *c*1850; and in directories 1856–66 is listed as Sharland & Co, wholesale jewellers, watchmakers, bookbinders, booksellers, printers, stationers and hardwaremen. Tiverton Museum has 8-day wall timepiece, 'Sharland & Co Tiverton', in black papier-mâché case inlaid with mother of pearl.

SHEA, JAMES: Plymouth. Watchmaker, 16 Old Town St, 1830 (D). Reported as bankrupt (EFP, 21 Feb 1833).

SHEPERD, W: Plymouth, 2 Russell St, 1856 (D).

SHEPHEARD, JABEZ: Plymouth, 9 George St, 1844–66 (D). Watchpaper noted.

SHEPHEARD, JAMES: Plymouth, 2 Whimple St, 1838 (D).

SHEPHERD, JESSE JOHN: Honiton, High St, 1893–1902 and later (D).

SHEPHERD, JOHN: He was paid 2s by the St John's, Exeter, churchwardens in 1590 'for his fee for mendyinge the clock this yeare'.

SHEPPARD, THOMAS: Devonport. 14 Morice St, 1850; 11 King St, 1856 (D).

SHERRY, F: Stonehouse, Plymouth, 38 Edgcumbe St, 1830; 4 Union St, 1856 (D).

SHORT & BAILEY (late Mallett): Watchmakers, jewellers and silversmiths, 4 High St, Barnstaple, and High St, Bideford, 1866 (D).

SHORT, S: Barnstaple. Advertisement in *North Devon Journal*, 13 May 1869: '4 High Street, Barnstaple. Established 1811. S. Short (late Mallett), watchmaker and jeweller etc. Being about to leave Barnstaple will offer for a short time previous to the disposal of the business the whole of his valuable stock of gold and silver watches'.

SHORT, SAMUEL: Barnstaple, 14 Boutport St, 1873 (D).

SHORT, SAMUEL: Bideford, Grenville St, 1850; High St, 1856 (D).

SHORT, WILLIAM MARSHALL: Holsworthy. Watch and clockmaker and jeweller; listed in 1866; at Higher Square, 1878–89; Station Rd, 1897 (D).

SHORTO, EDWARD HENRY HAYES (1809–81): Exeter. 'Watchmaker, jeweller & engraver, 2 North St . . . Upwards of 20 years with Mr Adams, 38 High St' (EFP, 22 Dec 1853). Supplied clock for St David's Church (now at Bradninch), *c*1852. Served for fifty years as parish clerk of St Petrock, looked after the clock there and wrote a history of the parish. A son, Edward Henry Shorto, was headmaster of Hele's School, Exeter, 1889–1905; another was town clerk of Exeter.

SIMONS, A: Bideford. Watch and clockmaker. Silvered-dial mahogany longcase clock *c*1800. Watch at Exeter Museum, 'A Simons Biddeford'.

SIMONS, MOSES: Bideford. Watch and clockmaker, High St, 1823 (D).

SIMPKINS, W: High Bickington, 1889 (D).

SKERRETT, GEORGE: Devonport, 89 Fore St, 1838; 37 Fore St, 1844–50 (D). His wife Emily died, aged 36, on 1 Jan 1846 (EFP). Longcase clock and watch (Bellchambers).

SKINNER, JAMES: Exeter. Watchmaker; son of the elder John Skinner; worked in the family business, and died, aged 34, in Jan 1818.

SKINNER, JOHN: Exeter. Watchmaker, goldsmith and jeweller; apprenticed in Honiton to Jerom Murch for a premium of £6 16s 6d paid in 1763. Married Susanna Mortimore at St Paul's Church, Exeter, Nov 1774. Was an active member of St Stephen's Church, regularly attending meetings at which the churchwardens' accounts were presented, his signature appearing along with those of other parishioners signifying they were in order, from 1773 onwards. At first he signs as John Skinner Junr. and then from 1780 onwards adopts the signature Jno. Skinner, and some of his clocks are inscribed in this manner. His wife, who died in 1784, appears to have been a nonconformist and their children were baptised at George's Meeting in South St. The main premises of the business were in High St, but he also owned leasehold property in Bartholomew St West, part of which was let as a public house – the Exeter Arms. In 1813 he acquired 20 High St for £1,700. Died 23 Apr 1818, aged 71, and was buried in St Sidwell's churchyard. His estate was later sworn in 'at under £7,000'. He looked after St Stephen's Church clock for about forty years. Numerous bracket and longcase clocks, wall timepieces, regulators, turret clocks, and watches. One of his suppliers was Thwaites of London (see Chapter 1). Several newspaper references to watches by him: No 56 pawned at Salisbury by a suspected highwayman in 1781; No 596, gold-cased, lost from a gentleman's coat pocket (EFP, 28 Apr 1808).

SKINNER, JOHN (ii): Exeter. Watchmaker, silversmith and jeweller, 20 High St. Worked in the family business, which he continued after his father's death in 1818. 'Married. On Sunday last, at St Stephen's Church, by the Rev W. Stabback, Mr John Skinner, watchmaker, to Miss Catherine Tooze, both of this city' (EFP, 7 Nov 1822). In 1824 his shop window was raided and four gold chains and three seals, value £30, stolen. Was a member of Exeter City Council on its formation in 1836 after the abolition of the old Chamber of Exeter. Died 25 Nov 1846, aged 70, and was buried in St Sidwell's. The property passed to his son, John Skinner, a surgeon, whose address is given in a deed of 1855 as 17 Leighton Villas, Kentish Town, Middlesex. When the new Bedford St was cut through to High St in the 1870s, No 20 was demolished, and the road junction (now pedestrianised) occupies the place where some of Exeter's most attractive clocks were sold.

SKINNER, THOMAS: Exeter. Lost watch so inscribed (*Hampshire Chronicle*, 10 July 1780 – Buckley). Probably a misprint or mistake for John Skinner.

SLEE, FRANCIS (FRANK): Holsworthy, Lower Square, 1873–1902 and later (D).

SLOMAN, ——: Exeter. Robbed of watches (EFP, 17 Oct 1811).

SLY & CO: Barnstaple, 89 High St, 1889–1902 and later (D). Successors to Charles Ford.

SMALE, JOHN: Exeter. Clock and watchmaker, Mary Arches St, 1828–67 (EFP). Recorded in 1841 Census, aged 40.

SMALE, TIMOTHY: Kingsbridge. Watch and clockmaker, silversmith and working jeweller, Fore St, 1830–56 (D).

SMERDON, EDWARD: Newton Abbot, 4 Wolborough St, 1889–1902 and later (D).

SMERDON, JOHN: Newton Abbot, Wolborough St, 1850–83 (D).

SMITH, JOHN: South Molton. Death notice: 'Jan. 29, at Southmolton, after a lingering illness and deeply lamented, Mr John Smith, watchmaker, aged 29' (*Exeter Gazette*, 4 Feb 1837).

SMITH, JOHN: Totnes, Bridgetown, 1856–78 (D).

SMITH, MARK: Totnes, 25 Bridgetown, 1893–1902 and later (D).

SMITH, REUBEN: Totnes, Bridgetown, 1883–9 (D).

SMITH, RICHARD: Plymouth, 1748 (Baillie).

SMITH, THOMAS C: Torquay, 167 Union St, 1902 (D).

SMITH, THOMAS CLAYTON: Brixham, 31 Fore St, 1878–89 (D).

SMITH, WILLIAM: Plymouth. Clock and watchmaker, Higher Broad St, 1812–14 (D).

SMITH, WILLIAM: Torquay, 28 Higher Fleet St, 1878 (D).

SNELL, EDWARD: Barnstaple. Watch and clockmaker and silversmith, listed in 1823 (D); died Apr 1827, aged 33 (EFP). Will in DRO, dated 24 Mar 1827, mentions wife Elizabeth and his freehold house in High St; John Snell of Tiverton, saddler, and William Young of Barnstaple, plumber, appointed as executors; witnesses Charles Trist, Jos W. Hunt and Samuel Gillard (qv); estate valued at 'under £1,500'. Longcase clock seen at hotel (Tapley-Soper).

SNELL, JAMES: Barnstaple. Watch and clockmaker and silversmith, High St, 1830 (D).

SNELL, JOHN HOLDER: Tavistock. Watchmaker, Market St, 1823; Fore St, 1830–8; Higher Market St, 1844–66 (D). Thirty-hour longcase clock (Bellchambers).

SNELL, WILLIAM: Honiton. Took out licence 11 Dec 1813 to marry Mary Vowler. 'W. Snell, clock and watchmaker, silversmith, jeweller, &c. is in immediate want of an apprentice, he will be treated as one of the family. A premium is expected' (EFP, 9 Mar 1815). Thirty-hour longcase clock reported (Bellchambers).

SNELL, WILLIAM: South Molton. Watch and clockmaker, Market Place, 1823 (D).

SNELL, WILLIAM: Exeter. Clock and watchmaker, 112 Fore St Hill. Goods stolen from kitchen by thieves who tried unsuccessfully to break into his shop (EFP, 17 May 1827).

SNELL, WILLIAM: Crediton. Watch and clockmaker, 1838–44 (D).

SNELL, WILLIAM THOMAS: Salcombe, 27 Union St, 1889–93; 1 Fore St, 1897; 73 Fore St, 1902 and later (D).

SNELLING, JAMES: Newton Abbot. Watch and clockmaker and agent to the Yorkshire Fire & Life Insurance, Wolborough St, 1830–44; Bridge St, 1850–6 (D).

SNOW, ROBERT: South Molton, 4 South St, 1873–8 (D).

SOADY, Mrs ELIZABETH: Devonport. Albert Rd, Morice Town, 1883 (D).

SOADY, WILLIAM: Devonport. Albert Rd, Morice Town, 1878 (D).

SOBEY, HENRY: Bere Alston, 1889–93; & Son, 1897–1902 (D).

SOBEY, WILLIAM RAWLINS: Exeter. Watchmaker, silversmith and

jeweller; admitted as freeman of Exeter Goldsmiths' Company 1835. Listed at various addresses from 1832; at Queen St, 1848. In an advertisement dated 13 Sept 1852 in *Exeter Gazette* he gave his address as 1 Queen St, and announced he was selling off his stock owing to long continuance of ill health; described as 'the late', 28 Mar 1855 (Assay records).

SOLOMAN, M: Exeter, 60 High St, 1866 (D).

SOLOMON, MOSES: Plymouth. Wholesale watch manufacturer, Frankfort St, 1836 (D).

SOMMERS, ALFRED: Plymouth, 7 Notte St, 1897–1902 and later (D).

SOMMERS, F. R: Plymouth, 34 Whimple St, 1893 (D).

SOMMERS, GEORGE jun: Ilfracombe, 4 High St, 1856 (D).

SOMMERS, GEORGE: Ilfracombe, High St, 1850 (D). Longcase clock at Ilfracombe Museum.

SOMMERS, GEORGE: Plymouth, 17 Southside St, 1873; 7 Notte St, 1883–93 (D).

SOMMERS, W. H: Plymouth. 60 Union St, Stonehouse, 1893 (D).

SOPER, ARTHUR THOMAS: Torquay, 47 Market St, 1893 (D).

SOPER, JOHN: Torquay. Tormohun baptisms include: Lewis Walter, son of John and Susan Soper, of Torquay, watchmaker, born 14 Oct, baptised 1 Nov 1829 (register).

SOPER, THOMAS: Moretonhampstead. Watchmaker, Market Place, 1823 (D).

SOPER, THOMAS: Torquay. Watch and clockmaker, 8 Braddons Row, 1830 (D).

SOPER, THOMAS: Salcombe. Watchmaker, 1838 (D).

SOUTHCOTT, HENRY: Hatherleigh, Bridge St, 1866–83 (D).

SOUTHEY, JOHN: Exeter. Black and gilt Japanned tavern clock-case with octagonal dial, signed beneath, 'Jno. Southey Exeter', the trunk with a painted river scene, 1.3m (53in) high, mid-eighteenth century.

SOUTHWORTH, PETER: Exeter. Watchmaker; gave evidence at the city Quarter Sessions 18 Jan 1705/6 about watches taken from his shop.

SPARKE (SPARK), ANGEL: Plymouth. Watch and clockmaker; listed in the *Universal British Directory*, 1798; at Market St in 1814 (D). 'Sunday morning early some villains broke open the shop of Mr A. Sparke, watchmaker, of Plymouth, and plundered it of several articles' (*Exeter Gazette*, 22 Mar 1792). St Andrew's, Plymouth, churchwardens' accounts have these items: 1786–7, 'Angel Sparke for repairing the clock & chymes 13s'; 1796–7, 'Mr A. Spark for repairing the church clock 17s'. A maker of excellent 8-day arched-dial longcase clocks; one fetched £1,100 at auction in Plymouth in 1979. A number of watches also noted.

SPARKE, CHRISTOPHER: Exeter. In 1644–5 'Paid Christopher Sparke for mending of the clock' (St Mary Steps, Exeter, churchwarden's accounts). He lived in St Mary Major parish (Protestation Return).

SPARKE, WALTER: Exeter. Locksmith; admitted as Exeter freeman by apprenticeship 31 Dec 1688; buried in St Lawrence's 2 Dec 1705. Carried out work on the St Stephen's Church clock and was possibly the maker of a clock and chimes for St Sidwell's Church in 1691. A Mr Sparke was paid £10 'for puting up ye cheemes' at St Mary Steps Church in 1691. These chimes have since been removed.

SPIEGELHALDER, ALBERT: Exeter. Watchmaker, South St; voted in 1864 election.

SPIEGELHALDER, ANSELM: Exeter. Watch and clockmaker specialising in German clocks, South St, 1844–75 (EPJ). Recorded in 1861 Census, aged 48, born Germany; together with wife Grace, 39, born Devon; and children Albert, 17, Walter, 15, Felix, 10, John, 1 month, all born Exeter. Weight-driven wall timepiece at Exeter Museum; longcase clock with broad mahogany case and arched painted dial with picture of ruins.

SPIEGELHALDER, WALTER: Exeter. Watch and clockmaker, jeweller etc; was in partnership for three years with his father, Anselm, and after his death continued the business at 'the Old Established Shop, 83 South St, opposite Bear St' (*Devon Weekly Times*). Listed there 1876–8 (D).

SPILLER, HENRY: Bideford. Watchmaker; listed in the *Universal British Directory*, 1793.

SPILLER, HENRY: Exeter. Clockmaker and glazier's vice-maker and repairer, 'from London'. Son of the Wellington, Somerset, clockmaker Joel Spiller; announced in Sept 1795 that he had taken a house in St Thomas 'near the New Bridge'. In 1804 he removed to the sign of the dial in Alphington St, where he remained until 1822 (EPJ). Carried out improvements to the stables clock at Killerton House in 1811, and the following year made a clock for St David's Church, Exeter, 'on a new and improved principle – it continues going during the time of winding up, and is altogether one of the best specimens of that kind of mechanism that has ever been seen in this part of the kingdom' (EFP, 12 Mar 1812). In 1814 he advertised for 'a lad, of a tractable disposition, and a good mechanical genius' as an apprentice. Longcase clocks.

SPILLER, HENRY: Plymouth. Watch and clockmaker, Stonehouse Row, 1822; 86 King St, 1830 (D).

SPILLER, JOHN: Alphington, Exeter. Clockmaker; married Elizabeth Bond at Nynehead, Somerset, 3 May 1802 (register).

SPRAKE, JOHN: Axminster. Watch No 12116 *c*1775 (Buckley).

SPRY, SHADRACK: Black Torrington, 1878 (D).

SQUIRE, C: Buckland Brewer. Clock reported (DNQ, vol 8).

SQUIRE, ROBERT: Buckland Brewer. Clockmaker, 1856 (D).

SQUIRE, ROBERT & SON: Bideford. Watch and clockmakers, 12 High St, 1866–1902 and later (D). Church clock at St Giles in the Wood has cast inscription: 'R. SQUIRE & SON BIDEFORD 1879'. Another, three-train, noted at Bideford.

STACEY, GEORGE: Bampton, Luke St, 1889 (D).

STAFFORD, TIMOTHY: Exeter. Watch and clockmaker from Coventry; worked for John Tucker at Thorverton in 1816 and then for Henry Ellis in Exeter. On 8 June 1821 took out licence to marry Mary Snow. Was in business in 1822 in Cowick St (EPJ). See Chapter 5.

STANBURY, ELISHA: Exeter. Watchmaker, St Lawrence parish (1804 jurors' list); New Market, 1816; Paul St, 1834–42 (EPJ). Longcase clock, 'E. Stanbury Exeter' (Tapley-Soper).

STANBURY, JOHN: Exeter. Watchmaker, North St, 1833 (EPJ). Recorded in 1841 Census, aged 60.

STAPLETON, ROBERT: Exeter. Recorded at Rack St in 1841 Census, clockmaker, aged 40.

STARTUDGE, ROGER: Modbury and Aveton Gifford. Took out licence 14 Oct 1740 to marry Mary Philips of Loddiswell. Numerous payments recorded to him in Modbury churchwardens' accounts between 1732 and 1765 for work about the parish clock, bells and locks. He also repaired Ugborough Church clock. Longcase clock, 'Roger Startudge Aveton Gifford' – 'lovely brass face and also tells the date' (Tapley-Soper).

STEER, WILLIAM: Brixham, The Quay, 1856–78; Fish St, 1883 (D).

STENLAKE, BENJAMIN COUCH: Tavistock. Watch and clockmaker and stationer, Market St. Last examination at Exeter Bankruptcy Court (EFP, 7 June 1849).

STEPHENS, C. F: Torquay, 21 Brunswick Square, 1893 (D).

STEPHENS, FARINTON: Torquay, 17 Abbey Rd, 1878 (D). See also Farinton, Stephen.

STEPHENS, GEORGE: Devonport, 39 Cornwall St, 1889; Tavistock St, 1893–1902 and later (D).

STEPHENS, N: Ilfracombe, 40b Fore St, 1902 and later (D).

STEPHENS (STEVENS), THOMAS: Plymouth. Watch and clockmaker; listed in the *Universal British Directory*, 1798; at 23 Treville St in 1822 (D). Described as a watchmaker, St Andrew's parish, when he took out licence 9 Aug 1790 to marry Mary Leigh of Charles, Plymouth. The *Exeter Gazette*, 24 Jan 1824, reported his death.

STEVENS, ——: Exeter. 'Stevens for clock &c £4 9s 4d' (St Mary Steps, Exeter, churchwarden's accounts, 1800).

STEVENS, HENRY JOHN: Bradninch, 2 High St, 1902 and later (D). Supplied Bradninch Guildhall clock *c*1904.

STIDWORTHY, FREDERICK: Kingsbridge, Fore St, 1873–93 (D).

STINCHCOMBE, EDWARD: Okehampton, Fore St, 1893–1902 and later (D).

STOCKER, WILLIAM: Honiton. Thirty-hour, one-piece brass-dial long-case clock seen at Devon inn in 1983. Another, mahogany cased, had long round-headed door to trunk and a white dial, 'Stocker Honiton', with moving figure in oval of Old Father Time cutting field of corn with his scythe (Tapley-Soper). A Wm Stocker, laceman was appointed as an executor of the will of Francis Pile jun, watchmaker, Honiton.

STONE, Mrs F. E: Exeter, 36 High St, 1878 (D).

STONE, HENRY: Exeter. Watchmaker, St Martin's St, 1867; 36 High St, 1869 (EPJ).

STONE, JOHN: Exeter. Manufacturing silversmith, jeweller, watch and clockmaker; moved to 36 High St in 1854 after twenty years at 30 New Bridge St. Was at one time the owner of Lovelace's Exeter Clock and displayed it in his shop. Died 24 Mar 1868: 'The funeral of Mr Stone took place on Monday at the new cemetery, and was attended, in addition to the members of his family and intimate friends, by a number of tradesmen and others who had known and respected "Honest John" for years. Throughout the High St most of the shops were partially closed during the day' (EFP, 1 Apr 1868).

STONE, JNO: Exeter. Watchmaker, Martin St, 1863 (EPJ).

STONE, THOMAS HART: Exeter, 36 High St. Died 30 May 1873, aged 36; 'Mrs Stone to continue her late husband's business' (EFP, 11 June 1873).

STONEMAN & GRAF: Exeter. Watchmakers and jewellers, 60 Longbrook St, 1889–1902 and later (D).

STRANG, JAMES W: Plymouth, 30 & 31 Treville St, 1878–83 (D).

STRICKLAND, GEORGE: Dawlish, 15a King St, 1883 (D).

STRINGER, ARTHUR T: Axminster, West St, 1889 (D).

STROWBRIDGE (STRAWBRIDGE), HENRY: Chudleigh. Watch and clockmaker, 1823–44 (D). White-dial longcase clocks.

STROWBRIDGE, HENRY: Dawlish and Exeter. Watch and clockmaker and jeweller, Mill Place, Dawlish, 1830; Strand, Dawlish, 1844–56; Bedford St, Exeter, from 1862 (D). Death notice in *Exeter Gazette*, 18 Feb 1881: 'Feb. 10 at 259, High St, Mr Henry Strowbridge (formerly of Dawlish), aged 86'. Watch at Exeter Museum.

STROWBRIDGE, W. T: Exeter, 16 Castle St, 1873 (D).

STUMBELS (STUMBLES), WILLIAM: Aveton Gifford and Totnes. Married in 1716 and died in 1769. His best clocks are superb.*

SUMMER, RALPH: From Blundell's School, Tiverton, accounts: '1651 Paid Ralph Summer for mending of the clocke 7s 1d'.

SUMMERS, HENRY: Plymouth. 22a Union St, Stonehouse, 1889 (D).

SUMMERS, WILLIAM H: Barnstaple, 12 Strand, 1902 (D).

SWEETNAM, GEORGE F: Plymouth, 31 Saltash St, 1893–1902 and later (D).

SYMONDS, ——: Plymouth. Longcase clock *c*1780 (Bellchambers).

SYMONS, HENRY HAWKES: Devonport. 8 Albert Rd, Morice Town, 1873 (D).

SYMONS, HERBERT G: Plymouth, 104 Union St, 1889–1902 and later (D).

SYMONS, JAMES: Totnes, Castle St, 1844–50 (D).

SYMONS, JOHN E: Tavistock, 10 Duke St, 1889–1902 and later (D).

SYMONS, WILLIAM: Exeter. Watchmaker and jeweller, Goldsmith St, 1868 (EPJ); 226 High St, *c*1876–1907; then at Alphington.

TARDREW, RICHARD: Holsworthy. Clock reported *c*1780.

TASKER, ARTHUR F: Bideford, 85 High St, 1902 (D).

TAYLOR, ALLAN: Torrington. Watch and clockmaker, High St, 1866 (D).

TAYLOR, Mrs FANNY: Torrington, High St, 1873–8 (D).

TAYLOR, JOHN: Plymouth, 2 Vennal St, 1838 (D).

TAYLOR, JOHN P: Exeter. Paris St, 1859. Recorded in 1861 Census at Harris Place, watchmaker, aged 26, born Exeter.

TAYLOR, RICHARD: Exeter. Watch and clockmaker, Old Bridge, St Edmund's, 1824; at Cowick St later; then at Thomas Pollard's former premises in Alphington St, *c*1841. Was appointed inspector of weights and measures; and voted in the 1864 election, being described as watchmaker.

TAYLOR, WILLIAM: Exeter. Watchmaker, Alphington St; voted in 1864 election.

TEMPLER, RICHARD: Plymouth. Watchmaker, St Andrew's parish; took out licence 14 Dec 1759 to marry Mary Marshall of Charles, widow; and second licence, 4 Jan 1764, to marry Elizabeth Ham of Charles, spinster.

TEMPLER, RICHARD: Exeter. Watchmaker; son of Richard Templer; admitted as Exeter freeman by succession, 2 Nov 1776.

TEMPLER, THOMAS: Barnstaple. Watchmaker; son of John Templer, brazier, admitted as Exeter freeman by succession, 9 Sept 1780. 'A caution. Whereas, on Tuesday the 13th instant, absconded from their master, Mr John Oatway, in Great Torrington . . . Thomas Templer and son, his journeymen clock and watchmakers, and supposed to have carried off with them various articles in the clock and haberdashery line, the property of their said master. This is to request the public in general, and those persons in the above branches in particular, to stop all such goods as may be offered to sale by the aforesaid Thomas Templer and son. N.B. They are supposed to have taken their direction towards Exeter' (EFP, 29 Aug 1793).

TEMPLER, THOMAS: Chard, Somerset. Watchmaker; son of Thomas Templer, watchmaker; admitted as Exeter freeman by succession, 6 July 1802.

TEPPER, JOSHUA TURNER: South Molton. Watch and clockmaker and jeweller, Broad St, 1830–8 (D). 'On Tuesday night, the shop of Mr J.

Tepper, watchmaker, South Molton, was broken open by means of holes bored in the shutters, and a pane of glass being broken, four watches were stolen therefrom. There were other watches, and a quantity of jewellery within reach, which remained untouched' (EFP, 25 Mar 1830).

THOMAS, clockmaker of Exeter: church clock repairer, 1483.*

THOMAS, HARRY F: Bovey Tracey, Fore St, 1883 (D).

THOMAS, H. F: Chudleigh, 1889 (D).

THOMAS, HENRY: Exeter. Practical watch and clockmaker and working jeweller; at Broadgate in Dec 1857 (advertisement); then at 189 High St. Died 1865. In June 1864 he advertised: 'A jewelled duplex watch (by Grimaldi and Johnson), with compensating regulator; the movement taken from gold cases; cost, originally, sixty guineas, to be sold for eight guineas, a great bargain' (EFP).

THOMAS, HENRY: Moretonhampstead, Ivy Cottages, 1883 (D).

THOMAS, JAMES: Newton Abbot. Watch and clockmaker, jeweller and general dealer, Wolborough St, 1838–78 (D). Eight-day white-dial longcase clock sold at auction in 1974 for £150.

THOMAS, Mrs MARY: Newton Abbot. She continued James Thomas's business at 16 Wolborough St; listed there in 1883; at 20 Courtenay St, 1889–93 (D).

THOMAS, JOHN: Ilfracombe. Watch and clockmaker, silversmith, jeweller and optician, 113 High St, 1866 (D).

THOMAS, JOHN C: Ilfracombe, 31 High St, 1878; & Son, 1889–1902 and later. Also at 6 Arcade (D).

THOMAS, MARTIN: Plymouth Dock. Watchmaker; married Elizabeth Gill at Stoke Damerel parish church, 23 Feb 1786 (register). Bracket clock in ebonised case with silvered dial and verge escapement, signed 'Martin Thomas Plymouth Dock' on the dial and on the engraved backplate.

THOMAS, WILLIAM: Salcombe, 1873; Lower St, 1878; Fore St, 1883 (D).

THOMAS, WILLIAM: Exeter. Recorded at St Thomas in 1851 Census, aged 17, apprentice to a watchmaker, born Exeter.

THOMAS, WILLIAM HENRY: Exeter. Watchmaker, High St; voted in 1864 election.

THORN, THOMAS MEMORY: Exeter. Watchmaker; took out licence 6 Sept 1742 to marry Dorothy Butler.

THORNE, ABRAHAM: Tiverton. Born 1 May 1730. Was paid 10s 6d in 1756 for keeping and cleaning the church clock at Huntsham (churchwarden's accounts). In a letter to his sister, Mrs French, dated 14 Feb 1759, Martin Dunsford, author of the *History of Tiverton*, writes: 'Abraham, son of Simon Thorne, Clockmaker at the lower end of our street, ridded house last Sunday night and is gone off, having taken in a number of watches to righting, many of which it is like have been privately sold. And Mr Smale the Landlord has found means to come at some. Mr Stone next to Mr Terry and my Cousin Martin who are two of ye sufferers can neither of 'em tell ye time of ye day and tell me has asked several others without information and laugh at one another whilst others are more gravely condoling on the like occasion' (DNQ, vol 9). Mahogany bracket clock, 8-day quarter repeater with silvered dial and original bracket.

THORNE, MICHAEL jun: South Molton. Watchmaker; took out licence in 1791 to marry Sarah Thorne of South Molton. Listed in the *Universal British Directory*, 1798, as watchmaker, together with Michael Thorne sen, freeman and glass merchant. Longcase clock in fruitwood case (Tapley-Soper); another, in oak case, with brass dial sold at farmhouse auction in 1977.

THORNE, RICHARD: South Molton. Clock reported (Tapley-Soper).

THORNE, SIMON: Late of Tiverton, insolvent (*London Gazette*, 27 Aug 1720 – Buckley). Buried at St Peter's, Tiverton, 21 Sept 1722.

THORNE, SIMON (ii): Tiverton. Watch and clockmaker; James Greenway was apprenticed to him *c*1717, premium £7. Married firstly Garthrud Mills in 1723 and, secondly, Mary Rice of Exeter, in 1740; buried at St Peter's, Tiverton, 26 Feb 1761. His house and shop, which he rented from 1745 onwards, now forms part of the Midland Bank in Fore St. Kept the church clocks at Tiverton in repair for many years. In *Brice's Weekly Journal* of 9 Oct 1730 there is mention of Mr Thorne, clockmaker, in Tiverton, from whom catalogues could be obtained for an auction of books in Exeter. Numerous longcase clocks, mostly in cases of walnut, oak or fruitwood; an exceptional 8-day example with subsidiary dials illustrated in C. N. Ponsford *et al*, *Clocks & Clockmakers of Tiverton*. Gold pair-cased verge watch with embossed design of St George and the Dragon, hallmarked 1736, sold at Christie's in 1977 for £1,000.

THORNE, WILLIAM: Tiverton, 104 West Exe South, 1910 (D).

TICKELL, CHARLES P: Kingsbridge. Watch and clockmaker, gold and silversmith, jeweller and engraver, Fore St, 1838–50 (D). Longcase clocks.

TICKELL, JOHN: Crediton. Clockmaker and gunsmith; took on three apprentices around the year 1737: Nathaniel Harris and William White as clockmakers, and John Maunder as a gunsmith. Made a large mural timepiece for Crediton Unitarian Church in 1727 (Plate 46); also a three-train church clock for Thorverton (1751) and a church clock now at Exminster, but made originally for Crediton parish church in 1748 at a cost of £20. Longcase clocks reported, including an early example, *c*1720–30, with single hand, spandrels of twin cherub and large crown pattern, and 30-hour movement with steel corner posts; another, 8-day, arched dial, with walnut case, 2.5m (8ft 4in) high, was sold at Sotheby's.

TICKELL, JOHN NATHAN: Crediton. Clockmaker and gunsmith; John Lang was apprenticed to him *c*1756, premium £3. Crediton baptisms include: Mary, daughter of John Nathan Tickel and Elizabeth, 16 June 1756 (register).

TICKELL, JOHN N: Kingsbridge. Watch and clockmaker and jeweller, Fore St; in business from *c*1770. Died 7 Oct 1831, aged 80, and was described in the *Exeter Gazette* as 'the oldest tradesman of the town, having conducted his business with integrity and honour during a period of nearly 60 years; and has left a character equalled by few and excelled by none'. Many longcase clocks (Bellchambers). White-dial example noted, signed 'Jno. Tickell Kingsbridge', with dial and falseplate supplied by Osborne's Manufactory, Birmingham; good quality movement.

TICKLE (TICKELL), JOHN: Widecombe in the Moor. Clockmaker; took out licence on 1 Aug 1727 to marry Mary Battishill of Spreyton. A clockmaker of this name worked at Crediton.

TOKER, ——: Described as clockmaker of Ashburton in the Exeter Cathedral fabric roll for 1444–5 when he was paid 10s for repairing the great clock (*magno horilog.*), staying four days in the city.

TOLMAN, JOHN: Colyton, High St, 1889–1902 and later; also listed at Queen St, Seaton, in 1906 (D).

TOMKINS, GEORGE: Axminster. Watchmaker; took out licence 7 May 1731 to marry Martha Torr of Lyme Regis.

TOVEY, HENRY: Honiton. Silversmith, jeweller and watch and clock manufacturer, High St, 1850–89 (D). He and Matthew Murch were contractors for the St Paul's Church, Honiton, clock in 1851. He also had a clock outside his shop in 1848 showing Honiton local time and Greenwich Mean Time (see Murch, Matthew).

TOVEY, WILLIAM ALFRED: Honiton, High St, 1893–1902 and later (D).

TOWAN, JOSEPH: Plymouth Dock. Watch and clockmaker; arched-dial mahogany longcase clock, the 305mm (12in) silvered dial having phases of the moon and showing 'High Water at Plymouth Dock', the case with rope-twist quarter mouldings and hood pillars. Date estimated as *c*1780, but it could be later. Listed at 11 Duke St in 1822 (D).

TOWILL, FREDERICK G: Exeter, Eastgate Arcade, High St, 1883–1902 and later (D).

TOWSON, JOHN GAY: Plymouth Dock (Devonport). Watch and chronometer maker; listed in the *Universal British Directory*, 1798. The *Exeter Flying Post* in 1803 reported that 'on the night of Friday the 20th of May, the shop of Mr John Towson, watchmaker, No 69, Fore St, Plymouth Dock, was broke open'. Articles stolen included a metal gilt watch, maker's name John Towson, Plymouth Dock, other watches including two silver 'stop and second' watches, ten pairs of tea-tongs, engraved burnished teaspoons, standard gold rings marked 'SN', bracket candlesticks, and two old boatswain's calls. Business listed at same address in 1838 as Towson & Son (D). By 1850 he had retired to 12 Trafalgar Place, Stoke Damerel. Died 17 Apr 1857, leaving estate valued at under £1,500. Will in DRO mentions son John Thomas Towson and daughters; one of the witnesses was Charles Henry Kallensee, 71 Fore St, Devonport, watchmaker.

TOWSON, JOHN THOMAS (1804–81): Devonport. Watch and chronometer maker, scientific writer, inventor of various processes in photography, and 'well known for being the first to direct attention to the advantages of sailing on the great circle' across the Atlantic.*

TOZER, JAMES: Devonport. 32 Charlotte St, Morice Town, 1878 (D).

TOZER, WILLIAM HENRY: Torquay, 10 Victoria Parade, 1878 (D).

TRAIES, JAMES: Dartmouth. Watchmaker, silversmith, jeweller and engraver; came from Crediton, but was admitted as a freeman of Exeter on 21 Aug 1790 after serving his apprenticeship there with Francis Trobridge. Married Betsy Masters at St Saviour's, Dartmouth, 21 Nov 1790. 'Trays, ——, Dartmouth' recorded as watchmaker in the *Universal British Directory*, 1793.

TRAIES, WILLIAM: Watchmaker, of Dartmouth, son of James Traies, of Crediton, jeweller and engraver, admitted as Exeter freeman by succession, 8 Feb 1817. Served his apprenticeship in Exeter with John Upjohn, being a witness to a lease concerning the Upjohn premises in South St in June 1813. Later went to London and worked as a silversmith (see Chapter 5).

TRATT, ——: Colyton. White-dial clock with flowers in dial corners, applewood case (Tapley-Soper). '1816. May 14th. Mr Trott for keeping the clock in order £1 1s' (Colyton churchwardens' accounts). A Mr Tratt mentioned in later accounts.

TREADWIN, JOHN: Exeter. Chronometer, watch and clockmaker, Cathedral Yard, 'Has commenced business . . . duplex, lever, horizontal, and all kinds of English and foreign watches, accurately repaired' (EFP, 28 Jan 1847). Listed until 1854 (EPJ). Watches etc worth £300, some signed Treadwin Exeter, stolen from his shop (EFP, 23 Nov 1848). Born *c*1820, his father, John Treadwin, an Exeter bookseller, died when he was a boy and he was brought up by Nathaniel Tucker, who made clock cases for Henry Ellis. After Tucker's death in 1833, Ellis and a Mr Stoneman took over the guardianship, Treadwin attending a school at Mount Radford. He was then apprenticed to Exeter watchmaker Walter Tucker, and on completing his term went to London to improve himself in the watchmaking line before returning to his native city. Married Charlotte Elizabeth Dobbs on 28 Jan 1850 at Gulliford Chapel, Lympstone. She was a lacemaker and for many years ran a manufactory at Cathedral Yard. Examples of her work can be seen at Exeter Museum.

TREGENA, ANDREW: Crediton. Was paid 5s for mending Crediton Church clock and dial in 1704, and 10s for similar work in 1705 (governors' accounts).

TREHANE, SAMPSON: Exeter. Jeweller, silversmith and watchmaker, St Stephen's parish, 'willing to serve' (1803 Militia List); High St, 1812–36 (EPJ). Churchwarden of St Stephen's, 1812–13. Bankrupt 1820 (papers in DRO); John Broom, of St Thomas, watchmaker, appointed as assignee.

TRELEAVEN, SILVESTER: Moretonhampstead. Watchmaker, Cross St, 1823–30 (D).

TRELEAVEN, SILVESTER jun: Moretonhampstead. Watch and clockmaker, 1838–97 (D). Also sub-distributor of stamps and postmaster, and agent to the Royal Insurance Company, New St.

TRELEAVEN, WALTER: Plymouth, 11 Cornwall St, 1873–8 (D).

TRELIVING, JOHN: Devonport, 40 St Aubyn St, 1838; 5 Catherine St, 1844–50 (D).

TRELIVING, JOHN: Plymouth. Watchmaker, 36 Buckwell St, 1866 (D).

TRESCOTT, JOHN: Crediton. Working there in 1797, according to Venn's MS history of Crediton.

TRESSIE, THOMAS: Plymouth, 6 Regent St, 1878 (D).

TREVERTON, JAMES: Plymouth. Mahogany longcase clock, 8-day, brass dial.

TREVOR, JOHN: Topsham. Watch and clockmaker, Fore St. Married Louisa McQuarie, a widow, at Topsham parish church, 25 June 1818; their family included four sons and two daughters. Sowton churchwardens' receipts include: 'J. Trevor Jan 8, 1841: To repairing the church clock 7s 6d'; and a payment of 12s to him for 'three journeys to Sowton to put the church clock in repair' in 1843. Verge watch illustrated in C. N. Ponsford, *Time in Exeter.*

TREVOR, JOHN (ii): Topsham, Fore St; son of John and Louisa, born 21 Dec 1820. Recorded in 1851 Census together with his brother, James, 28, also a watchmaker. Still at Fore St in 1866; a John Trevor later recorded in Shapter St, 1893–1902 (D).

TREVOR, RICHARD (*c*1763–1827): Topsham. Clock and watchmaker; son of a maltmaker; married Margaret Elliott at Topsham parish church 17 Feb 1791 and was buried in the churchyard there 18 Jan 1827 (registers). 'Paid Richd. Trevor Clockmaker his bill & stamp £4 17s 8½d' (Topsham churchwarden's accounts, 1795). He subsequently had the care of the clock, his salary for keeping it in repair being agreed at £2 a year in 1810. Examples of his work at Topsham Museum, including 8-day silvered-dial, mahogany, longcase clock, showing 'High Water at Topsham Bar'. One of his watchpapers is pasted on the inside of the trunk door.

TREWIN, WILLIAM: Plymouth. 120 Union St, East Stonehouse, 1883–93 (D).

TRICK, WILLIAM: Bideford. Watch and clockmaker, Grenville St, 1844 (D). In Dec 1845 filed a petition for protection from process in the Court of Bankruptcy for the Exeter district (EFP).

TRIGGER, RICHARD: Dolton, 1878 (D).

TRIGGS, JOHN RICHARD: Cullompton, Fore St, 1893 (D).

TRIMBLE, SON & BROOKING: Exeter. Watch and plate dealers, 155 Fore St, 1852 (D).

TRIST, BENJAMIN: Brixham, Fore St, 1830–44; the business listed under Sarah Trist in 1850 (D).

TRIST, JAMES: Exeter. Watchmaker, St Sidwell's (1802 jurors' list).

TRIST, JAMES: Torquay, Braddons Row, 1844; Braddons Place, 1850 (D). Longcase clocks, including 8-day example with white-painted arch dial and inlaid oak case.

TRIST, JOSEPH C: Exeter, 30 New Bridge St, 1873–83 (D). Clock reported.

TROBRIDGE, JOHN: Apprenticed *c*1770 to Jas. Bucknell, watchmaker, Crediton, premium £8 8s.

TROUT (TROWT), WALTER: Exeter. Clockmaker; he submitted evidence at the city Quarter Sessions, 15 Jan 1697/8.

TRUGARD, JAMES: Plymouth Dock. A maker of longcase clocks, early to mid eighteenth century. Thirty-hour example with cherub-head spandrels *c*1720 in oak case; an arch dial with centre seconds *c*1760; another, in green lacquer case, sold at auction in 1979.

TRUGARD, MOSES: Dartmouth. Watchmaker; listed in the *Universal British Directory*, 1793. Eight-day longcase clock (Bellchambers).

TRUSCOTT, JOHN: Tavistock. Watchmaker; took out licence 5 June 1809 to marry Mary Welsford of Crediton. Premises for sale (EFP, 15 Mar 1810).

TUCKER, ——: He was paid £2 2s in 1736 for work on Newton St Cyres Church clock (churchwarden's accounts).

TUCKER, Miss HELEN JANE: Tiverton, Fore St, 1852–73 (D). Recorded in the 1861 Census as watchmaker and jeweller, aged 41, born Tiverton. She continued J. W. T. Tucker's business after his eyesight failed, and was succeeded by James Grason. Blundell's accounts include several payments to her between 1855 and 1860 for repairs to the school clock.

TUCKER, JAMES H: Exeter. Watchmaker, parish of Allhallows, Goldsmith St (1835 polling list).

TUCKER, JOHN: Dartmouth, Lower St, 1823 (D).

TUCKER, JOHN: Crediton, High St, 1883; 32 East St, 1889–1902 and later (D). Watchpapers noted.

TUCKER, JOHN: Tiverton. Watch and clockmaker; took over from Simon Thorne in 1760 as keeper of the town's church clocks and in 1771 made a new clock with additional quarter striking for St Peter's Church (replaced 1883). Was the son of another John Tucker, a slea and shuttle maker at Thorverton, and worked as a clock and watchmaker at Tiverton for upwards of forty-five years; his shop was on the north side of Fore St (Land Tax assessments). Was succeeded by his son John in 1805 and died, aged 77, on 3 Mar 1807 (EFP). His widow died in 1809. Apprentices included Henry Marsh (*c*1761), Samuel White (1771), and Benjamin Bowring. Pair-cased watch No 782 in Exeter Museum.*

TUCKER, JOHN (ii): Exeter and Tiverton. Born *c*1760; died 15 Aug 1829.*

TUCKER, JOHN WALTER TOTHILL: Tiverton. Watchmaker, jeweller and silversmith, Fore St; son of the above, with whom he served his apprenticeship; set up in business *c*1815. In Jan 1823 his wife Elizabeth, daughter of John Rossiter of Tiverton, died, leaving him with five young children. In 1826 his shop was burgled, and the next year was burgled yet again, when sixty-two watches and a quantity of plate were stolen. Two men, George Champion, 21, and John James, 22, were subsequently found guilty of the second burglary and publicly hanged on 24 Aug 1827 at the drop in front of the Devon County Gaol in Exeter. He made the 1830 clock at Tiverton pannier market; and served the town as mayor in 1847. Miss Helen Jane Tucker took over the business in the 1850s when his eyesight began to fail. He was recorded in the 1861 Census as retired watchmaker, aged 68, born Exeter; together with Mary A. Tucker, wife, 71, born London. Died 1 Feb 1881, aged 89, and was buried in St Peter's churchyard, where his headstone is by the path just to the south-east of the chancel window. Tiverton Museum has white-dial wall clock, also a number of billheads and receipts (Fig 12).*

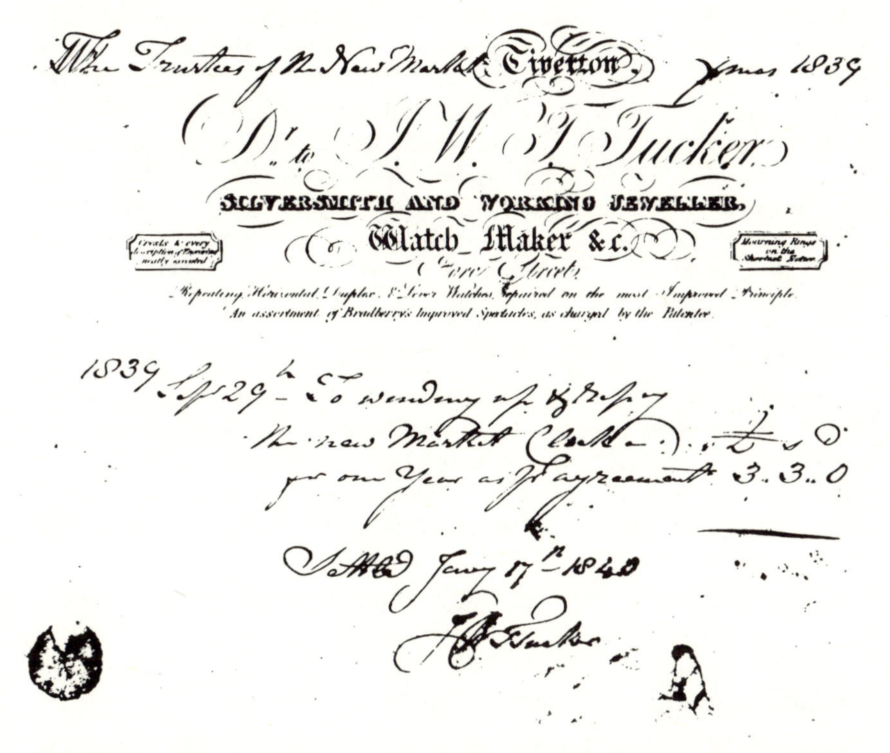

The Trustees of the New Market Tiverton Xmas 1839

Dr. to J. W. T. Tucker

SILVERSMITH AND WORKING JEWELLER,

Watch Maker &c.

Fore Street

1839 Sep 29th To winding up & keeping the new Market Clock for one Year as per agreement 3.3.0

Settled Jany 17th 1840

J W T Tucker

Fig 12 John W. T. Tucker, maker of the clock at Tiverton pannier market, submitted this bill for its maintenance at Christmas 1839

TUCKER, R. & M: Honiton, High St, 1897–1902 and later (D).

TUCKER, RICHARD WOODGATES: Honiton. Silversmith, watchmaker, jeweller, optician and circulating library, High St, 1866–93 (D).

TUCKER, ROBERT: Bideford, High St, 1844 (D). Witnessed codicil dated 30 Nov 1841 to will of Samuel White (ii), silversmith and watchmaker, Bideford (DRO). 'Married . . . May 18, at St Peter's Church, Tiverton, by the Rev Mr Kirwan, Mr Robert Tucker, silversmith, of Bideford, to Ellen Mary, eldest daughter of Mr James Sellick, relieving officer, Tiverton' (EFP, 6 June 1844). Watch movement at Exeter Museum, 'R. Tucker Bideford', verge converted to lever.

TUCKER, WALTER JAMES KELLAND: Exeter. Clock and watchmaker, silversmith and jeweller; youngest son of John Tucker (ii), born *c*1801. In business in the High St from 1826–48; listed at 9 Belmont Terrace in 1850. Apprentices included John Treadwin. White-dial longcase clocks.

TUCKER, WILLIAM: Chulmleigh, East St, 1902 and later (D).

TUCKER, WILLIAM GOLDSWORTHY: Exeter. Watchmaker and jeweller, High St; married Henrietta, daughter of Jacob Bricknell, of Exeter, Jan 1826 and died at York 27 Mar 1852. The St Petrock's, Exeter, churchwarden's accounts for 1827–8 record payment of a guinea to him for keeping the church clock in repair. Bankrupt in 1829, stock-in-trade and household furniture at 64 High St offered for sale including '5 gold watches, 27 silver watches. One regulator, one 30-hour, 1 round, and 3 bracket 8-day clocks; works of an 8-day clock, 12 dials . . . Turning lathe, quadrant, three vices, clock & watch tools, two working benches, lead weights, grinding stone, vice box with clock work' (EFP, 25 June 1829).

TURNER, ——: He is said to have made verge watches at Totnes *c*1810.

TURNER, GEORGE: Honiton. Married Miss J. Lee in Dec 1815 (EFP). Listed in 1838 as watchmaker, registrar of births, deaths and marriages, and agent for the Clerical & Medical General Life Assurance Office, High St (D). Longcase clock reported. His name noted on gallery clock at St Paul's Church, Honiton. 'Died – February 19, at Honiton, Mr George Turner, formerly silversmith, aged 77' (EFP, 24 Feb 1869).

TURNER, GEORGE: Torquay, Lower Union St, 1844–73 (D).

TURNER, H: Plymouth. 67 Union St, Stonehouse, 1856 (D).

TURNER, JOHN: Ottery St Mary, Silver St, 1889 (D).

TURNER, WILLIAM MILLS: Ilfracombe, 24 High St, 1878; 13 High St, 1883–1902 and later.

TURTLIFF (TURTLIEF), THOMAS: Of Plymouth; apprenticed as clock-maker *c*1715 to Jacob Lovelace, Exeter.

TUTTLE, ROBERT SHEPHERD: Plymouth. 121 Union St, East Stone-house, 1902 (D).

TWINER, WILLIAM M: Ilfracombe (see Turner, W. M.).

TYACKE, GEORGE: Torrington, South St, 1883 (D).

UGLOW, A: Ridgway, Plympton, 1856 (D).

UGLOW, A: Exeter. 'Clock and watch manufacturer, 61 High St, Exeter, opposite the Guildhall' (EFP, 3 Feb 1859). Selling off stock of watches and jewellery 'at a great reduction previous to retiring from business' (*Western Times*, 8 Dec 1860).

UGLOW, ABEL: Watchmaker; married Agnes Perkin at Milton Damerel, 21 Nov 1799 (Miles Brown).

UGLOW, ABEL: Tavistock. Watchmaker, West St, 1838 (D).

UGLOW, NICHOLAS JOHN: Sidmouth, Fore St, 1866–97; business listed under Mrs N. J. Uglow in 1902 (D).

UGLOW, WILLIAM: Cullompton. Listed in 1823 (D). 'A burglary was effected a few nights since at the shop of Mr William Uglow, clock & watchmaker at Collumpton and a quantity of goods carried off' (EFP, 19 Apr 1827). He and wife, Mary, had three children who were baptised at Cullompton Methodist Church: Kerenhappuch, born 1821, Henretta, 1824, and Keziah, 1826 (register). Eight-day white-dial longcase clocks.

UGLOW, WILLIAM: Newton Abbot. Watchmaker and jeweller, Court-enay St, 1866–73 (D). The frame of Combeinteignhead Church clock has repairer's mark, 'W. Uglow Feb 9th 1866', and others dated 1870 and 1873.

UGLOW, Mrs: Newton Abbot, 8 Bank St, 1878 (D). 'Executors of William Uglow, watchmaker, optician and agent for Medical & General Insurance Co.'

UMBER, THOMAS: Kept the St Peter's Church, Tiverton, clock in repair for £2 annually in 1738, 1739 and for six months of 1740 (churchwardens' accounts).

UPCOTT, CHARLES HAYNES: Lympstone. Watchmaker, 1838–66 (D).

UPHAM, RICHARD: Ottery St Mary. Clock and watchmaker; kept East Budleigh Church clock in repair between 1772 and 1778 (churchwardens' accounts). Insolvent and imprisoned for debt (EFP, 1 July 1774).

UPJOHN & COMPANY: Exeter, 39 High Street. 'Messrs Upjohn and Co Beg to call the attention of the Nobility, Gentry, and inhabitants of Exeter and its vicinity to their celebrated manufacture of watches and clocks, English and Foreign; also to their well-assorted and elegant selection of jewellery' (*Devon Weekly Times*, 18 June 1875).

UPJOHN & MICHIE: Exeter, 39 High St, 1868–75 (EPJ).

UPJOHN, EDWARD: Founder of the Devon branch of the family; at Topsham in 1726; in Exeter *c*1740; died 1764.*

UPJOHN, EDWARD (ii): Exeter. Watch engraver; son of the above, died Jan 1741/2.*

UPJOHN, HENRY: Bideford. Watch and clockmaker; listed in the *Universal British Directory*, 1793. Son of Peter and Elizabeth Upjohn; baptised 4 June 1769; married Mary Bowen at Bideford, 2 Jan 1791. Cleaned Northam Church clock in 1815 (churchwarden's accounts). Longcase clocks include one in a massive carved case. A wall clock, 'Hy Upjohn Bideford', has 305mm (12in) wooden dial, 5-pillar movement with tapered plates, and bevelled pendulum with heavy bob.

UPJOHN, JAMES: Exeter. Watchmaker; apprentice of John Upjohn, St Mary Major parish (1803 Militia List). Was probably Thomas Upjohn's son James, baptised 1783. From Cheriton Fitzpaine marriage register: 'James Upjohn, of St Martin, Exeter, and Anna Catherine Upjohn, by licence 21 Oct 1806. Witnesses W. Upjohn, John Upjohn'.

UPJOHN, JAMES ROUSE: Exeter. Son of John and Maria; was briefly in partnership with his brother Robert after their father died in 1848; recorded in St Thomas in 1851 Census, described as 35, unmarried, visitor, watchmaker, born Exeter.

UPJOHN, JOHN: Exeter. Watchmaker. Thomas Day was apprenticed to him *c*1762. A John Upjohn can later be traced in Clerkenwell.

UPJOHN, JOHN (ii): Exeter. Clock and watchmaker; son of Thomas and Charlotte, born 3 Jan 1771; married Maria Wittingham and had a family of thirteen children. Worked in London for twelve years, then in 1803 took over William Upjohn's business in Exeter. Churchwarden of St Mary Major 1816–17; mentioned as a surviving trustee of the parish in 1842 (DRO). The business in South St became known as Upjohn & Sons in the 1840s and he retired to Cowick St, where he died 2 Dec 1848 'after a long and painful illness, borne with christian fortitude'. A silvered brass dial wall timepiece, 'J. Upjohn Exeter' illustrated in *English Dial Clocks* by Ronald E. Rose (p134). Turret clock in Exeter Museum collection.*

UPJOHN, JOHN WILLIAM HENRY: 'Died. On the 13th inst., at his father's house in this city [Exeter], universally beloved by all who knew him, aged 41, John, eldest son of Mr J. Upjohn, and of the firm of J. and T. Upjohn, Chandos St, London' (*Western Luminary*, 13 June 1836). Burial recorded in St Mary Major parish.

UPJOHN, NATHANIEL: Plymouth. Watch and clockmaker, St Andrew's parish. Son of Edward and Mary; baptised at Shaftesbury, Dorset, in 1717; took out licence on 29 May 1749 to marry Martha Brown, a widow. William Harry was apprenticed to him *c*1761, premium £68 5s. Watches and longcase clocks noted; one of the latter fetched £560 at auction in Kingsbridge in 1976. Died in December 1782; his stepsons, John and William Brown, were both in the trade (Loomes).

UPJOHN, PETER: Bideford. Son of Edward and Mary; baptised at Shaftesbury, Dorset, in 1714. Twice married, firstly to Elizabeth Smith at Bideford, 17 Aug 1745, then to Elizabeth Isaac at Barnstaple, 10 Mar 1754 (registers). He served as Bideford churchwarden in 1774 and was buried at the parish church 30 Jan 1795. Longcase clocks include a silvered arched dial example in mahogany case; another, 30-hour, is signed on a curved strip affixed beneath the top of the chapter ring, dial centre engraved with coastal scene. He or his son Peter listed at Bideford in the *Universal British Directory*, 1793, as watchmaker, silversmith, maltster and common brewer. A daughter Elizabeth, by his first marriage, married a Shepton Mallet draper called Moore (Loomes). There were six children by his second marriage; the eldest son, Peter, was described in Nov 1819 as 'late of Bideford, decd., maltster and brewer' (*Exeter Gazette*); the youngest, Henry, was a watchmaker. Exeter Museum has verge watch by Peter Upjohn, Bideford, No 213, case hallmarked London 1780–1.

UPJOHN, RICHARD: Exeter. Clock and watchmaker; son of Edward and Mary; baptised at Topsham 7 Feb 1727/8; died Exeter, May 1778.*

UPJOHN, ROBERT WITTINGHAM: Exeter. Chronometer, watch and clockmaker; son of John and Maria; married a Miss Ellis of St James St, Exeter, at Bradninch, in 1846; died, aged 59, on 1 July 1866. Took over from his father as keeper of the Exeter Cathedral clock; his initials 'RU 1824' are carved in the clockroom. 'Removed from 16 South St to 39 High St, opposite Queen St' (*Western Times*, 8 Sept 1855).*

UPJOHN, THOMAS: Launceston and Exeter. Son of William and Anna; died 1783.*

UPJOHN, THOMAS NATHANIEL JAMES: 'Died: Aug. 21, at Exbourne, in this county [Devon], aged 40, Thomas, second son of Mr J. Upjohn, of this city [Exeter], and of the firm of J. and T. Upjohn, watch manufacturers, Chandos Street, London' (*Exeter Gazette*, 26 Aug 1837). Married Ann Westlake at Exbourne in 1827.

UPJOHN, WILLIAM: Exeter. Son of Edward (i); married 1736; died 1768.*

UPJOHN, WILLIAM (ii): Exeter. Watch and clockmaker and silversmith; son of the above; baptised St Mary Major 29 Mar 1754; completed his apprenticeship with his elder brother Thomas in Exeter, a premium of £6 being paid in 1772. Set up in business in Fore St, his name appearing in Quarter Sessions jury lists from 1775, being described as of the parish of St John in Oct 1779 and of St Mary Arches in Jan 1782. On night of 5 Feb 1781 his shop was broken open and robbed of several watches, rings, and other valuables. The stolen watches included 'one gold chas'd watch, name W. Upjohn, Exeter, device on the case Telemachus and Calipso'. A man was subsequently brought to trial and sentenced 'to be sent on the River Thames to hard labour for three years'. His premises in 1784 were 'opposite the Corn Market in the Fore Street' (EFP); he then moved to the Cathedral Close, but in 1800 advertised for an apprentice and stated he had removed from there 'to a house three doors above Bear Lane, Southgate St, where he continues to manufacture and repair all sorts of clocks and watches' (EFP). Was succeeded there in 1803 by his nephew, John Upjohn, and went to London, where he died in 1812. By his marriage to Elizabeth Glanville he had a large family; his son William John Upjohn was a watch manufacturer at 11 St John's Square, Clerkenwell; a daughter, Henrietta, married a watchmaker named Bird, from Bristol; and another daughter married John Mallett, the Barnstaple watch manufacturer. His widow, Elizabeth, lived on to her ninetieth year, dying at the house of her son-in-law in Brentford 24 May 1841. William Upjohn made Plymtree Church clock.

VALLACK, THOMAS E: Plymouth, 151 North Rd, 1873; 32 High St, 1878 (D).

VANSTONE, JAMES: Holsworthy. Watchmaker, 1850 (D).

VEALES, EDWARD: Cullompton. Watch and clockmaker, Fore St, 1830 (D).

VEALS, JAMES: Exmouth. Watchmaker and jeweller, Wellington Place, 1838–56; Sheppards Walk, 1866; 45 Strand, 1873–93 (D).

VELVIN, J: Plymouth, 33 Buckwell St, 1866 (D).

VOYSEY, JOHN: Kenton. Watchmaker and clock cleaner, South Town, 1850–73 (D). Tendered successfully in 1846 to repair Thorverton Church clock (vestry book). A daughter was baptised at Kenton in 1826; three sons, Henry, William and John, received £150 between them under the will of Abraham Payne, of Kenton, gent, *c*1846 (DRO).

WADELTON, CHARLES: Plymouth. 20 Chapel St, Stonehouse, 1889 (D).

WADELTON, JOHN: Devonport. Clock and watchmaker, 6 Duke St, 1830 (D).

WADELTON, JOHN: Plymouth, 23 Frankfort St, 1873–83 (D).

WAKELIN, JOHN R: Exeter, 26 Holloway St, 1870; South St, 1873–8 (D).

WALDRON, JOHN: Tiverton. 'Paid Mr Waldren for righting ye clock 7s 6d' (Huntsham churchwarden's accounts, 1721). Took out licence 22 Nov 1723 to marry Miss Elizabeth Gale of Tiverton, and after her death took out a further licence 28 Sept 1732 to marry Sarah Parkhouse of Uplowman, a widow. The *London Gazette* of 23 July 1737 lists him as insolvent. A lantern clock noted, also several 30-hour longcase clocks and a good quality 8-day clock with engraved dial border and crown and cherub spandrels, housed in a carved oak case.

WALKER, JOSIAS: Ottery St Mary. Watch and clockmaker; took out licence 20 Nov 1783 to marry Mary Ashford of Ottery St Mary. Listed in several directories; at Silver St in 1830. Longcase clocks, including 30-hour silvered brass dial example (Tapley-Soper).

WALTER, MICHAEL: Plymouth, Frankfort St, 1850 (D).

WALTER, S: Devonport. 12 William St, Morice Town, 1856 (D).

WALTERS, ——: Received £1 10s in 1720 'for mending & cleansing' Gittisham Church clock and at the same time promised to keep it in repair for seven years (churchwarden's accounts).

WALTERS, EDWARD: Exmouth, Lower Parade, 1883–1902 and later (D).

WALTERS, MICHAEL: Plymouth. 'Wooden clock maker', Old Town St, 1836 (D). A Michael Walter listed later.

WALTON, PHILIP: Holsworthy. Longcase clock, 'Phil. Walton Holsworthy'.

WARD, SAMUEL: Plymouth, 36 Tavistock Rd, 1893–7; at 45 Old Town St later (D).

WARNER, ——: 'Dartmouth, England. 1844. Curious clock with clock dial and hands and works of a watch in a watchcase at the back to make it go. Owned in Boston, Mass' (N. Hudson Moore: *The Old Clock Book*, 1912).

WAY, JOHN: Exeter. Pawnbroker, plate and watch dealer; recorded in the 1851 Census at 155 Fore St, aged 30, married, born Axmouth.

WAYCOTT, PETER: Holne, Ashburton and Totnes. Clockmaker. In 1916–17 a correspondent reported: 'There are many of his clocks in Devon today and some by his son Robert, of Paignton and Torquay. Peter lived at Ashburton in 1799 and made clocks then or soon after. He had a machine for cutting the wheels. He was a "jack of all trades", so made the clock cases, the works and painted the dials. His son Robert continued to manufacture until the forties, when these tall clocks appear to have been eclipsed by the Dutch and American article. Peter had two brothers, William and Richard, who sailed from Dartmouth somewhere about the year of Waterloo – 1815. They went to Nova Scotia and made clocks there, many of which are to be found today' (DNQ, vol 9).

WAYCOTT, PETER: Plymouth. Clock and watchmaker; granted a patent, No 6126, in conjunction with John Lee, auctioneer, 22 June 1831 for improvements in mangles (DNQ, vol 9).

WAYCOTT, PETER: Watchmaker; signed petition in 1836 'for the making of the Torbay Road' along Torquay seafront.

WAYCOTT, ROBERT: Paignton and Torquay. Watch and clockmaker; son of Peter Waycott. Recorded at Paignton in 1838; at 9 Braddons Row, Torquay, in 1844 (D).

WAYCOTT, WILLIAM: Ottery St Mary. Watchmaker, 1823 (D).

WEBB, CHARLES: Exeter. Watchmaker; provided sureties, at Exeter Quarter Sessions 4 May 1722, and for a marriage licence application, 18 Sept 1728.

WEBB, JOSEPH: 'Ran away from his master, Mr John Mallett, Clock & Watch Maker, Barnstaple, on the 24th day of June last, Joseph Webb, his apprentice. He is about 5 feet 6 inches high; wore away a blue coat and trowsers.' Whoever harbours or employs him after this public notice, will be prosecuted to the utmost severity of the law' (EFP, 27 July 1820).

WEBBER, AUGUSTUS: Ilfracombe, 13 Portland St, 1873; 35 High St, 1878; and at 57 High St, 1889–97 (D).

WEBBER, C. P: Plymouth, 96 Union St, 1897–1902 and later (D).

WEBBER, GEORGE: Exeter. Watchmaker, North St, 1828–30 (EPJ). The same name listed at Crediton later.

WEBBER, GEORGE: Crediton. Watch and clockmaker, engraver, jeweller and silversmith. The Exeter Flying Post has a notice dated 9 Nov 1833, announcing that Mr Bucknell, watchmaker and silversmith, Market Place, Crediton, has resigned his business in favour of Mr George Webber, 'a person well qualified to give every satisfaction in the above line'. He married Fanny, youngest daughter of John Mogridge, at Crediton on 13 Mar 1839, and is listed in directories, 1838–56, at High St. Several high quality clocks noted, including a regulator in Gothic-style case (Bellchambers); an imposing bracket clock with silvered convex dial with Greek key pattern around the edge; and a wall timepiece in carved mahogany case with inlaid brass decoration and moulding; also a watchpaper, 'WEBBER Watch & Clock MAKER Engraver & Jeweller Successor to Mr Bucknell CREDITON'.

WEBBER, HENRY: Ilfracombe. Clock and watchmaker, 1823–66 (D); address given as 35 High St.

WEBBER, HENRY: Bideford, Mill St, 1850 (D).

WEBBER, JOHN G: South Molton. The Antiquarian Horological Society publication *Horology in Provincial and Rural Museums* gives his dates as 1856–1934 and states that South Molton Museum has eighteenth-century wheel-cutting engine last used by him (p30). Listed at 10 Barnstaple St in 1897; at 25 East St, 1902 and later (D).

WEBBER, RICHARD: Pilton, Barnstaple. Reputed to have made Pilton Church clock in 1713. Braunton churchwardens' accounts have these items: 1714, 'pd Mr Webber for setting up the clock £1'; 1715, 'pd Richard Webber for keeping and repairing the clock 5s'. Further payments recorded in 1716 and 1717.

WEEKS, JOHN: Plymouth. Watchmaker. The *Exeter Flying Post*, 17–24 Jan 1772, has an interesting notice: 'An eloped wife. Whereas Esther, the wife of me Anthony Grmielne, of Plympton, in the county of Devon, left my house on the 31st day of December last, under pretence of paying a visit to her mother at Chepstow, in Monmouthshire, instead of which she has been lately discovered to cohabit with one John Weeks, of Plymouth, watchmaker, and gives out and pretends she is not my wife. These are therefore to caution all persons whatsoever not to credit her on my account, as I am determined not to pay or be accountable for any debts that she may contract; and I do hereby caution the clergy in general, within the diocese of Exon, from being imposed on by her pretending to be a single woman. Witness my hand the 17th day of January, 1772. A. Grmielne. NB. She is now about the age of 18 years, her maiden name was Hodges, and was born at Bristol. Witness, E. Hodges, her mother'.

WELCH, GEORGE: Items in Crediton governors' accounts include: '1687 Sept. 10th pd then and formerly to George Welch for mending the church clock and making two wheeles & clensing the clock £2'. Further payments recorded to him up to 1699, including one for mending muskets 'in the tyme of Monmouth's rebellion'.

WELCH, WILLIAM: Plymouth Dock. Primarily a silversmith, flourishing 1758–1800 (*Catalogue of Exeter Silver* in Exeter Museum, p58). His watch No 7207 reported lost (*Public Advertiser*, 4 Oct 1771 – Buckley). William Welch jun was a witness at the wedding of the jeweller John Colman Hearle at Stoke Damerel on 26 Nov 1791 (register).

WELLER, GEORGE: Exeter. Clockmaker; kept the Exeter Cathedral clock in repair from 1718 to about 1742 (fabric accounts), and also looked after the St Thomas, Exeter, Church clock and chimes. A 'Mr Wellers' was paid 10s at Ashburton in 1715 'for fitting ye Clock & Chaimes'. Worked on the Kenton Church clock in 1720 and, at South Molton in 1726, was paid £15 for 'new makeing the Clock & Chimes' (churchwarden's accounts). His signature is among those approving the St Thomas parish accounts for 1732 and 1733. Longcase clocks noted, including one with a very fine dial inscribed 'Geo. Weller Exon' (Tapley-Soper). A George Weller repaired Wells Cathedral clock in the late seventeenth century.

WELLINGTON, A. F: Kingsbridge, 10 Fore St, 1902 and later (D).

WELLINGTON, CHARLES H: Totnes, 15 High St, 1897–1902 and later (D).

WELLINGTON, CHARLES HAMILTON: South Brent, Station Rd, 1902 (D).

WELLINGTON, WILLIAM THOMAS: Kingsbridge. Ebrington Place, Dodbrooke, 1873–8; Market Place, Dodbrooke, 1883; Duke St, 1889–97; at Loddiswell, 1902 (D).

WELSFORD (WELSHFORD), WILLIAM: West Teignmouth. Watchmaker; took out licence 16 Dec 1802 to marry Elizabeth Taylor. John Bulkeley was apprenticed to him *c*1802–3. Moved to Plymouth *c*1807.

WEST, JAMES: Sidbury. Clockmaker; entry in Colyton burial register reads: '1766. Crago, Joseph, carpenter, who married Rachel, daughter of James West of the parish of Sidbury, clockmaker, 15 Apr. (died) 22 Apr. (buried)'. The Salcombe Regis clockmaker Richard Pinn married an Anne West, of Sidbury, in 1763.

WESTAWAY, JOHN: Exeter and Teignmouth. Apprentice of Thomas Pollard, St Thomas, Exeter. 'John Westaway, watch and clock manufacturer, begs leave to inform . . . he has just commenced business at 167 Fore St, Exeter, opposite the Corn Market . . . He is lately from Burnett's, Fleet St, London, where he had for some years past, the sole management of the concern. Foreign and English watches and clocks of every description cleaned and repaired' (EFP, 22 Jan 1818). He also opened a shop 'during the season' at Teignmouth, but went bankrupt in 1821; after which he continued to work in Exeter for some years, moving to a house near the Old London Inn, St Sidwell's, in 1826. His wife's name was Jane and two children were baptised at St Petrock's, Exeter, in Oct 1823 (see Chapter 5).

WESTLAKE, ——: Exmouth. Watch and clockmaker, 1823 (D).

WESTLAKE, GEORGE: Devonport. 118 Navy Row, Morice Town, 1838 (D).

WESTLAKE, JOHN CAREW: Exeter, 9 Alphington St, St Thomas, 1883 (D).

WESTLAKE (WESLAKE), ROBERT: Plymouth, early nineteenth century (Guildhall Index, London).

WESTLAKE, S. J: Plymouth, 29 Ebrington St, 1902 and later (D).

WHEELER, JAMES: Torquay. Watch and clockmaker, Higher Union St, 1850; Fore St, St Marychurch, 1866 (D).

WHERLY, PHILIP: Barnstaple, Boutport St, 1850 (D). German clocks.

WHITBY, WILLIAM: Cullompton. Watch and clockmaker; listed in the *Universal British Directory*, 1793, and in other directories, at Fore St, 1823–30. Was owner of two new-built houses at Cullompton, advertised to let in June 1812 (EFP). Numerous payments to him were recorded in the Cullompton churchwardens' accounts for keeping the clock and chimes in repair from 1798 to *c*1823. Longcase clock with painting of local militia on dial.

WHITE & CROOK: Morchard Bishop, 1873 (D).

WHITE, ROBERT: Morchard Bishop. Watch and clockmaker, 1850–78 (D). Painted-dial longcase clock, 'White Morchard'.

WHITE, SAMUEL: Bideford and Torrington. Son of a Torrington saddler, he served six years of his apprenticeship at Tiverton with the watchmaker John Tucker, commencing 11 Mar 1771. Later in business at Bideford, at the Quay and then in Allhalland St, on the corner of Bridge St; he also had a shop in Torrington which he opened on market days. Married Mary Green – at Torrington parish church 5 Apr 1781 – by whom he had three sons and six daughters; died, aged 81, at Bideford on 12 Oct 1837, being described in the *Exeter Gazette* as 'an affectionate husband, a kind father, and universally esteemed by all who knew him'. The business was listed in directories, 1823–30, as S. White & Son. A silver watch, Samuel White, Torrington, No 75,860, reported lost (*Exeter Gazette*, 25 Dec 1824).*

WHITE, SAMUEL (ii): Bideford. Watchmaker and silversmith; son of the above. Born 13 Dec 1793, he married the daughter of a Bideford innkeeper named Pim in March 1823, and died 8 Apr 1843. Worked in the family business, which he continued after his father's death in 1837. Will in DRO; estate valued at under £800; codicil dated 30 Nov 1841 is signed with his mark.*

WHITE, WILLIAM: Son of Daniel; apprenticed *c*1737 to Jn. Tickell, of Crediton, clockmaker, premium £5.

WILBY, HENRY: Plymouth, 5 York St, 1893; Lower Compton Rd, 1897 (D).

WILKINSON, EDWIN: Cullompton, Fore St, 1873 (D).

WILKINSON, ROBERT: Exeter, Sidwell St, 1873–83 (D). 'From London, and 12 years with Ellis Brothers [Exeter], watch jobber to the trade' (*Western Times*, 8 Dec 1876).

WILLCOCK, J: Plymouth, 2 Bank St, 1856 (D).

WILLIAMS, ALFRED: Plymouth, 88 Old Town St, 1878 (D).

WILLIAMS, CHARLES: Exeter, 35 South St, 1844 (D).

WILLIAMS, GEORGE: In 1799–1800 he supplied 'a clock of 2 dials' (since replaced) for Bideford parish church and was paid £61 9s 6d (churchwarden's accounts); he carried out repairs in 1804. There was a maker of this name at Bristol.

WILLIAMS, JAMES: Plymouth. Watchmaker, Charles parish; took out licence 28 Dec 1769 to marry Elizabeth Lymbery.

WILLIAMS, JOHN: Bideford. Clockmaker, Allhalland St, 1838; Mill St, 1844–56 (D). Henry Ellis, of Exeter, relates that in the early 1830s 'my sale of clocks not being very extensive, I caused them to be manufactured at Bideford by John Williams, a person in the employ of Mr White'. Mahogany longcase clock, 'J. Williams Bideford', noted at Bateman's, the former home of Rudyard Kipling at Burwash, Sussex.

WILLIAMS, JOHN: Clockmaker; married Jane Squance at Torrington parish church 17 Sept 1820 (register).

WILLIAMS, JOHN: Torrington, Potacre St, 1844–50 (D). Death notice:

'March 14, at Torrington, aged 33, Eliza, wife of Mr John Williams, clock and watchmaker, of that town' (EFP, 30 Mar 1848).

WILLIAMSON, ——: Was paid £1 8s 10d in 1768 for cleaning and repairing the St Sidwell's, Exeter, church clock (churchwarden's accounts).

WILLIS, WILLIAM: Exeter. Watchmaker, Sidwell St, 1835 (D).

WILLMOTT, GEORGE: Axminster, Chard St and Victoria Place, 1873–1902 (D).

WILLMOT, JOHN: Plymouth Dock. Watchmaker; took out a licence 28 Sept 1807 to marry Grace Goodridge of Stoke Damerel.

WILLS, GEORGE H: South Molton, 7 King St, 1889; at Romansleigh, 1897–1902 (D).

WILLS, JOSEPH: South Molton and Philadelphia, USA. South Molton churchwardens' accounts have these items: 1722, 'Pd Joseph Wills towards makeing the clock & hand £1 1s. In my hand to pay Joseph Wills when the work is finished £10 19s' (Clock was new made in 1726 by 'Mr Weller'). American correspondent reports that he went to Philadelphia (*c*1725–30) where he made many 30-hour and 8-day clocks, and died in 1759. He left monetary portion of his estate to Giles and Grace Tucker, son and daughter of Grace Tucker of South Molton.

WINDEATT, RICHARD: Exeter, 7 Catherine St, 1873–8 (D).

WINGATE, G. M: Plymouth, 86 Treville St, 1893 (D).

WINSOR, Mrs J. & SON: Paignton, Winner St, 1883 (D).

WINSOR, THOMAS: Paignton. Watchmaker; listed in directories from 1850; at Winner St, 1878.

WINSTANLEY, JOSEPH: Barnstaple. Clockmaker; said to have made new chimes for the parish church in 1710 for £19. His name noted as surety on document dated 3 July 1730 (Tapley-Soper).

WINTERHALTER, FERDINAND: Torquay, Fleet St, 1878–93; 6 Braddons Rd West, 1897 (D).

WITHINGTON, WILLIAM B: Axminster, West St, 1893; Lyme St, 1897–1902; and at West St again later (D).

WOLF, AARON: Plymouth, 20 Whimple St, 1866–78 (D).

WOOD, DAVID: Rackenford. Clock and watchmaker, 1866–73 (D).

WOOD, H: Dartmouth, Higher St, 1856 (D).

WOOD, HENRY: Dartmouth. Watch and clockmaker, Lower St, 1830; Spithead, 1838 (D).

WOOD, HUBERT: Tiverton, St Peter St, 1878–89 (D).

WOOD, JOHN: Exeter. Watchmaker; a son, James, baptised at Bow Meeting House, Exeter, in 1727 (register). An 8-day, square-dial longcase clock by this maker was sold at Exeter in 1976. Several other longcase clocks also recorded.

WOOD, JOHN BOURGOYNE: Torquay. Fore St, St Marychurch, 1883–1902 and later (D).

WOODE, WILLIAM: 'Ite paide to William Woode for mendying of the clocke 20d' (South Tawton churchwarden's accounts, 1600).

WOOLLEN, FRANCIS: Barnstaple, Anchor Lane and Holland St, 1878 (D).

WOOLLEN, FRANCIS J: Barnstaple, Queen St, 1893 (D).

WOOLLEN, FREDERICK: Barnstaple, Holland St, 1889 (D).

WOTTON, GEORGE: Torquay, 12 Lucius St, 1883–93 (D).

WOTTON, GEORGE: Paignton, 9 Victoria St, 1897–1902 and later (D).

WOTTON, SILVANUS: Paignton, 2 Church St, 1889; 9 Victoria St, 1893 (D).

WYATT, EDWARD WILLIAM: Exeter. Watchmaker; listed 1832–78; at Commercial Rd from 1848 (EPJ).

WYATT, ELIZA: Plymouth. Watchmaker, 7 Kinterbury St, 1866 (D).

WYATT, JOHN SAMUEL: Devonport, 65 St Aubyn St, 1883–1902 (D).

WYATT, R. jun: Plymouth, 34 Whimple St, 1856 (D).

WYATT, ROBERT: Plymouth. Clock and watchmaker, 27 Treville St, 1822; 27 Kinterbury St, 1830; 7 Kinterbury St, 1838–57; also at 3 Buckwell St in 1850 (D). St Andrew's, Plymouth, churchwarden's accounts have these items: 1815–16, 'Paid Mr Wyatt repairing the clock £10'; 1820, 'June. Paid Wyatt his bill for the chimes by order of vestry £14 4s 6d'. He further repaired the clock in 1825–6 and kept it wound at least until 1833. Died 22 May 1857, leaving estate valued at under £450. Will in DRO, mentions daughters Eliza and Mary Ann.

YELLAND, ARTHUR JOHN: Exeter, 46 Paul St, 1889–93 (D).

YELLAND, JOHN CHAPMAN: Exeter. Working jeweller, watchmaker

and engraver, Market St (1864 list of voters); 4 North St, 1866–73 (D).

YEO, THOMAS: Holsworthy. Clock and watchmaker and ironmonger, 1823–44 (D).

YOUNG, JOHN & CO: Exeter (see Depree, F.T.).

ZARINGER, ALBERT: Plymouth, 55 Union St, 1878 (D).

ZIEGLSBAUER (ZIEGELBAUER), CHARLES: Torquay, Higher Union St, 1866–78 (D).

7

# MAKERS BY TOWN AND VILLAGE

*Appledore*
Burnecle, William
Lemon, Hubert
Oatway, Alfred
Parkhouse, William

*Ashburton*
Bidlake, Thomas
Burston, John
Conneybear, James
Conneybear, Samuel
Dunsford, Martin
Ezekiel, Eleazer
Foster, James
Hamlyn, Thomas jun
Harding, Charles
Harding, John
Harding, John (ii)
Harding, Richard
Hays & Parsons
Hays, Adrian
Hudge, William
Osborn, W. G.
Pearse, James
Perryman, John
Toker, ——
Waycott, Peter

*Aveton Gifford*
Startudge, Roger
Stumbels, William

*Axminster*
Bannister, Richard
Bishop, Thomas
Davis, Adolphus E.
Elford, Joseph
Hawkins, Joseph
Jones, Eli
Kenneton, James
Legg, John
Marsh, E. & J.
Masters, Philip D. F.
Pearse, Samuel
Pearse, William
Peream, George
Pulman, Philip
Pulman, William
Scherzinger, Joseph
Sprake, John
Stringer, Arthur T.
Tomkins, George
Willmott, George
Withington, William B.

*Bampton*
Hobbs, Elias
Hobbs, Mrs Elias
Hobbs, Elias (ii)
Stacey, George

*Barnstaple*
Alexander, Albert
Alexander, Charles
Beare, John
Coffin, James
Cole, John
Cook, John
Darke, John
Darke, John (ii)
Davis, J.
Delve, John
Dyer, William
Easton, James & Co
Feurier, Leon
Ford, Charles
Fox, J. F.
Gaydon, Frederick
Gaydon, John
Gent, John R.
Gillard, Samuel
Green, John
Hill, Benjamin
Hodge, John
Hunt, Harry P.
Hunt, John
Huntley, William
Jones, Arthur & Thomas
Jones, Edmund
Jones, John William
Jones, Thomas Henry
Joyce, John
Kent, William
Kistler, A.

Snell, William
Tickell, John
Tickell, John Nathan
Tregena, Andrew
Trescott, John
Trobridge, John
Tucker, John
Webber, George
Welch, George
White, William

*Cullompton*
Baker, John Henry
Barber, Benjamin
Benham, Mrs Jane
Benham, John
Bidgood, George
Bilbie, Thomas
Bilbie, Thomas Castleman
Box, Frederick
Bussell, Thomas
Davis, Arthur
Hart, C.
Hill, William jun
Hornsey, Charles
Hutchings, William
Long & Co
Marshall, John
Martin, John
Masne, Thomas
Mayne, Mrs Jane
Mayne, William
Mills, John
Parrott, E.
Richards, John
Salter, Isaac
Triggs, John Richard
Uglow, William
Veales, Edward
Whitby, William
Wilkinson, Edwin

*Dartmouth*
Adams, Harold
Ash, Henry
Braithwaite, William
Burgoine, Richard
Dundass, Robert
Follett, Richard
Ford, Thomas
Haddon, Charles
Harvey, George
Hodge, John
Hole, William Henry
Land, Thomas
Langmead, George
Lidstone, George
Lidstone, John
Lidstone, Robert
Manley, John jun
Masters, Elizabeth
Masters, John
Mortimore, Robert
Mortimore, Robert M.
Newcomen, Thomas
Roberts, John
Rogers, John
Sanders, William
Traies, James
Traies, William
Trugard, Moses
Tucker, John
Warner, ——
Wood, Henry
Wood, H.

*Dawlish*
Barnes, Reuben Thomas
Bolt, Mrs Emily
Bolt, William
Bulkeley, John
Chasty, Charles
Cornelius, William
Easterbrook, Henry
Hoefler, R.
Lear, John
Strickland, George
Strowbridge, Henry

*Dolton*
Langbridge, John
Trigger, Richard

*East Anstey*
Bradford, John
Bradford, W.

*Exeter*
Adams, Ann & Son
Adams, Edward Hewish
Adams, John
Adams, William
Adamson, William
Allin, John
Angel, Francis
Ashford, James
Avent, Thomas
Aviolet, Samuel Anthony
Bake, Robert
Baker, John
Balle, John
Bartlett, Moses
Bayly, Richard
Bealey, Bertram C.
Bedford, Stewart H.
Beirley Bros.
Blackbeard, H. J.
Blackbeard, John
Blackbeard, William J.
Blackmore, John
Bloodworth, John
Bloodworth, Richard
Boteler, Henry

Bowden, John
Bowden, Matthew
Bowring, Benjamin
Bradford, William
Brierly, Joseph
Bristow, William George
Brooking & Son
Broom, John
Brown, H. A.
Brown, John
Brown, John & Anthony
Brown, Mathias & Co
Brown, William
Bruford, William & Son
Burnett, Abraham Filmore
Burrington, John
Burrington, Thomas
Bussey, Giles
Caine, John
Calle, Samuel
Canon, C. E.
Cavill, William John
Celler, ——
Chamberlain, William
Chambers, William Henry
Charlton, Cornelius
Chaunter, John
Chaunter, Thomas
Cheriton, George
Clampitt, Charles H.
Clarke, Joseph
Cleak, Ezekiel
Cleak, Ezekiel (ii)
Cleak, Ezekiel (iii)
Clement, Edward
Clockmaker of Exeter
Clockmaker, Peter
Clode, Harold
Cockren, Edward
Coffin, James
Cohen, Hyam
Coles, William
Colleber, John
Collier, John
Conibear, George Henry
Coombe, William
Cottey, Abell
Cottey, John
Couldridge, William Henry
Courti, Paul
Cross, Charles
Cross, Charles (ii)
Cross, Henry John
Crosse, Edmund
Curtis, William
Dashwood, John
Davy, Robert
Davyes, John
Day, Thomas
Depree, Frederick Templer
Depree, Raeburn & Young
Depree & Young
Dietrich, J. G.
Donovan, John
Dufner, Leopold
Early, W.
Eastcott, Richard
Eastman, Henry
Edwards, Thomas
Ellis, Henry
Ellis, Henry Samuel
Ellis, William Horton
Ellis & Co
Ellis & Son
Ellis, Depree & Tucker
Eustace, Thomas
Evans, David
Evans, David (ii)
Ezekiel, Abraham
Ezekiel, Ezekiel Abraham
Ezekiel, C. & A.
Ezekiel, Henry
Ferenbach, Joseph & Co
Finnimore, William
Flashman, George
Flood, ——
Folland, William
Folland, William John
Fontana, Baptista
Ford, John jun
Freeman, Richard
Frost, Jonathan
Frost, William
Frost, William (ii)
Frost & Johns
Gant, Robert Drake
Gard, Henry
Gard, William
Gifford, John
Gilpin, Robert John
Gouldsworthy, Edward
Grant Brothers
Green, James
Gregory, William Arthur
Gregory, W. T.
Grinking, Robert
Grinking, Thomas
Groves, J.
Groves & Leyman
Guillaume, Guillaume
Guillaume, William
Hall, John
Hall, John James
Halstaffe, Peter
Halstaffe, Peter (ii)

Harris, Israel
Harrison, Robert
Hart, Moses
Harvey, Robert
Hawkins, Ambrose
Hawkins, John
Haydon, John
Hearn, Albert
Herbert, Sydney T.
Herbert, Thomas
Hettish, A.
Hettish & Co
Hettish, Gordon
Hettish, S. & F.
Hicks, Joseph
Hill, Henry
Hine, James
Hock, Primus
Hoppin, Charles
Hoppin, Matthew
Hoppin, William
Hornsey, Charles
Hornsey, William
Howard, William
Howe, Mrs C. & F.
Howe, Frederick
Howe, Joseph
Hunt, William
Hunter, Thomas James
Hurlstone, John
Isle, John
Jacobs, Morris
Johns, Richard
Johns, Samuel
Jole, John
Jonas, B.
Jonas, Samuel
Joseph, B. & B.
Josias
Kallendar, Orlando
Kerfutt, Walter
Keys, Joseph
Keys, William
Kingdon, Samuel
Lake, Henry
Lake, John Ellett
Lakeman, John
Lakeman, William
Lamby, John
Langworthy, William
Larmuth, Thomas
Lazarus, Isaac
Lazarus, Moses
Lear, James
Levy, Emanuel
Levy, Jonas
Levy, Joseph & Son
Lewis, Henry
Leyman, Frank
Linton, George
Lisle, William
Lisle, William Richard
Lovelace, Jacob
Luke, William
Lyons & Sons
McNay, A. & Son
Manner, John
Manning, G. & Co
Markes, Thomas
Markeys, John
Marks, Charles
Marks, Sidney
Marks & Elsey
Mather, Edward
Matthews, Edward
Maunder, John
Maynard, William Thomas
Midgley, Richard
Miller, Jacob
Miller, Richard
Mole, William
Morgan, John
Morgan, Robert William
Morle, John
Moss, John
Moxey, Milford Kemp
Murch, John
Mutton, Robert
Needs, Thomas
Newcombe, Henry
Newton, George
Nichols, James
Nix, B. W.
Pace, Henry
Palmer, Thomas
Pare, ——
Parkin, Edward A.
Parrish, John
Passmore, Hercules
Pearce, H. J.
Pfaff & Brown
Pfaff, John
Pfaff, Joseph
Piper, Edward
Plumleigh, Thomas
Plumley, William
Poling, Abraham
Poling, Antony
Poling, John
Poling, John (ii)
Pollard, T. & J.
Pollard, Thomas
Purchase, Joseph
Pye, John Elliott
Pyne & Sons
Reddaway, Robert
Reid, Frederick Treeby
Richards, William Jenkin
Rickard, Hercules
Rickard, Hercules (ii)
Rickard, John
Rickard, John (ii)

Rickard, John (iii)
Rihll, William
Robertson, Joseph
Roof, Charles
Rose, George
Rose, Richard
Ross, Daniel
Rosse, James
Ruff, Conrad
Runder, Francis
Sanderson, George
Savage, Abraham
Savage, John
Savage, John (ii)
Savage, Peter
Savage, Thomas
Savage, Thomas (ii)
Sayer, Matthew
Scott, Frederick G.
Shepherd, John
Shorto, Edward H. H.
Skinner, James
Skinner, John
Skinner, John (ii)
Sloman, ——
Smale, John
Snell, William
Sobey, William Rawlins
Soloman, M.
Southey, John
Southworth, Peter
Sparke, Christopher
Sparke, Walter
Spiegelhalder, Albert
Spiegelhalder, Anselm
Spiegelhalder, Walter
Spiller, Henry
Spiller, John
Stafford, Timothy
Stanbury, Elisha
Stanbury, John
Stapleton, Robert
Stevens, ——
Stone, Mrs F. E.
Stone, Henry
Stone, John
Stone, Jno.
Stone, Thomas Hart
Stoneman & Graf
Strowbridge, Henry
Strowbridge, W. T.
Symons, William
Taylor, John P.
Taylor, Richard
Taylor, William
Templer, Richard
Templer, Thomas
Templer, Thomas (ii)
Thomas, clockmaker of Exeter
Thomas, Henry
Thomas, William
Thomas, William Henry
Thorn, Thomas Memory
Towill, Frederick G.
Traies, William
Treadwin, John
Trehane, Sampson
Trimble, Son & Brooking
Trist, James
Trist, Joseph C.
Trout, Walter
Tucker, James H.
Tucker, John
Tucker, Walter J. K.
Tucker, William G.
Turtliff, Thomas
Uglow, A.
Upjohn & Company
Upjohn & Michie
Upjohn, Edward
Upjohn, Edward (ii)
Upjohn, James
Upjohn, James Rouse
Upjohn, John
Upjohn, John (ii)
Upjohn, Richard
Upjohn, Robert W.
Upjohn, Thomas
Upjohn, William
Upjohn, William (ii)
Wakelin, John R.
Way, John
Webb, Charles
Webber, George
Weller, George
Westaway, John
Westlake, John Carew
Wilkinson, Robert
Williams, Charles
Williamson, ——
Willis, William
Windeatt, Richard
Wood, John
Wyatt, Edward W.
Yelland, Arthur John
Yelland, John C.
Young, John & Co

*Exmouth*

Boyce, Edmund
Carter, Edward
Coombes, George
Ferris, William
Hill, William W.
Lazarus, J.
Maynard, George
Maynard, William
Mortimore, Harry
Nicks, John
Palmer, Charles
Pinn, Richard

Richards, Francis H.
Veals, James
Walters, Edward
Westlake, ——

*George Nympton*
Gould, George

*Hartland*
Bright, James
Chope, William
Heard, William

*Hatherleigh*
Braund, John
Chasty, Robert
Ditchet, Samuel
Downing, John
Groves, J.
Harris, William
Hill, Henry H.
Hurford, John
Merryfield, John
Southcott, Henry

*High Bickington*
Manning, John
Simpkins, W.

*Hockworthy*
Perry, William

*Holne*
Waycott, Peter

*Holsworthy*
Badcock, William
Bate, S.
Darke, John
Friend, Walter
Mill, John O.
Nottle, George
Nottle, John
Reynolds, John
Short, William Marshall
Slee, Francis
Tardrew, Richard
Vanstone, James
Walton, Philip
Yeo, Thomas

*Honiton*
Brown, John
Bryan, Thomas
Chubb, John
Clark, Alfred
Compton, John
Coombe, Edmund John
Deeme, Henry
Dent, George
Hurd, Thomas
Land, Thomas
Lott, George
Murch, Jerom
Murch, John
Murch, Matthew
Pearce, Samuel
Pile, Francis jun
Richards, John
Salter, Thomas T.
Saunders, John Palmer
Shepherd, Jesse John
Snell, William
Stocker, William
Tovey, Henry
Tovey, William Alfred
Tucker, R. & M.
Tucker, Richard W.
Turner, George

*Ilfracombe*
Barns, William Fry
Clemmow, Charles
Grant, John D.
Hall, Thomas
Harris, B. S.
Harris, Francis
Harris, John
Hobbs, Elias
Organ, William
Raw, Henry
Sommers, George
Sommers, George jun
Stephens, N.
Thomas, John
Thomas, John C.
Turner, William Mills
Webber, Augustus
Webber, Henry

*Instow*
Dobbs, Alfred Reginald

*Ivybridge*
Bidgood, William
Letcher, Henry

*Kentisbeare*
Davis, Arthur

*Kenton*
Voysey, John

*Kingsbridge*
Bennett, James
Bennett, Joseph
Bentley, Samuel
Brockedon, Philip
Brown, Nicholas
Ching, William Thomas
Easterbrook, William
Frazer, Alexander
Haddon, John Charles
Hughes, David

Iverson Bros
Luckham, Henry
Osborn, William
Paddon, George
Partridge, James T.
Randle, John
Smale, Timothy
Stidworthy, Frederick
Tickell, Charles P.
Tickell, John N.
Wellington, A. F.
Wellington, William Thomas

*Landscore*
Billing, John

*Lifton*
Miller, Richard

*Loddiswell*
Wellington, William T.

*Luppitt*
Cleak, Adam
Deeme, Henry

*Lydford*
Routleigh, George

*Lympstone*
Coventon, John G.
Upcott, Charles H.

*Lynton & Lynmouth*
Blackford, Sydney
Burnell, William
Chant, L. J.
Harris, Francis
Hurcombe, Harry
Kennelly, Samuel

*Meeth*
Bowden, Samuel

*Membury*
Harner, ——

*Merton*
Elliott, Joseph

*Milton Abbot*
Hicks, William

*Modbury*
Brown, John
Ellory, Thomas
Fesant, John
Flashman, Joseph
Paddon, George
Paddon, William
Startudge, Roger

*Morchard Bishop*
Cann, Stephen
Passmore, John
Passmore, Richard
Pugsley, Mark
White & Crook
White, Robert

*Moretonhampstead*
Boyce, John H.
Bucklow, Alexander
Davies, Henry Charles
Lang, John
Laskey, Samuel
Miles, Albert
Miles, Thomas
Rihll, Albert John
Rihll, William
Soper, Thomas
Thomas, Henry
Treleaven, Silvester
Treleaven, Silvester jun

*Morteboe*
Easton, William

*Newton Abbot*
Balhatchet, Francis J.
Beyer, George C.
Blackbeard, William J.
Bradford, James
Brooking, John
Bulkeley, John
Burge, Arthur Henry
Buston, John
Chudleigh, William J.
Clarke, ——
Coombe, Edmund John
Cross, Charles
Crossman, Frederick
Davis, Henry C.
Durckheim, Isaac Moses
Ezekiel, Eleazer
Friend, Walter
Gerry, Frederick W.
Harley, John
Huxtable, Edmund
Huxtable, Mrs Sarah J.
Hyne, Charles
Jeffard, ——
Jonas, Jonas
Lombardini, Francis
Maisters, John
Michell, Richard
Pickford, W.
Pike, James
Pike, James (ii)
Prout, Joseph
Ridgway, Harry White
Smerdon, Edward
Smerdon, John
Snelling, James

Thomas, James
Thomas, Mrs Mary
Uglow, William
Uglow, Mrs

*Northam*
Clark, Edward

*North Molton*
Boon, ——

*North Tawton*
Bond, C.
Day, Herbert
Foster, T.
Gill, John
Gould, Thomas
Pike, John
Saunders, E. Jenkin

*Okehampton*
Bassett, William
Bevan, Thomas
Bond, C.
Bridges, James Colman
Cornish, William H.
Drew, S. R.
Hooper, Humphrey
Lee, William
Lugg Brothers
Lugg, Thomas J.
Miles, John
Nottle, John
Perrien, John H.
Philp, Richard
Rowe, Thomas Squire
Russell, George P.
Seymour & Bond
Stinchcombe, Edward

*Ottery St Mary*
Adams, William
Allford, George
Capon, Francis W.
Gillham, William
Godfrey, Henry
Lancey, W. J. O.
Mayer, Joshua H.
Mills, John
Peream, John
Piper, F. J.
Scherzinger, Silas
Scobell, John Spurway
Turner, John
Upham, Richard
Walker, Josias
Waycott, William

*Paignton*
Bowden, William
Jago, William Henry
Lidstone, John F.
Parker, Henry May
Rodgers, F. J.
Rodgers, F. R.
Waycott, Robert
Winsor, Mrs J. & Son
Winsor, Thomas
Wotton, George
Wotton, Silvanus

*Plymouth & Plymouth Dock (Devonport)*
Abraham, Abraham
Abrahams, Aaron
Abrahams, M.
Adams, Charles
Adams, Thomas
Almond, Richard P.
Antill, Samuel
Atwill, William
Bake, Robert
Baker, John
Baker, W.
Banks, George
Banks, Ralph
Barnett, John
Bartlett, George
Bartlett, William
Basch, Emanuel
Bastow, John
Bastow, Thomas
Bate, Albert
Beer, Benjamin
Beer, John R.
Bennett, John
Bennett, Joseph
Bernstein, Philip
Bickle, Thomas
Bidgood, William
Blackford, John
Blackmore, William H.
Blight, Edward G.
Bone, George
Boney, Caleb
Boney, F. H.
Boney, John
Borelli, Dominick
Bowden, Ambrose
Bowden, Frederick E.
Bradden, Thomas
Bradey, John
Brissington, W.
Brock, Charles
Brock, John
Brown, John
Brown, John (ii)
Brown, N. E.
Browne, Thomas
Buckingham, John
Bulford, John Rogers
Burt, Robert Henry
Burt, William Henry
Butland, Ben
Buttall, John

Capps, William George
Carne, George
Carpenter, G.
Chenhall, ——
Chenhall, J.
Ching, William H.
Clarke, George P.
Clutterbuck, H.
Cohen, D.
Cohen, L.
Cohen, Mrs
Cohin, Samuel
Collins, Henry
Copplestone, William
Cornelius, W. J.
Cornelius, W. J. (ii)
Cornish, C. H.
Coryndon, William
Coryton, George
Cox, William C.
Coxworthy, Thomas
Crews, Charles F.
Croft, John
Croft, Robert
Crook, Henry
Crowther, Henry W.
Croydon, Charles
Crunkhorn, W. J.
Daniel, Abraham
Daniell, Robert
Davey, ——
Davies, Francis
Davis, Richard F.
Daws, Richard
Dawson, Matthew
Dietschey, Charles
Dilke, Edward
Dillin, Robert
Dodd, Edward
Dominy, R.
Dorning, Robert
Down, Albert
Down, James
Dugdale, James R.
Dugdale, L. H.
Dunsford, William
Dupuy, Adrian
Durant, Walter
Dyer, John
Dymond, Charles
Edgcombe, James
Edgcumbe, Samuel
Edgcumbe, Stephen Luke
Edwards, Joseph
Edwards, William
Efford, Charles
Egbert, William H.
Ellett, Charles
Elliott, John
Elliott, Mary
Elliott, Walter
Ellis, Thomas
Ellis, William James
Elms, Joseph
Emden, Gompert Michael
Emdon, Mark
Evans, William
Ezekiel, Philip
Fawkes, Charles
Fehrenbak & Ketterer
Ferenbach, Xavier
Fenn, J. S.
Ford, W. C.
Ford, William L.
Forrest, Jonathan
Frost, Charles
Fulford, Robert
Gard, Edward
Garland, Richard
Gawman, Edwin
Gendle, Thomas
Gibbs, William
Gibbs, Mrs
Giles, George
Gilmore, George
Godfrey, Thomas Dove
Godfrey, William
Gorfin, Henry
Gorfin, William
Gormully, Philip
Gossier, Daniel
Goulding, Frank H.
Granville, Samuel
Granville & Linde
Green, John
Greenleaf, William
Griffin, Henry
Griffin, John
Grigg, Humphrey
Gudridge, Percy
Guillaume, Benjamin
Haddy, William
Haley, Samuel B.
Hall, Edward
Halse, John
Halse, William Henry
Harry, John
Harry, William
Hart, Emanuel
Hart, Henry
Hart, Moses
Hart, Samuel
Harvey, William
Harvie, J. C.
Harvie, William C.
Hawkins, Henry
Hayhurst, James
Hayward, Frederick
Healy, Thomas
Heard, William
Heart, Henry
Heine & Co
Heles, Richard
Hellyer, Richard

Hill, Albert
Hill, Richard
Hill, William
Hillman, William
Hoare, William Henry
Hocking, Henry R.
Hodge, Frederick
Hodge, John
Hoefler, Fidel
Hoefler, Severin
Hoppen, William Vosper
Hornbrook, Aaron
Huet, John W.
Hutton, John
Hyman, Abraham
Hyman, H.
Jackson, Joseph
Jago, H.
James, Nicholas
Jarvis, Henry
Jeffery, Alfred Byron
Jeffery, F.
Jeffery, William Abbott
Jerger, John
Jonas, Benjamin
Joseph, A. & Sons
Kallensee, Charles H.
Kammerer, Seraphin
Kaufman, Bernard
Kennett, Arthur W.
Kent, Soloman
Kent, William
Ketterer, M. & Co
Kneebone, Arnold
Kressinger, Christian
Lakeman, George
Lakeman, Robert
Lamport, Henry
Lamport, Henry (ii)
Lamport, William
Lancaster, Thomas
Lancaster, William
Lang, Henry
Lang, Richard
Lashmore, Charles E.
Lear, Richard
Leddra, John
Ledum, John
Lempert, Saul
Levi & Co
Levi, Lion
Levy, A.
Levy, Barnett
Levy, Marks
Lewis, William Charles
Lewis, William Thomas
Libbey, Daniel
Libbey, Daniel Harson Pearce
Libby, J.
Lock, Charles
Loveguard, Mrs Flora
Luxmore, Henry
Luxton, J. H.
Lyon, Francis
Lyon, Judah
Macey, Robert
March, H.
Markes, Charles
Maunder, Michael
Mayne, Thomas
Mends, Benjamin
Mends, John
Metherell, Josiah
Michell, Walter
Millman, Arthur J. G.
Mitchell, Barnet
Mitchell, James
Mitchell, Richard John
Mitchell, Thomas
Monk, Richard R.
Morshead, Robert
Mortimer, William
Moser, B.
Moser & Wehrle
Mudge, Thomas
Murray, Robert
Mutton, Samuel
Myers, T.
Nathan, ——
Neame, William John
Neugard, John
Newey, Joshua
Nicholson, Benjamin
Northcote, James
Northcote, Samuel
Northcote, Samuel (ii)
Oliver, Thomas
Oliver, W.
Osborne, Henry
Osbourn, Henry J.
Osbourn, John
Owen, A. W.
Oyens, Peter jun
Page, Keen & Page
Parsons, William
Pasmore, S.
Patterson, Thomas
Pearce, James
Pearce, Joseph W.
Pearce, William
Pearce, W. P.
Pearse, Frederick
Penberthy, Frederick
Penny, P. L.
Pepperell, Albert
Petherick, J.
Phillips, ——
Pilson, Abraham
Pinhey, Richard Lear
Pleace, William H.
Pollard, John
Pope, William

Proctor, William
Quinlan, John P.
Ramsey, Edward
Ramsey, Jonathan
Rawling, Alfred
Rawlinson & Deacon
Rawlinson, John
Raworth, Henry
Raworth, Samuel
Reed, B.
Reepe, Joseph
Reynolds, Mrs Elizabeth
Reynolds, T.
Reynolds, William
Riedlinger, Ludwig
Roberts, William
Rodd, William
Rohrer, Andrew
Row, William Harris
Rowe, Henry
Rudall Brothers
Rudd, John William
Rundell, Joseph
Rupe, ——
Rutt, Richard
Ryder, John W.
Ryder, Richard J.
Sampson, John
Sampson, Thomas
Sandford, Frederick
Saunders, T. G.
Saxby, Harry
Searle, G. Elliott
Shea, James
Sheperd, W.
Shepheard, Jabez
Shepheard, James
Sheppard, Thomas
Sherry, F.
Skerrett, George
Smith, Richard
Smith, William
Soady, Mrs Elizabeth
Soady, William
Solomon, Moses
Sommers, Alfred
Sommers, F. R.
Sommers, George
Sommers, W. H.
Sparke, Angel
Spiller, Henry
Stephens, George
Stephens, Thomas
Strang, James W.
Summers, Henry
Sweetnam, George F.
Symonds, ——
Symons, Henry Hawkes
Symons, Herbert G.
Taylor, John
Templer, Richard
Thomas, Martin
Towan, Joseph
Towson, John Gay
Towson, John Thomas
Tozer, James
Treleaven, Walter
Treliving, John
Tressie, Thomas
Treverton, James
Trewin, William
Trugard, James
Turner, H.
Tuttle, Robert Shepherd
Upjohn, Nathaniel
Valleck, Thomas E.
Velvin, J.
Wadelton, Charles
Wadelton, John
Walter, Michael
Walter, S.
Walters, Michael
Ward, Samuel
Waycott, Peter
Webber, C. P.
Weeks, John
Welch, William
Welsford, William
Westlake, George
Westlake, Robert
Westlake, S. J.
Wilby, Henry
Willcock, J.
Williams, Alfred
Williams, James
Willmot, John
Wingate, G. M.
Wolf, Aaron
Wyatt, Eliza
Wyatt, John Samuel
Wyatt, Robert
Wyatt, R. jun
Zaringer, Albert

*Plympton*
Batho, Thomas
Eveleigh, John Pearse
Hillson, Richard
Nicholls, John
Ramsey, Frederick A.
Reynolds, James
Reynolds, William
Uglow, A.

*Plymstock*
Ball, Alfred Ernest

*Rackenford*
Wood, David

*Rockbeare*
Reynolds, James

*Salcombe*
Snell, William Thomas
Soper, Thomas
Thomas, William

*Salcombe Regis*
Pinn, Richard

*Sampford Courtenay*
Huxtable, John

*Sandford*
Pridham, Lewis

*Seaton*
Good, Edward Dare
Good, Samuel

*Shebbear*
Hearn, H.

*Sidbury*
West, James

*Sidmouth*
Blackmore, James
Blackmore, James (ii)
Denby, William
Follett, ——
Gibbs, James
Harding, Charles W.
Holwill, Robert
Jones, James
Moore, Henry
Passmore, Richard
Paul, Thomas
Pinn, Richard
Porter, Henry
Scadding, James
Uglow, Nicholas J.
Uglow, Mrs N. J.

*South Brent*
Ball, Christopher
Bolitho, Samuel
Foot, Richard
Mason, Charles
Salmon, William
Wellington, Charles H.

*South Molton*
Allen, Peter
Bickell, John
Bickell, Richard John
Bradford, Philip
Clarke, Charles
Cruwys, ——
Day, Christopher
Delve, John
Foster, Henry
Gould, George
Gould, Thomas
Hammond, Frederick S.
Hellier, John
Huxtable, William
Lyddon, William
McKenzie, Joseph
Mills, Samuel G.
Potter, John
Smith, John
Snell, William
Snow, Robert
Tepper, Joshua Turner
Thorne, Michael jun
Thorne, Richard
Webber, John G.
Wills, George H.
Wills, Joseph

*South Tawton*
Marks, Richard

*Starcross*
Bulkeley, James
Bulkeley, John
Eastmond, John

*Stokeinteignhead*
Sare, John

*Tavistock*
Alford, George
Alford, James
Barnett, John
Bennett, Edwin
Chenhall, Mrs Grace & Sons
Chenhall, James
Davy, William
Dingle, William
Fehrenbach, J.
Ford, Robert
Graf, F.
Hoidge, John
Jackson, John
Jackson, John & Eastcott, William
Jessop, Peter
Jessop, Thomas
Jessop & Jackson
Jewel, Leonard
Joseph, ——
Lidstone, George Henry
Miles, Mrs C.
Miles, Christian
Miles, John
Miles, Mrs Mary Jane Vigars
Miles, Samuel
Miles, Thomas O.
Parnell, John
Pearse & Bounsell
Pearse, John

Pearse, William
Percy, William
Richards, William
Snell, John Holder
Stenlake, Benjamin C.
Symons, John E.
Truscott, John
Uglow, Abel

*Teignmouth*

Adam, John Richard
Adams, John
Bolt, Richard
Bolt, William
Boyce, Arthur
Boyce, John
Boyce, William
Bradford, Edwin G.
Burston, John
Chasty, Charles Heard
Chasty, Mrs Susan
Chasty, William
Coe, Alfred I.
Coe, Mrs Alfred I.
Gilpin, William R.
Hele, W. W.
Hele, William R.
Hill, John
Jonas, Benjamin
Loosemore, John
Maynard, J.
Rossiter, E.
Rossiter, George
Rossiter, Mrs S.
Welsford, William
Westaway, John

*Thorverton*

Commings, James
Commings, James Levil
Commings, Mrs Jane
Crosse, Edmund
Cummings, James Richard
Cummings, William Lipscombe

*Tiverton*

Alwood, William
Arch, Samuel
Archibald, Alexander
Beer, William
Besley, Thomas
Boyne, Charles H.
Bradford, J.
Bradford, William
Broom, Charles
Chamberlain, Henry
Chamberlain, T. F.
Chorley, John
Collins, John
Cookson, William
Cottrell, Mrs Mary
Cottrell, Thomas
Cross, Henry
Crow, T. Leonard
Curwood, Miss Rhoda J.
Davis, Arthur
Debnam, John
Downing, R. E.
Eames, George
Eames, William
Early, Henry
Early, William Henry
Evans, P.
Foster, H.
Foster, Henry
Foster, Thomas
Franke, Alexander
Grason, James
Greenway, James
Hooper, Richard
Jarman, Allan
Jarman, Henry
Jones, William
Kerridge, Frederick
Land, Thomas
Long, John
Long, Samuel
Long, Thomas
Long, William
Marsh, Henry
Maunder, William Dail
Minifee, John
Mudford, Frederick
Munford, Edward H.
Pack, Isaac
Pack, Jacob
Parker, Henry May
Payne, John Ellis
Pinkstone, Mrs Charlotte M.
Pinkstone, George
Pinkstone, Richard
Pinkstone, Thomas
Prickman, John
Rendell, Henry
Rendell, Henry jun
Rendell, James
Rendell, John O.
Richards, John
Sharland, Mrs Emma
Sharland, Herbert J.
Sharland, John
Sharland & Co
Summer, Ralph
Thorne, Abraham
Thorne, Simon
Thorne, Simon (ii)
Thorne, William
Tucker, Miss Helen Jane
Tucker, John
Tucker, John (ii)
Tucker, John W. T.

Umber, Thomas
Waldron, John
Wood, Hubert

*Topsham*

Burnett, Abraham Filmore
Gale, J. B.
Grignion, Daniel
Gubb, Theophilus
Gubb, T. & H.
Hunn, Henry H.
Manley, J.
Manley, William
Margrie, S.
Newman, George
Salter, Robert
Trevor, John
Trevor, John (ii)
Trevor, John (iii)
Trevor, Richard
Upjohn, Edward

*Torquay*

Adams, Charles R. S.
Babidge, W. H.
Bailey, Frederick
Beyer, George C.
Bradford Brothers
Bradford, Dennis
Bradford, Edwin G.
Bradford, James
Braham, James
Braham, John
Burton, Henry
Carleton, William
Cary, Richard
Cawdle, Henry
Cawdle, William
Chasty, William
Couch, Conroy
Crofts, John
Cumming, John
Dashper, Alfred
Davey, John
Easterbrook, Frank
Easterbrook, George P.
Easterbrook, William
Farinton, Stephen
Galley, Arthur R.
Gerry, Nicholas
Goodfellow, E. H.
Grant Brothers
Honey, John B.
Hutt, Henry
Jacobs, Alexander
Jeffery, William Henry
Jonas, Henry
Kellond, George Albert
Kramm, Charles
Kramm, John
Lear, John
Lear, Samuel
Lombardini, F.
Mabin, William S.
Netherway, Albert
Parker, Harry
Parker, Henry M.
Pearce, Christopher
Peters, William James
Pratt, Albert J.
Pratt, William & Son
Pridham, William
Rawle, Frederick
Rodgers, William Henry
Sermon, Edward
Smith, Thomas C.
Smith, William
Soper, Arthur Thomas
Soper, John
Soper, Thomas
Stephens, C. F.
Stephens, Farinton
Tozer, William Henry
Trist, James
Turner, George
Waycott, Robert
Wheeler, James
Winterhalter, Ferdinand
Wood, John B.
Wotton, George
Zieglsbauer, Charles

*Torrington*

Barrow, Walter Philip
Bartlett, Hubert
Blight, John
Bloodworth, John
Collins, Augustus
Dennis, John
Eplett, Joseph
Feevings, John
Gaydon, Frederick
Green, Thomas
Hawkings, ——
Langbridge, James
Oatway, John
Passmore, Richard
Quick, Mrs M. J.
Rowe, John Passmore
Saunders, William
Taylor, Allan
Taylor, Mrs Fanny
Tyacke, George
White, Samuel
Williams, John

*Totnes*

Bennett, Charles
Bennett, Joseph
Bradford, James
Brockedon, Philip
Brockedon, William
Burrow, Samuel E.

Carter, John
Chasty, William
Chenhall, John
Clement, William
Codner, Phillip
Couldridge, William H.
Evans, Evan
Evens, Nicholas
Ford, John
Friend, John Walter
Friend, Robert
Good, Jacob
Grute, William
Harvie, J. C.
Jacob, J.
Jacobs, Mrs Jane
Jacobs, Lewis
Lombardine, Francis
Luscombe, Richard
Luscombe, Robert
Martin, Edward Braker
Pike, William
Pinkstone, William R.
Preddy, William
Reed, Robert
Richards, John
Roberts, Robert
Roucklieffe, John
Smith, John
Smith, Mark
Smith, Reuben
Stumbels, William
Symons, James
Turner, ——
Waycott, Peter
Wellington, Charles H.

*Uffculme*
Burrow, James
Long, Speccott Thomas
Long, Walter & James
Long & Sons

*Upottery*
Cleak, Adam

*Westleigh*
Davis, Arthur

*Widecombe in the Moor*
Tickle, John

*Winkleigh*
Bartlett, Albert
Francis, Edward
Linington, William Richard
Saunders, Edward

*Witheridge*
Bradford, William
Downing, Robert Edward
Downing, William

*Place unknown*
Ball, Thomas Beckenham
Blackwell, Robert
Cottie, ——
Gould, William
Hoppin, John
Hume, William
Isack, Mabell
Jerwood, John
Jerwood, Thomas
John Clockmaker
Kymm, ——
Mallett, Stephen
Mitchell, Hezekiah
Passmore, William
Peryam, Clement
Peryman, Richard
Philips, William
Phillips, Richard
Pyke, H.
Tucker, ——
Walters, ——
Williams, George
Woode, William

# APPENDIX: Tools and Materials of Abell Cottey (1711)

Two large bench Vises at 15s each
one ditto Small
one ditto Smaller
one Large Turn bench
one fusie Screw
one Watch lathe
one Small ditto
two Iron Braces & seven Bitts
two Sweiges
two pair of large hand Vises
two pair smaller Ditto one 4s ye other 2s
four Smaller ditto
three pair Clock plyers
five pair Watch Plyers at 2/6 each
two pair of slideing Tongs
four Skrew Plates wth tops
one beakIron
two small watch Ditto & one stake
three borax boxes two brass 1/6 each one tinn
three blow pipes
one plate for drawing weyer
three vise stakes
two pair of beam compases
two pair of sweep Compase & one pair skew dividers
one pair of brass Shares
two Oiles Stones one 4s the other 2s
two hones one 1/6 the other 6d
two Anvill hand hammers
three Clock hamers
two Squares
one burnisher & two Scrapers

one and a half dozen of larg files for Iron at 7d per doz
four Equalling Files
five small Ditto at 5d each
three small Watch hamers
one more Ditto
one spring Sawe
four dozen of Watch files at 2/6 per doz
sixteen new file handles at 2d per peice
one freizing Tool
two small gages
four setting tongs
two pair of Callipers & pair of turning tongs
one drill bitt & six drills
six screw arbors
one screw for adjusting a fusie
four old scratch brushes
one Calliper plate & two pinion plate guages
three gravers at 4d each & Counter borer 6d
two Stakes
four brushes & a skrew driver for watch work
nine figures & Stamps at 3d each
one Steel drill bow two brest drivers one skrew driver
one Watch dividing plate
several small drills & square broches
three new Rubbers
two dozen of new half rounds files
one hand ditto
one half round smooth ditto
twenty-one new files 5d each
twenty-three old worn files
two scratch brushes
three new files at 5d each
twenty-two doz of Watch files new
one Anvill
one Clock Stake
one pair of Old Bellows
one sand box two pair of screws & severall old Flails
one grind stone Iron Axeltree
a parcell of old Iron
a parcell of shruff brass
a spoon Mould
two Raisors at 1s each

a parcel of Cast Clock Work
four and a half dozen of Watch keys at 8s per doz
two dozen better sort at 12s per doz
five dozen of Watch Strings
three Watch Chains at 6s each
two dozen & two Pollised Glasses at 12s per doz
six dozen & ten Comon watch Ditto at 6s per doz
nine Watch Springs at 2/6 each
one oz. of Borax
two bobbins of pendlum Weyer.

# BIBLIOGRAPHY


Allix, Charles. 'Mudge Milestones (Watch Dates)', *Antiquarian Horology*, Summer 1981, 627–34; Addenda by A. F. Cliborne, December 1981, 144–5

Baillie, G. H. *Watchmakers and Clockmakers of the World*, second edition (NAG Press Ltd, 1947)

Bellchambers, J. K. 'Some Westcountry Clockmakers', *Antiquarian Horology*, September 1958, 144–5, 150

——. 'The Mudges in the Westcountry with a few notes on Samuel Northcote, Watch and Clockmaker', *Antiquarian Horology*, March 1960, 45–7

——. *Devonshire Clockmakers* (Devonshire Press Ltd, 1962)

——. *Somerset Clockmakers*, Antiquarian Horological Society, 1968

Britten, F. J. *Old Clocks and Watches and their Makers*, second edition (B. T. Batsford, 1904)

Brown, H. Miles. *Cornish Clocks and Clockmakers*, third edition (David & Charles, 1980)

Cescinsky, Herbert and Webster, Malcolm R. *English Domestic Clocks* (George Routledge & Sons, Ltd, 1913)

Chandlee, Edward E. *Six Quaker Clockmakers*, Historical Society of Pennsylvania, 1943 (reprinted New England Publishing Co, 1975)

Clutton, Cecil. 'The Second Mudge lever Watch?' [The 'Flint' Mudge], *Antiquarian Horology*, March 1960, 47–8

Cuss, T. P. C. 'The Huber-Mudge Timepiece with Constant Force Escapement', *Pioneers of Precision Timekeeping*, Antiquarian Horological Society Monograph No 3, 1965

Daniels, George. 'Thomas Mudge, the Complete Horologist', *Antiquarian Horology*, December 1981, 150–173

Ellis, Henry S. 'Description of Shepherd's Galvano Magnetic Regulator', Monthly Notices of the Royal Astronomical Society, Vol XV No 3, 1854 (reprinted in C. N. Ponsford, *Time in Exeter*)

Flint, Stamford Raffles. *Mudge Memoirs: being a record of Zachariah Mudge and some members of his family* (printed by Netherton & Worth, Truro, 1883)

Good, Richard. Descriptions of 'The Mudge Marine Timekeeper' and 'The

First Lever Watch made by Thomas Mudge', *Pioneers of Precision Timekeeping*, Antiquarian Horological Society Monograph No 3, 1965
——. 'Watch by Thomas Mudge, London, No. 574, with perpetual calendar mechanism', *Antiquarian Horology*, December 1981, 178–187
Gould, Rupert T. *The Marine Chronometer* (J. D. Potter, 1923)
Gwynn, Stephen. *Memorials of an Eighteenth Century Painter – James Northcote* (T. Fisher Unwin, 1898)
Hall, John James. 'The Story of Bishop Grandisson's Clock in the Church of St Mary at Ottery', *Horological Journal*, 1907
——. 'Historical and other Records of an Ancient Astronomical Clock in the Cathedral Church of St Peter in Exeter', *Horological Journal*, 1913
——. 'The Story of the Life and Work of Jacob Lovelace of Exeter', *Horological Journal*, 1931–2
Keir, David. *The Bowring Story* (The Bodley Head, 1962)
L——. 'Memoirs of the Life and Mechanical Labours of the late Mr Thomas Mudge', *Universal Magazine*, July 1795
Loomes, Brian. *Watchmakers and Clockmakers of the World – Volume 2* (NAG Press Ltd, 1976)
——. 'The amazing life of James Upjohn', *Clocks*, November 1983, 12–13
Maskelyne, Nevil. *An answer to a pamphlet entitled 'A Narrative of Facts', lately published by Mr Thomas Mudge, Junior* (London, 1792)
Mudge, Thomas jun. *A Narrative of Facts relating to some Timekeepers constructed by Mr Thomas Mudge* (London, 1792)
——. *A Reply to the Answer of the Rev Dr Maskelyne, Astronomer Royal, to A Narrative of Facts* (London, 1792)
——. *A description with plates of the timekeeper invented by the late Mr Thomas Mudge; to which is prefixed a narrative . . . a republication of a tract by the late Mr Mudge on the improvement of timekeepers; and a series of letters written by him to His Excellency Count Bruhl* (London, 1799; reprinted Turner & Devereux, 1977)
Parliament, House of Commons. *Select Committee Report on the Petition of Thomas Mudge*, printed 11 June 1793
Pearson, the Reverend W. *Rees's Clocks Watches and Chronometers* (1819–20; reprinted David & Charles, 1970)
Ponsford, Clive N. *Time in Exeter; a History of 700 years of Clocks and Clockmaking in an English provincial city* (Headwell Vale Books, 1978)
——. 'The Matthew the Miller Clock at the Parish Church of St Mary Steps, Exeter', *Antiquarian Horology*, September 1981, 52–66
Ponsford, C. N., Scott, J. G. M., and Authers, W. P. *Clocks and Clockmakers of Tiverton*, second edition (W. P. Authers, 1982)
Robinson, Tom. *The Longcase Clock* (Antique Collectors' Club Ltd, 1981)
Rogers, Inkerman. 'Early Clocks and Clockmakers of Barnstaple', *The North Devon Journal*, 16 June 1932

——. 'Horology Ancient and Modern', volume in manuscript, North Devon Athenaeum, Barnstaple

Rowe, Margery M., and Jackson, Andrew M. *Exeter Freemen 1266–1967*, Devon & Cornwall Record Society, 1973

Scott, J. G. M. 'Time to Feed the Navy' [The Vulliamy clock at the Royal William Victualling Yard, Plymouth], *Clocks*, March 1979, 20–2

——. 'Some Dating Features for Turret Clocks', *Antiquarian Horology*, Spring 1981, 536–8

Smith, Alan. 'The Exeter Lovelace Clock', *Antiquarian Horology*, June 1966, 78–85

Tribe, Tom, and Whatmoor, Philip. *Dorset Clocks and Clockmakers with a Supplement on the Channel Islands* (Tanat Books, 1981)

Wardle, Arthur C. *Benjamin Bowring and his Descendants* (Hodder & Stoughton, 1938)

Way, Stanley. 'Giant Act of Parliament Clock' [by John Tickell, Crediton, 1727], *Clocks*, May 1980, 57

# ACKNOWLEDGEMENTS

The author and publishers wish to thank all those who have provided photographs or have permitted clocks and related items to be photographed for use in this book, namely: Mrs Lorna Bellchambers and James F. Bellchambers, Plates 20 (*Newswork Ltd*), 27 (*Eric M. Morrison*), 36, 38, 39, 40; the Worshipful Company of Clockmakers, Plate 9; the Devon and Exeter Institution library, Exeter, Plate 5; the Devon Record Office, Exeter, Figs 1, 2, 3, 10, 11; Roger Ellis, Plates 41, 42, and Fig 8; Exeter Cathedral library, Fig 5; the Royal Albert Memorial Museum, Exeter, and Jeremy Pearson, Plates 15, 44; Charles S. Hadfield, Plate 14 (*L. H. Hildyard*); R. A. Hallet, Plate 47; A. F. Lilley, Plate 8 (*Douglas West Associates*); the Reverend Michael Moreton, Plate 3; Gilbert Partridge, Plate 16; Plymouth Museum and Art Gallery, Plate 37 (*Robert Chapman Photography*); Mrs H. R. Read, Plates 10, 11, 12; Prebendary J. G. M. Scott, Plate 7; Tom Tribe, Plate 4; the Victoria and Albert Museum, London, Plate 35; B. J. Wareham, Plate 13; Günther von Waskowski, Plates 21, 22, 23, 24, 25, 26 (*Nicholas Horne Associates*), 28, 29, 30 (*E. R. H. Francis*), 31, 32, 33, 34 (*Searle Photographic*), 45, and Fig 6; Stanley Way, Plate 46; Tiverton Museum, Fig 12; the Western Australia Museum, Perth, and David E. Hutchison, Plate 43; the Westcountry Studies Library, Exeter, Fig 9; Philip Whatmoor, Plates 1 and 2. Plates 5, 6, 10, 11, 12, 16, 17, 18, 19, 41, 42, and Figs 3, 7, 8, 9, 10, 11 and 12 are by the author.

# INDEX

NOTE – If particulars of a maker are required, refer first to Alphabetical List, which begins on page 169.